Lecture Notes in Mathematics

Edited by A. Dold and B. Eckmann

430

Constructive and Computational Methods for Differential and Integral Equations

Symposium, Indiana University
February 17–20, 1974

Edited by D. L. Colton and R. P. Gilbert

Springer-Verlag
Berlin · Heidelberg · New York 1974

Prof. Dr. David Lem Colton
Department of Mathematics
Indiana University
Bloomington, IN 47401/USA

Prof. Dr. Robert Pertsch Gilbert
Department of Mathematics
and Research Center for Applied Science
Indiana University
Bloomington, IN 47401/USA

Library of Congress Cataloging in Publication Data

Symposium on Constructive and Computational Methods for Differential and Integral Equations, Indiana University, 1974.
Constructive and computational methods for differential and integral equations.

(Lecture notes in mathematics ; 430)
"Organized through the Research Center for Applied Science at Indiana University under the sponsorship of the Air Force Office of Scientific Research grant no. 74-2674."
Bibliography: p.
Includes index.
1. Differential equations--Numerical solutions--Congresses. 2. Differential equations, Partial--Numerical solutions--Congresses. 3. Integral equations--Numerical solutions--Congresses. I. Colton, David L., 1943- ed. II. Gilbert, Robert P., 1932- ed. III. Indiana. University. Research Center for Applied Science. IV. Title. V. Series: Lectures notes in mathematics (Berlin) ; 430.
QA3.L28 no. 430 [QA372] 510'.8s [515'.623]
74-28189

AMS Subject Classifications (1970): 34A50, 34E15, 35A20, 35A40, 35C15, 35G15, 35G30, 35J25, 35J35, 35K20, 35L20, 45B05, 45L10, 49A20, 65L10, 65N05, 65N30, 65P05, 65R05

ISBN 3-540-07021-4 Springer-Verlag Berlin · Heidelberg · New York
ISBN 0-387-07021-4 Springer-Verlag New York · Heidelberg · Berlin

Offsetdruck: Julius Beltz, Hemsbach/Bergstr.

PREFACE

The following articles represent the contributed and invited lectures given at the Symposium on Constructive and Computational Methods for Differential and Integral Equations held at Indiana University, Bloomington, Indiana, from February 17-20, 1974. The Symposium was organized through the Research Center for Applied Science at Indiana University under the sponsorship of the Air Force Office of Scientific Research Grant No. 74-2674.

One of the main objects of this Symposium was to collect together those mathematicians working in the general area of constructive and computational methods for solving differential and integral equations in order to prepare a survey of recent developments in this important area of applied mathematical research. Such a survey would have the aim of not only coordinating the work of the active researchers in this area, but would also supply a means through which applied mathematicians and engineers could become more acquainted with the new methods now available for solving problems they may be presently grappling with unsuccessfully. The following collection of addresses therefore contains both reviews and discussions of current problems and it is hoped they will provide a beginning towards accomplishing the long range objectives of the Symposium.

It was decided by the Editors that it also would be interesting to hear if the new generation of computers such as the ILLIAC IV would pose new techniques and perhaps introduce new mathematical problems in the study of the numerical solution of differential and integral equations. As it is apparent from the lecture by Robert Wilhelmson, parallel computation is still somewhat in its infancy and even though it shows great promise, it is difficult to assess at this time how much it will influence the development of numerical methods for solving differential and integral equations.

The organizers take this opportunity of thanking the Air Force Office of Scientific Research for making the Symposium possible. They would also like to gratefully acknowledge the assistance of many members of the Systems Analysis Institute and Computer Science Department of Indiana University whose help contributed immeasurably to the success of the Symposium.

David L. Colton

R. Pertsch Gilbert

Editors

CONTENTS

CONFERENCE PARTICIPANTS

P. M. Anselone
G. Auchmuty
Michael P. Benson
Harald Bergstrom
S. Bhatnagar
James Bramble
Richard Calhoun
John Chadam
Y. F. Chang
David L. Colton
Joel Davis
Todd Dupont
S. C. Eisenstat
John Gibbs
R. P. Gilbert
Robert Glassey
Charles Groetsch
Ken Harrison
Dirk Hofsommer
Eberhard Hopf
Elias Houstis
George Hsiao
J. Thomas King
David Korn
Philip LaFollette
Peter Langhoff
Andrew Lenard
Peter Linz
William Ludescher
S. K. Mitra
Bill Moss
Zuhair Nashed
Edward Newberger
Charles Newman
Bill Odefey
R. E. O'Malley, Jr.
T. Papatheodorov
Frank Prosser
Madan L. Puri
Lt. Col. Enrique Ramirez
B. E. Rhoades
William Rundell
Ken Shafer
Quentin Stout
Maynard Thompson
Wolfgang Walter
R. J. Weinacht
Wolfgang Wendland
Robert Wilhelmson

Convergence of the Discrete Ordinates Method for the Transport Equation

P. M. Anselone* and A. G. Gibbs**

1. Introduction

The transport equation (transfer equation, linear Boltzmann equation) governs the distribution of neutrons in a nuclear reactor or of radiation in a stellar or planetary atmosphere. It is an integro-differential equation of some complexity in its more general forms. Primary independent variables are position and direction, speed (alternatively energy, frequency or wavelength), plus possibly time. The differential operator in the equation describes attenuation due to both absorption and scattering away from a particular direction. The integral operator describes scattering into a particular direction from other directions and production by fission. Source terms may be present. On any boundary surface there are boundary conditions corresponding to incoming directions. For time-dependent problems there are initial conditions.

---

* Oregon State University, Corvallis, Oregon 97331

** Battelle, Pacific Northwest Laboratory, Richland, Washington 99352

Supported by the U. S. Atomic Energy Commission at the Pacific Northwest Laboratory, Richland, Washington, and the Union Carbide Nuclear Division, Oak Ridge, Tenn.

Since the transport equation can be solved explicitly only in a few very special cases, approximation methods are commonly used. The discrete ordinates method involves the replacement of the integral operator by a numerical integral operator in order to obtain differential equations for approximate solutions. Empirically, this method yields very satisfactory results for a large class of transport problems. However, a complete and rigorous convergence and error analysis is presently available only in some rather idealized cases.

The discrete ordinates method was first applied by Wick [18] and by Chandrasekhar [7, 8] to particular steady-state problems from astrophysics with homogeneous media bounded by single planes or parallel planes (half-spaces or slabs), usually under conditions of isotropic scattering and isotropic sources. Convergence theorems for the approximations were obtained for various problems of this type with isotropic scattering by Anselone [1-5], Kofink [13], Keller [10, 11] and Wendroff [17]. Subsequently, convergence theorems have been obtained for certain transport problems involving anisotropic scattering, other geometries, and time-dependence, by Keller [12], Nestell [16], Madsen [14] and Wilson [20].

We shall survey these contributions from an abstract and unifying point of view in order to isolate what is essential to the convergence of the discrete ordinates approximations. In the process, methods of analysis will be developed which appear to be applicable to very general transport problems and to other integro-differential equations. The discussion is in terms of a sequence of model problems of increasing complexity. All functions considered will be real. For mathematical convenience and physical relevence specified functions are continuous and usually nonnegative. We remark that the problems surveyed here involve discrete ordinates approximations only for the angular integrals, and thus do not represent the most general case of discretization as practiced, say, in many nuclear reactor applications today, where the spatial variable is also discretized. However, as noted below, the powerful analytical tools developed in applications to angular discrete ordinates approximations also show promise for investigations of the more general case.

2a. An Isotropic Transport Problem in a Finite Slab

Consider the problem for $\phi(x, \mu)$ with $a \leq x \leq b$, $-1 \leq \mu \leq 1$, given by

$$\mu \frac{\partial \phi(x, \mu)}{\partial x} + \phi(x, \mu) = \frac{c}{2} \int_{-1}^{1} \phi(x, \mu')d\mu' + g(x),$$

(2.1)

$$\phi(a, \mu) = 0 \quad \text{for} \quad \mu > 0, \ \phi(b, \mu) = 0 \quad \text{for} \quad \mu < 0,$$

where $\phi(x, \mu)$ is the one-speed angular flux at depth x in a homogeneous slab in a direction making an angle $\theta = \cos^{-1}\mu$ with the positive x-axis. The positive number c is of order 1 and represents the mean number of secondary neutrons per collision. In the above equation distance has been measured in units of neutron mean free paths, i.e., $x = \sigma\tau$, where σ is the total cross section and τ is the actual distance.

The boundary conditions correspond to a situation with no reflection of neutrons at the surface and no incident neutrons. Actually, a non-zero incident flux boundary condition could have been assumed. However, such a problem can be reduced to (2.1) for $\phi_1 = \phi - \phi_0$, where ϕ_0 satisfies the homogeneous equation

$$\mu \frac{\partial \phi_0(x, \mu)}{\partial x} + \phi_0(x, \mu) = 0$$

and the given boundary conditions. If this reduction is not made, then an unnecessary approximation is introduced for ϕ_0, which can be found explicitly. For the same reason, zero boundary conditions can and should be assumed also in applications of the discrete ordinates method to more general transport problems.

To proceed, assume a numerical integration rule such that, for any continuous function $h(\mu)$, $-1 \leq \mu \leq 1$,

$$\sum_{j=\pm 1}^{\pm n} w_{nj}\, h(\mu_{nj}) \to \int_{-1}^{1} h(\mu)\, d\mu, \quad n \to \infty,$$

$$w_{n,-j} = w_{n,j} > 0, \quad \mu_{n,-j} = -\mu_{n,j},$$

$$0 < \mu_{n1} < \mu_{n2} \cdots < \mu_{nn} \leq 1.$$

The Gauss quadrature formula is an example. The symmetry conditions on the w_{nj} and μ_{nj} are a convenience rather than a necessity.

Discrete ordinates approximations $\phi_n(x, \mu)$, $n = 1, 2, \ldots,$ satisfy

$$\mu \frac{\partial \phi_n(x, \mu)}{\partial x} + \phi_n(x, \mu) = \frac{c}{2} \sum_{j=\pm 1}^{\pm n} w_{nj}\, \phi_n(x, \mu_{nj}) + g(x),$$

(2.2)

$$\phi_n(a, \mu) = 0 \quad \text{for} \quad \mu > 0, \ \phi_n(b, \mu) = 0 \quad \text{for} \quad \mu < 0.$$

For $\mu = \mu_{ni}$, $i = \pm 1, \ldots, \pm n$, this is a system of ordinary differential equations for $\phi_n(x, \mu_{ni})$. Then (2.2) yields $\phi_n(x, \mu)$.

The problems for ϕ and ϕ_n have equivalent integral equation formulations which are more convenient for the convergence analysis. Let

$$f(x) = \frac{c}{2} \int_{-1}^{1} \phi(x, \mu)\, d\mu + g(x), \tag{2.3}$$

$$f_n(x) = \frac{c}{2} \sum_{j=\pm 1}^{\pm n} w_{nj}\, \phi(x, \mu_{nj}) + g(x). \tag{2.4}$$

From (2.1) and (2.3), it follows that

$$\phi(x, 0) = f(x),$$

(2.5)

$$\phi(x, \mu) = \frac{1}{\mu} \int_{z}^{x} e^{(y - x)/\mu} f(y)\, dy, \qquad z = a, \ \mu > 0, \quad z = b, \ \mu < 0.$$

The equations for ϕ_n in terms of f_n are identical. From (2.3) and (2.5),

$$(2.6)\qquad f(x) - \frac{c}{2}\int_a^b E_1(|x - y|)\, f(y)\,dy = g(x),$$

where E_1 is the exponential integral function of order one,

$$(2.7)\qquad E_1(s) = \int_0^1 e^{-s/\mu}\,\mu^{-1}\,d\mu, \quad s > 0,$$

which has a logarithmic singularity at $s = 0$. Similarly,

$$(2.8)\qquad f_n(x) - \frac{c}{2}\int_a^b E_{n1}(|x - y|)\, f_n(y)\,dy = g(x),$$

where E_{n1} is a numerical integration approximation to E_1:

$$(2.9)\qquad E_{n1}(s) = \sum_{j=1}^{n} w_{nj}\,\frac{e^{-s/\mu_{nj}}}{\mu_{nj}}, \quad s \geq 0,$$

$$(2.10)\qquad E_{n1}(s) \to E_1(s) \quad \text{uniformly for } s \geq \varepsilon \text{ as } n \to \infty$$

for each $\varepsilon > 0$. Thus, the discrete ordinates method is seen to be equivalent to the approximation of the integral operator in (2.6) by the integral operator in (2.8).

Express (2.6) and (2.8) in operator form

$$(2.11)\qquad (I - K)\,f = g, \qquad (I - K_n)f_n = g$$

on the space $C[a, b]$ with the max norm. Then K and K_n are Fredholm integral operators with kernels

$$k(x, y) = \frac{c}{2}E_1(|x - y|), \qquad k_n(x, y) = \frac{c}{2}E_{n1}(|x - y|).$$

These are bounded linear operators on $C[a, b]$ and for c not too large (the case of a subcritical medium, assumed here) we have

$$\|K\| = \max_x \int_a^b k(x, y)dy < 1.$$

Hence, $(I - K)^{-1}$ exists,

$$\|(I - K)^{-1}\| \leq (1 - \|K\|)^{-1},$$

and f and ϕ are uniquely determined.

From (2.10),

$$\|K_n - K\| \to 0, \qquad \|K_n\| \to \|K\| \quad \text{as} \quad n \to \infty.$$

Thus, we are in the realm of standard operator approximation theory. For n sufficiently large, $\|K_n\| < 1$, $(I - K_n)^{-1}$ exists and is bounded uniformly in n, and

$$(2.12) \qquad (I-K_n)^{-1} - (I-K)^{-1} = (I-K_n)^{-1}(K_n-K)(I-K)^{-1},$$

$$(2.13) \qquad \|(I-K_n)^{-1} - (I-K)^{-1}\| \leq \|(I-K_n)^{-1}\| \, \|K_n-K\| \, \|(I-K)^{-1}\|,$$

$$(2.14) \qquad \|(I-K_n)^{-1} - (I-K)^{-1}\| \leq \frac{\|(I-K)^{-1}\|^2 \, \|K_n-K\|}{1 - \|(I-K)^{-1}\| \, \|K_n-K\|},$$

$$(2.15) \qquad \|(I-K_n)^{-1} - (I-K)^{-1}\| \leq \frac{\|(I-K_n)^{-1}\|^2 \, \|K_n-K\|}{1 - \|(I-K_n)^{-1}\| \, \|K_n-K\|},$$

$$(2.16) \qquad \|(I-K_n)^{-1} - (I-K)^{-1}\| \to 0 \quad \text{as} \quad n \to \infty.$$

The error bound in (2.14) is "theoretical" in the sense

that it involves $(I - K)^{-1}$, whereas (2.15) is "practical" because it depends on $(I - K_n)^{-1}$, and can thus be computed. The bound (2.13) is of mixed type. It follows from (2.16) and (2.5)ff that $f_n \to f$ and $\phi_n \to \phi$ uniformly as $n \to \infty$.

This is essentially the path followed by Anselone [4, 5] and by Keller [11], although Keller's analysis was less abstract and he defined the discrete ordinates approximations only at the quadrature points. Previously, Keller [10] and Wendroff [17] treated the differential operators directly without inverting them. The results were less satisfactory: L^2 convergence on quadrature points, and uniform convergence on quadrature points under a restrictive assumption. Kofink [13] established L^2 convergence on quadrature points by exploiting an equivalence between the discrete ordinates method and the spherical harmonics method.

2b. An Isotropic Transport Problem in a Half Space

Anselone [1-4] also derived uniform convergence theorems for isotropic transport problems in the case of semi-infinite slabs $(0 \leq x \leq \infty)$. In particular, the classical Milne problem leads to the homogeneous equation $(I - K)f = 0$, where K is the same integral operator as above. The Wiener-Hopf method, based on Fourier transforms, was devised originally to solve this equation.

A more direct method, which anticipated later theories of positive and monotone operators, was given by Hopf [9] for the case $c = 1$. It involves a change of variable $f(x) = x + q(x)$. Then q satisfies the inhomogeneous equation $(I - K)q = E_3$, where E_3 is the exponential integral function of order 3,

$$E_3(s) = \int_0^1 e^{-s/\mu}\,\mu d\mu, \quad s \geqq 0.$$

A nonnegative solution q is sought in the space of bounded continuous functions on $[0, \infty)$. Although now $\|K\| = 1$, monotonicity considerations yield a unique solution q given by the uniformly convergent Neumann series

$$q = \sum_{m=0}^{\infty} K^m E_3.$$

The discrete ordinates approximation problem can be recast in the form $(I - K_n)f_n = 0$. Let $f_n(x) = x + q_n(x)$. Then $(I - K_n)q_n = E_{n3}$, where E_{n3} is a numerical integration approximation to E_3, and q_n is given by

$$q_n = \sum_{m=0}^{\infty} K_n{}^m E_{n3}.$$

A detailed term by term analysis yields $q_n \to q$ uniformly as $n \to \infty$. It follows that $f_n \to f$ and $\phi_n \to \phi$ uniformly.

3a. An Anisotropic Transport Problem in a Finite Slab

Consider the problem for $\phi(x, \mu)$ with $a \leq x \leq b$, $-1 \leq \mu \leq 1$, given by

$$\mu \frac{\partial \phi(x, \mu)}{\partial x} + \phi(x, \mu) = \frac{1}{2} \int_{-1}^{1} p(x,\mu,\mu')\phi(x,\mu')d\mu' + g(x,\mu$$

(3.1)

$$\phi(a, \mu) = 0 \quad \text{for} \quad \mu > 0, \quad \phi(b, \mu) = 0 \quad \text{for} \quad \mu < 0.$$

Here, x represents the optical depth measured from the plane of the origin, i.e.,

$$x(\tau) = \int_{0}^{\tau} \sigma(\tau')d\tau',$$

with $\sigma(\tau)$ the total cross section at position τ, and τ the distance measured from 0. The differential kernel $p(x, \mu, \mu')d\mu$ represents the average number of neutrons emerging with direction cosines in $d\mu$, following a collision by a neutron with direction cosine μ' at position x. The function g represents sources within the slab. We assume here that both p and g are continuous functions of x and μ.

Let

$$(3.2) \qquad f(x, \mu) = \frac{1}{2} \int_{-1}^{1} p(x, \mu, \mu')\phi(x, \mu')d\mu'.$$

Then the problem is expressed in operator form as

$$(3.3) \qquad D\phi = f, \qquad f = L\phi + g,$$

where $f,g \in C(X)$ with $X = [a, b] \times [-1, 1]$. Under reasonable conditions on ϕ, $M = D^{-1}$ exists as an operator on $C(X)$, and $\phi = Mf = D^{-1}f$ is given by

$$\phi(x, 0) = f(x, 0),$$

(3.4)

$$\phi(x, \mu) = \frac{1}{\mu}\int_z^x e^{(x' - x)/\mu} f(x', \mu)\,dx', \qquad \begin{array}{ll} z = a & \text{for } \mu > 0, \\ z = b & \text{for } \mu < 0. \end{array}$$

Therefore, an equivalent formulation of the problem is

$$\text{(3.5)} \qquad \phi = Mf, \qquad (I - K)f = g, \qquad K = LM.$$

It can be shown that K is a bounded integral operator which maps $C(X)$ into $C(X)$. Since K has a non-negative kernel

$$\|K\| = \|Ke\|, \qquad e \in C(X), \qquad e \equiv 1.$$

If p is not too large, then $\|K\| < 1$, which we assume in what follows. Then $(I - K)^{-1}$ exists as a bounded operator on $C(X)$, and f, ϕ are uniquely determined.

Assume a convergent, positive quadrature rule:

$$\sum_{j=1}^{n} w_{nj}\, h(\mu_{nj}) \to \int_0^1 h(\mu)\,d\mu \quad \text{as} \quad n \to \infty, \qquad h \in C[-1, 1],$$

$$w_{nj} > 0, \qquad 1 \le j \le n, \qquad n = 1, 2, \ldots .$$

In particular, for $h = e \equiv 1$,

$$\sum_{j=1}^{n} w_{nj} \to 1 \quad \text{as} \quad n \to \infty.$$

Hence, there exists $B < \infty$ such that

$$\sum_{j=1}^{n} w_{nj} \le B, \qquad n = 1, 2, \ldots .$$

The discrete ordinates approximations ϕ_n satisfy

$$\mu \frac{\partial\phi_n(x,\mu)}{\partial x} + \phi_n(x,\mu) = \frac{1}{2} \sum_{j=1}^{n} w_{nj}\, p(x,\mu,\mu_{nj})\phi(x,\mu_{nj}) + g(x,\mu)$$

(3.6)

$$\phi_n(a,\mu) = 0 \quad \text{for} \quad \mu > 0, \qquad \phi_n(b,\mu) = 0 \quad \text{for} \quad \mu < 0.$$

Equivalent formulations are

(3.7) $\quad D\phi_n = f_n, \qquad f_n = L_n\phi_n + g,$

(3.8) $\quad \phi_n = Mf_n, \qquad (I - K_n)f_n = g, \qquad K_n = L_nM,$

where K_n is a bounded linear operator on $C(X)$. Since the "kernel" of K_n is positive, $\|K_n\| = \|K_ne\|$, where $e \equiv 1$.

It follows from (3.5) that $K_n \to K$, i.e., $\|K_nh - Kh\| \to 0$ as $n \to \infty$ for each $h \in C(X)$. However, $\|K_n - K\| \not\to 0$ in the anisotropic case, so the standard operator approximation theory used in Section 2 above is not applicable. An alternative course of action is pursued here.

It follows from $K_n \to K$, $\|K\| = \|Ke\|$ and $\|K_n\| = \|K_ne\|$ that $\|K_n\| \to \|K\|$. Recall that $\|K\| < 1$. Hence, for n sufficiently large, $\|K_n\| < 1$, $(I - K_n)^{-1}$ exists and is bounded uniformly in n, and

(3.9) $\quad (I - K_n)^{-1} - (I - K)^{-1} = (I - K_n)^{-1}(K_n - K)(I - K)^{-1},$

(3.10) $\quad f_n - f = (I - K_n)^{-1}(K_nf - Kf),$

$$(3.11) \qquad \|f_n - f\| \leq \|(I - K_n)^{-1}\| \; \|K_n f - Kf\| .$$

It follows that $f_n \to f$ and $\phi_n \to \phi$ uniformly as $n \to \infty$. The error bound in (3.11) is of mixed type, neither purely theoretical nor purely practical.

The fact that K and K_n are positive operators is essential to the foregoing analysis, which is an abstract version of that carried out by Keller [12] in a more classical spirit.

3b. An Anisotropic Transport Problem in a Half Space

We remark that the case of anisotropic scattering problems in a half-space was treated by Nestell [16] by an adaptation of the positive operator theory of Hopf. That work is an extension of the isotropic scattering problem discussed in Section 2b above.

4. Collectively Compact Operator Approximation Theory

We shall consider the same problem as in 3 from another point of view, which yields both theoretical and practical (computable) error bounds. The analysis is based on collectively compact operator approximation theory (cf. Anselone [6]).

It can be shown that the operator K in (3.4) is compact: $\{Kh : \|h\| \leq 1\}$ is relatively compact or, equivalently, bounded and equicontinuous. The sequence $\{K_n : n \geq 1\}$ is collectively compact: $\{K_n h : \|h\| \leq 1, n \geq 1\}$ is relatively compact. Thus, we have

$$K_n \to K, \quad K \text{ compact}, \quad K_n \text{ collectively compact.}$$

These are the hypotheses for a general operator approximation theory in a Banach space setting. Some of the main conclusions of the theory are as follows.

The operator $(I - K)^{-1}$ exists iff for n sufficiently large $(I - K_n)^{-1}$ exists and is bounded uniformly in n, in which case

$$(I - K_n)^{-1} \to (I - K)^{-1}.$$

Let $f = (I - K)^{-1}g$ and $f_n = (I - K_n)^{-1}g$. Then $\|f_n - f\| \to 0$. Let

$$\Delta_n = \|(I - K)^{-1}\| \, \|(K_n - K)K_n\| ,$$

$$\Delta^n = \|(I - K_n)^{-1}\| \, \|(K_n - K)K\| .$$

Then

$$\Delta_n \to 0, \quad \Delta^n \to 0 \quad \text{as} \quad n \to \infty,$$

$$\|f_n - f\| \le \frac{\|(I - K)^{-1}\| \, \|K_n g - Kg\| + \Delta_n \|f\|}{1 - \Delta_n} \qquad \text{for} \quad \Delta_n < 1,$$

$$\|f_n - f\| \le \frac{\|(I - K_n)^{-1}\| \, \|K_n g - Kg\| + \Delta^n \|f_n\|}{1 - \Delta^n} \qquad \text{for} \quad \Delta^n < 1.$$

Moreover, $(I - K)^{-1}$ exists whenever $(I - K_n)^{-1}$ exists and $\Delta^n < 1$. Thus the existence of $f = (I - K)^{-1}g$ can be inferred from approximations.

The convergence of the discrete ordinates approximations was established by means of the collectively compact theory by Nestell [16] for an anisotropic transport problem in a finite slab with $p = p(\mu, \mu')$ and by Nelson [15] in greater generality. Both authors worked directly with explicit representations of K and K_n. Since K is an integral operator on functions of two variables, its kernel involves four variables. The kernel has a weak singularity. The definition of K_n is similar, with the μ-integral replaced by a sum. The complications of K and K_n, particularly the singularity, make it difficult to see what makes the analysis go through. We propose a different approach here, which is more transparent and extends more readily to a larger class of transport problems. By (3.5) and (3.7),

$$K = LM, \qquad K_n = L_n M$$

where M is defined in (3.4) and

$$(L\phi)(x, \mu) = \frac{1}{2}\int_{-1}^{1} p(x, \mu, \mu')\phi(x, \mu')d\mu',$$

$$(L_n\phi)(x, \mu) = \frac{1}{2}\sum_{j=1}^{n} w_{nj}\, p(x, \mu, \mu_{nj})\phi(x, \mu_{nj}).$$

The basis of the analysis is to take advantage of the fact that only L, which has a continuous kernel, is approximated in order to define $K_n = L_n M$. The singularity comes from M, which can be treated separately.

Consider $\phi = Mf$ for $f \in C(X)$. From (3.4),

$$\phi(x, 0) = f(x, 0),$$

$$\phi(x, \mu) = \int_0^{\frac{(x-z)}{\mu}} e^{-t} f(x - t\mu, \mu)\,dt, \qquad \begin{array}{l} z = a \text{ for } \mu > 0, \\ z = b \text{ for } \mu < 0. \end{array}$$

It follows that ϕ is continuous except perhaps at (a, 0) and (b, 0). The boundary conditions

$$\phi(a, \mu) = 0 \text{ for } \mu > 0, \quad \phi(b, \mu) = 0 \text{ for } \mu < 0,$$

are satisfied and

$$\phi(x,\mu) \to \phi(a,0) \qquad \text{as } x \to a, \ \mu \to 0-,$$

$$\phi(x,\mu) \to \phi(b,0) \qquad \text{as } x \to b, \ \mu \to 0+,$$

$$\phi(x,\mu) - (1 - e^{(a-x)/\mu})\phi(a,0) \to 0 \qquad \text{as } x \to a, \ \mu \to 0+,$$

$$\phi(x,\mu) - (1 - e^{(b-x)/\mu})\phi(b,0) \to 0 \qquad \text{as } x \to b, \ \mu \to 0-.$$

Thus, R(M), the range of M, consists of bounded functions which are continuous except perhaps at two points. By routine arguments, L and L_n map R(M) into C(X). Define the domains of L and L_n by $D(L) = D(L_n) = R(M)$. Then the operators $K = LM$ and $K_n = L_nM$ map C(X) into C(X).

It follows easily from $K_n - K = (L_n - L)M$ and the properties of R(M) that $K_n \to K$. To prove that K is compact and $\{K_n\}$ is collectively compact, consider $Kf = LMf$ and $K_nf = L_nMf$ for $\|f\| \leq 1$. As above, let $\phi = Mf$. Then for any $\varepsilon > 0$ and any $\nu > 0$ there exists $\delta = \delta(\varepsilon, \nu)$ independent of f such that

$$|\phi(x, \mu) - \phi(x', \mu)| < \varepsilon \qquad \text{for } |x - x'| < \delta, \ |\mu| > \nu.$$

This is an equicontinuity property in x, uniform for $|\mu| > \nu$. Essentially, M is a compact operator with respect to x, uniformly for each $\nu > 0$. Similarly, the operator L is compact with respect to μ uniformly for $a \leq x \leq b$. Examination of $(L\phi)(x, \mu) - (L\phi)(x, \mu')$ and $(L\phi)(x, \mu) - (L\phi)(x', \mu)$ reveals that the sets

$$\{Kf = LMf : \|f\| \leq 1\}, \qquad \{K_n f = L_n Mf : \|f\| \leq 1, \ n \geq 1\}$$

are bounded and equicontinuous. Therefore, K is compact and K_n is collectively compact.

The error bounds in the collectively compact theory depend particularly on

$$\|(K_n - K)g\| = \|(L_n - L)Mg\| ,$$

$$\|(K_n - K)K\| = \|(L_n - L)MLM\| ,$$

$$\|(K_n - K)K_n\| = \|(L_n - L)ML_nM\| .$$

Calculations based on these equations should yield sharper bounds than would be obtained otherwise.

5. A Three-Dimensional Anisotropic Transport Problem

Let Γ be a domain in R^3 with boundary $\partial\Gamma$ which, at least piece-wise, has a well-defined tangent plane at each point. Directions will be denoted by $\Omega \in S$, the unit sphere in R^3. Consider the problem for $\phi(x, \Omega)$, $x \in \Gamma$, $\Omega \in S$:

$$\Omega \cdot \nabla \phi(x, \Omega) + \sigma(x, \Omega)\phi(x, \Omega) = f(x, \Omega),$$

(5.1)

$$f(x, \Omega) = \int_S p(x, \Omega, \Omega')\phi(x, \Omega')d\Omega' + g(x, \Omega),$$

with $\phi(x, \Omega) = 0$ for $x \in \partial\Gamma$ and Ω an incoming direction. The gradient operator ∇ acts with respect to x. Assume that f, g and p are continuous and and that p is nonnegative. In operator form (5.1) becomes

(5.2) $$D\phi = f, \qquad f = L\phi + g.$$

Motivated by the fact that $\Omega \cdot \nabla$ is the directional derivative in the direction Ω, we introduce additional notation as illustrated in the following diagram.

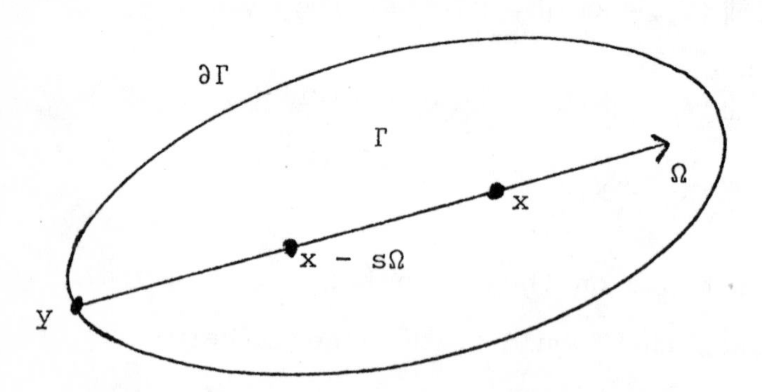

The ray through x in the direction $-\Omega$ is given by $x - s\Omega$, $s \geq 0$. Let $d(x, \Omega) = \|x - y\|_2$, the Euclidean distance from x to $\partial\Gamma$ along the ray.

An equivalent formulation of $D\phi = f$ is

$$(5.3) \quad -\frac{d}{ds}\phi(x - s\Omega,\Omega) + \sigma(x - s\Omega,\Omega)\phi(x - s\Omega,\Omega) = f(x - s\Omega,\Omega).$$

Solve (5.3) to obtain

$$(5.4) \quad \phi(x,\Omega) = \int_0^{d(x,\Omega)} f(x - s\Omega,\Omega)\, e^{-\int_0^s \sigma(x - r\Omega,\Omega)\,dr}\, ds.$$

Thus, $M = D^{-1}$ exists as an operator on $C(X)$, where $X = \overline{\Gamma} \times S$, and $\phi = Mf$ is given by (5.4). The original problem now becomes

$$(5.5) \quad \phi = Mf, \qquad (I - K)f = g, \qquad K = LM,$$

where L is the integral operator in (4.1).

Discrete ordinates approximations $\phi_n(x, \Omega)$ satisfy

$$(5.6) \quad D\phi_n = f_n, \qquad f_n = L_n\phi_n + g,$$

where L_n is a numerical integration approximation to L defined in terms of a positive convergent quadrature rule. Now-familiar reasoning yields the equivalent formulation of (5.6),

$$(5.7) \quad \phi_n = Mf_n, \qquad (I - K_n)f_n = g, \qquad K_n = L_nM.$$

The tasks are the same as before, namely to prove that $(I - K)^{-1}$ exists, $(I - K_n)^{-1}$ exists for n sufficiently large, and

$$(I - K_n)^{-1} \to (I - K)^{-1},$$

which implies that $f_n \to f$ and $\phi_n \to \phi$ uniformly. If $\|K\| < 1$ and $K_n \to K$, then these conclusions follow as in Section 3. If also K is compact and $\{K_n\}$ is collectively compact, then the error bounds of Section 4 are obtained. This program will be undertaken in a forthcoming sequel.

The collectively compact operator approximation theory may also be applicable to the case where the approximations are made with respect to the spatial variable, either by approximating the detailed variation of $\sigma(x, \Omega)$ and $p(x, \Omega, \Omega')$ with respect to position x by appropriate smoothed functions, and/or by carrying out the spatial integrations by some quadrature approximation. In this more general setting, we have M approximated by M_j as well as L approximated by L_n. Thus, it will be necessary to consider $L_n M_j \to LM$ in an appropriate sense.

6. A Positive Operator Approach

Explicit results for the three dimensional problem can be obtained rather easily if an additional assumption is made. Consider the problem of Section 5 in the form

$$(6.1) \qquad T\phi = g, \qquad T = D - L$$

under the restriction

$$(6.2) \qquad \sigma(x,\Omega) - \int_S p(x,\Omega,\Omega')d\Omega' > c > 0, \; x \in \Gamma, \; \Omega \in S.$$

Physically, this requirement means that absorption dominates production by fission, i.e., that $\sigma_a > (\nu - 1)\sigma_f$, where σ_a and σ_f are the absorption and fission cross sections, and ν is the mean number of neutrons produced produced per fission. This inequality is always satisfied in a non-multiplying medium $(\sigma_f = 0)$ as long as some absorption is present $(\sigma_a > 0)$; however, it may not be satisfied in some nuclear reactor applications.

Suppose that for each $g \in C(X)$ the equation $T\phi = g$ has a unique solution $\phi = T^{-1}g$ such that $\phi \geq 0$ if $g \geq 0$. Then T^{-1} is a positive operator from $C(X)$ to an appropriate solution space. Let $e(x, \Omega) \equiv 1$ on X. Then (6.2) is equivalent to

$$(6.3) \qquad Te > c > 0 \quad \text{on} \quad X.$$

By an elementary argument, $\phi = T^{-1}g$ is bounded for each $g \in C(X)$ and $\|\phi\| < c^{-1}\|g\|$ in terms of the sup norm. Thus, T^{-1} maps $C(X)$ onto a subspace of the bounded functions on X, T^{-1} is bounded, and $\|T^{-1}\| = \|T^{-1}e\| < 1/c$.

The discrete ordinates approximations ϕ_n satisfy

$$(6.4) \qquad T_n\phi_n = g, \qquad T_n = D - L_n.$$

As above, suppose that T_n^{-1} exists and is a positive operator on $C(X)$. Since $L_n \to L$, we have $T_n \to T$ and, in particular, $T_ne \to Te$. Therefore, for n sufficiently large,

$$(6.5) \qquad T_n e > c > 0 \quad \text{on} \quad X,$$

T_n^{-1} is bounded and $\|T_n^{-1}\| < 1/c$. Now,

$$T_n^{-1} - T^{-1} = T_n^{-1}(T - T_n)T^{-1}.$$

Let $\phi = T^{-1}g$ and $\phi_n = T_n^{-1}g$. Then

$$\phi_n - \phi = T_n^{-1}(T - T_n)\phi,$$

$$\|\phi_n - \phi\| \leq \|T_n^{-1}\| \; \|T_n\phi - T\phi\|,$$

$$\|\phi_n - \phi\| \to 0.$$

The error bound for $\|\phi_n - \phi\|$ is of mixed type, neither purely theoretical nor purely practical.

A special case of this scheme was carried out by Madsen [14] in a classical analysis setting.

7. A Time-Dependent Transport Problem

Wilson [20] investigated the discrete ordinates method for a generalized transport equation in several space dimensions:

$$(7.1) \qquad \frac{\partial\phi(t,x,v)}{\partial t} + v \cdot \nabla\phi(t,x,v) + \|v\|_2 \, \sigma(t,x,v)\phi(t,x,v)$$

$$= \int_{\mathcal{B}} p(t,x,v,v')\phi(t,x,v')dv' + g(t,x,v).$$

Independent variables are time $t \in (0, T)$, position $x \in \Gamma$, a rectangle in R^n, and velocity $v \in \mathcal{B}$, a ball in R^n. The gradient acts with respect to x. An initial condition specifies $\phi(0, x, v)$. Partial specular reflection on $\partial\Gamma$, ranging from full reflection to an exterior vacuum, was assumed by Wilson.

In operator form (7.1) becomes $D\phi = L\phi + g$. Discrete ordinates approximations ϕ_n satisfy equations $D\phi_n = L_n\phi_n + g$, where L_n is a numerical integration approximation to L with a convergent positive quadrature rule.

Wilson [19] obtained existence and uniqueness theorems for weak and strong solutions of a general problem which includes both $D\phi = L\phi + g$ and $D\phi_n = L_n\phi_n + g$ as special cases. The analysis involves an equivalent formulation of the problem as an integral equation which is of Volterra type in t. Direct iteration yields the existence and uniqueness of ϕ and ϕ_n, plus inequalities of the form

$$(7.2) \qquad \|\phi_n(t)\| \leq (a\|\phi_n(0)\| + b\|g\|)\, e^{(c\|\sigma\| + d\|L_n\|)t},$$

with max norms on the functions over undisplayed arguments.

Since $D(\phi_n - \phi) = L_n(\phi_n - \phi) + L_n\phi - L\phi$ with zero initial conditions, (7.2) implies

$$(7.3) \qquad \|\phi_n(t) - \phi(t)\| \leq b\|L_n\phi - L\phi\|\, e^{(c\|\sigma\| + d\|L_n\|)t},$$

$$(7.4) \qquad \phi_n \to \phi \quad \text{uniformly as} \quad n \to \infty.$$

The error bound in (7.3) is of mixed type.

8. Conclusions

In the preceding, we have attempted to survey developments in the investigation of convergence properties and error bounds for the discrete ordinates approximations to the transport equation. This summary has been carried out in an abstract setting which, we believe, serves to unify and clarify much previous work in this area. The direction of future investigations is also clearly indicated.

While the first studies of convergence properties of the discrete ordinates methods were carried out nearly fifteen years ago, in an astrophysical setting, they have generally not come to the attention of workers interested in practical applications of the method to nuclear reactor computational problems. In fact, as recently as 1968, a standard work on reactor computing methods refers to the convergence of the discrete ordinates method as an unsolved problem.

A number of practical implications follow from the results surveyed above. For example, the equivalence of the spherical harmonics (P_N) and Gauss-quadrature methods for slab geometry transport problems (in the sense that the solutions agree at the quadrature points) establishes the convergence of, and provides error bounds for, the P_N solutions at the quadrature points.

Hopefully, future extensions of the results surveyed here to include three-dimensional systems and discretization of the spatial variable will provide convergence proofs and practical error estimates which will be of use even in the most complicat practical applications of discrete ordinates methods.

References

Anselone, P. M. Integral Equations of the Schwarzschild-Milne Type. J. Math. and Mech., 7 (1958), pp. 557-570.

Anselone, P. M. Convergence of the Wick-Chandrasekhar Approximation Technique in Radiative Transfer. Astrophys. J., 128 (1958), pp. 124-129.

Anselone, P. M. Convergence of the Wick-Chandrasekhar Approximation Technique in Radiative Transfer, II. Astrophys. J., 130 (1959), pp. 881-883.

Anselone, P. M. Convergence of Chandrasekhar's Method for Inhomogeneous Transfer Problems. J. Math. and Mech., 10 (1961), pp. 537-546.

Anselone, P. M. Convergence of Chandrasekhar's Method for the Problem of Diffuse Reflection. Royal Astr. Soc. Monthly Notices, 120 (1960), pp. 498-503.

Anselone, P. M. Collectively Compact Operator Approximation Theory and Applications to Integral Equations. Prentice-Hall, Inc., Englewood Cliffs, N. J., 1971.

Chandrasekhar, S. On the Radiative Equilibrium of a Stellar Atmosphere, II. Astrophys. J., 100 (1944), pp. 76-86.

Chandrasekhar, S. Radiative Transfer. Oxford University Press, 1950.

Hopf, E. Mathematical Problems of Radiative Equilibrium. Cambridge University Press, 1934.

10. Keller, H. B. Approximate Solutions of Transport Problems. SIAM J. Appl. Math., 8 (1960), pp. 43-73.

11. Keller, H. B. On the Pointwise Convergence of the Discrete Ordinate Method. SIAM J. Appl. Math., 8 (1960), pp. 560-567.

12. Keller, H. B. Convergence of the Discrete Ordinate Method for the Anisotropic Scattering Transport Equation. Provisional Intl. Comp. Center, Rome Symposium, Birkhauser-Verlag, Basel/Stuttgart, 1960, pp. 292-299.

13. Kofink, W. Studies of the Spherical Harmonics Method in Neutron Transport Theory, Nuovo Cimento 10, 9 (1958), Supplemento, pp. 497-541.

14. Madsen, N. K. Pointwise Convergence of the Three-Dimensional Discrete Ordinate Method. SIAM J. Num. Anal., 8 (1971), pp. 266-269.

15. Nelson, P. Convergence of the Discrete Ordinates Method for Anisotropically Scattering Multiplying Particles in a Subcritical Slab. SIAM J. Num. Anal., 10 (1973), pp. 175-1

16. Nestell, M. K. The Convergence of the Discrete Ordinates Method for integral Equations of Anisotropic Radiative Transfer. Tech. Report 23, Math Dept., Oregon State University, 1965.

17. Wendroff, B. On the Convergence of the Discrete Ordinate Method. SIAM J. Appl. Math., 8 (1960), pp. 508-511.

18. Wick, G. C. Über Ebene Diffusions Problem. Z. für Physik 121 (1943), pp. 702-718.

19. Wilson, D. G. Time-Dependent Linear Transport, I., J. Math. Anal. Appl., to appear.

20. Wilson, D. G. Time Dependent Linear Transport, III; Convergence of Discrete Ordinate Method, submitted for publication.

The Numerical Solution of the Equations for Rotating Stars

J. F. G. Auchmuty

Department of Mathematics
Indiana University
Bloomington, Indiana 47401

1. Introduction.

In this paper, a method of numerically approximating the solutions of the equations for some models of rotating stars will be developed.

The simplest, realistic, model of a star is that of a compressible, inviscid fluid which is in equilibrium under self-gravitational and centrifugal forces. The equations are the Euler equations

$$f'(\rho)\,\text{grad}\,\rho = \rho\left[\text{grad}(B\rho) + \frac{L(m_\rho(r))}{r^3}\hat{i}_r\right]. \qquad (1.$$

Here we are using cylindrical polar coordinates (r,θ,z); $\hat{i}_r$ is the unit vector in the radial direction and

$$m_\rho(r) = \frac{1}{M}\int_0^r\int_{-\infty}^{\infty}\int_0^{2\pi} s\,\rho(s,\theta,z)\,d\theta\,dz\,ds\,. \tag{1.2}$$

is the proportion of the mass inside a cylinder of radius r.

The equation of state of the fluid is

$$p = f(\rho) \tag{1.3}$$

and the gravitational potential is

$$B\rho(x) = \int_{\mathbb{R}^3} \frac{\rho(y)}{|x-y|}\,dy \tag{1.4}$$

The function $L(m)$ represents the square of the angular momentum per unit mass. For a more detailed discussion of these equations see [3] or [8].

For our problem, the functions f and L are prescribed and one wants to find a density distribution $\rho(x)$ which satisfies (1.1). Equation (1.1) is an integro-differential equation and in the non-rotating case $(L \equiv 0)$, one can specify solutions of (1) by also giving $\rho(0)$ see Chandrasekhar [3] chapters 3 and 4.

When the star is rotating (i.e., $L \not\equiv 0$), the theory is much more complicated. A detailed exposition is given in Auchmuty and Beals [2] and a summary appears in [1]. In these papers, it is shown that the solutions of (1) are given by the local extrema of the functional

$$E(\rho) = \int_{\mathbb{R}^3} A(\rho(x))\,dx + \tfrac{1}{2}\int_{\mathbb{R}^3} \rho(x)\frac{L(m_\rho(r))}{r^2}\,dx - \tfrac{1}{2}\int_{\mathbb{R}^3} \rho(x)\,B\,\rho(x)\,dx \qquad (1$$

subject to the conditions

$$\rho(x) \geq 0 \qquad (1.6)$$

and

$$\int_{\mathbb{R}^3} \rho(x)\,dx = M \qquad (1.7)$$

Here $A(s) = s\int_0^s f(t)\,t^{-2}\,dt$.

The class of allowable functions ρ are those real-valued functions which are defined and measurable on $\mathbb{R}^3$, obey (1.6) and (1.7), are axially symmetric and obey

$$\int_{\mathbb{R}^3} A(\rho(x))\,dx < +\infty . \qquad (1.8)$$

One also requires f , L to satisfy certain conditions which will be given in the next section.

The problem which will be treated in this paper is the problem of finding local minima of the functional (1.5) subject to (1.6) and (1.7).

These problems have a long history, dating back to the time of Maclaurin. The first methods assumed that the equidensity and equipotential surfaces had special forms with spheroidal, ellipsoidal or toroidal symmetries. More recently, perturbation methods have been developed and methods of successive approximation (such as self-consistent-field methods) have been quite successful. For historical references and a description of the self-consistent field method see Ostriker and Mark [8]. Other recent methods are described and compared in Papaloizou and Whelan [9]. The method to be described here is quite different to all of these. Details of extensive computations on this problem will appear elsewhere.

The first section will summarize the variational characterization of the solutions of the equations for stellar structure. The conditions on the equation of state for there to be solutions will be given, and the properties of the solutions will be described. The second section is devoted to giving a discrete version of the problem and to describing an iterative method of solving it. Finally we show that the solutions of the discrete problems converge to those of the original problem.

The author is very grateful to Norman Lebovitz for many helpful discussions over a long period of time about this work.

2. The Continuous Problem.

This section is devoted to a description of the relevant mathematical results about our variational problem.

Assume that f and L obey the following conditio

P_1: $f(s)$ is non-negative, continuous and strictly increasing for $s > 0$.

P_2: $f(s) \geq c\, s^{\gamma}$ for $s \geq 1$ and some constants $c > 0$ and $\gamma > 6/5$.

P_3: $\int_0^1 f(s)\, s^{-2}\, ds < \infty$.

P_4: $L(m)$ is non-negative and absolutely continuous for $0 \leq m \leq 1$ and $L(0) = 0$.

P_5: $\lim_{s\to 0} f(s)\, s^{-4/3} = 0$ and $\lim_{s\to\infty} f(s)\, s^{-4/3} = +\infty$.

Let W be the set of all measurable, real-valued functions on $\mathbb{R}^3$ which are axially symmetric, obey (1.6), (1.7) and (1.8), and which are centered in that

$$\int_{z(x)<0} \rho(x)\, dx = \int_{z(x)>0} \rho(x)\, dx$$

Under these conditions one can prove (see [2]) the following result.

Theorem 2.1. Suppose $P_1 - P_5$ holds. Then there exists a function ρ_0 in W which minimizes $E(\rho)$ among all functions in W . Any such ρ_0 has compact support, is continuous on $\mathbb{R}^3$ and continuously differentiable wherever it is positive and satisfies the variational inequality

$$A'(\rho(x)) - B\rho(x) + \int_{r(x)}^{\infty} \frac{L(m_\rho(s))}{s^3} ds \geq \lambda \qquad x \text{ in } \mathbb{R}^3 \tag{2.1}$$

Equality holds in (2.1) at any point x where $\rho(x) > 0$. λ is a negative constant.

When one replaces P_5 by the condition

$$P_5': \quad \lim_{s\to 0} f(s)s^{-4/3} = 0 \qquad \lim_{s\to\infty} f(s)s^{-4/3} = K > 0$$

then theorem 1 no longer holds. Instead one has the followin result

<u>Theorem 2.2.</u> Suppose $P_1 - P_4$ and P_5' holds. Then there exists a constant $M_0 > 0$, depending only on K, such that if $M < M_o$ there is a function ρ_0 in W which minimizes $E(\rho)$ on W. This ρ_0 has all the properties of the minimizing function in Theorem 1.

The conditions $P_1 - P_5$ or P_5' allow many of the usual equations of state and angular momentum distributions.

The polytropic equation of state

$$p = K\rho^\gamma \tag{2.2}$$

obeys $P_1 - P_3$ and P_5 provided $\gamma > 4/3$, $K > 0$.

The equation of state for white dwarf stars, given parametrically by

$$\begin{aligned} p &= Au(2u^2 - 3)(u^2+1)^{\frac{1}{2}} + 3\sinh^{-1}u \,. \\ \rho &= Bu^3 \end{aligned} \tag{2.3}$$

obeys $P_1 - P_3$ and P_5' with $K = 2AB^{-4/3}$.

Theorems 2.1 and 2.2 are sharp. If $L = 0$ and the equations of state is given by (2.2) with $6/5 < \gamma < 4/3$, then the functional $E(\rho)$ is unbounded below on W. Similarly when $L = 0$ and the equation of state is given by (2.3) and $M > M_0$, the functional $E(\rho)$ is again unbounded below. In [2] it was shown that M_0 is the Chandrasekhar limiting mass for white dwarfs.

Using methods similar to those in [2] one can also show that for rotating polytropic stars, there is a minimum mass M_1. That is, there is no solution of the variational problem for these models if $M < M_1$.

To solve these problems numerically, one makes essential use of the fact that the solutions have compact support and have some regularity. This enables one to do all the calculations on bounded sets in $\mathbb{R}^3$ and to use Sobolev spaces of functions.

For simplicity, for the rest of this paper we shall only use the polytropic equation of state (2.2) with $\gamma > 4/3$. We shall also assume L is uniformly Lipshitz continuous on $[0,1]$ and L is differentiable at 0. As can be seen, analogous results hold if other equations of state are used, but the analysis becomes much more complicated.

Given f and L, Theorem 2.1 implies that one can find $R > 0$ such that

$$\rho_0(x) = 0 \qquad \text{for} \quad |x| \geq R$$

where ρ_0 is any minimizing function for the variational problem.

Let B_R be the ball of radius R in $\mathbb{R}^3$ and $L^p(B_R)$ be the usual Lebesgue space with the norm given by

$$\|u\|_p^p = \int_{B_R} |u(x)|^p \, dx$$

P.T.O

Similarly let $W_0^{1,p}(B_R)$ be the usual Sobolev space of L^p functions whose first destributional derivatives are in $L^p(B_R)$ and which are zero, in a generalized sense on the sphere of radius R. Then by an appropriate adaption of the regularity argument in [2] one can show that the solutions ρ_0 of the variational problem lie in $W_0^{1,p}(B_R$ for $1 \leq p \leq \infty$.

Let $W_R = \{u \in W\colon u(x) = 0 \text{ a.e. for } |x| \geq R\}$. Since

$$\int_{B_R} A(\rho)\, dx = \frac{K}{\gamma - 1} \int_{B_R} \rho^\gamma(X)\, dx = \frac{K}{\gamma - 1} \|\rho\|_\gamma^\gamma \tag{2.4}$$

when f is given by (2.2), one sees that W_R may be considered as a subset of $L^\gamma(B_R)$.

This may be summarized by the following theorem.

Theorem 2.3. Suppose the equation of state is given by (2.2) with $\gamma > 4/3$ and that L obeys P_4. Then there are solutions of each of the following problems

(I). Minimize E among all functions in W.

(II). Minimize E on the subset W_R of $L^\gamma(B_R)$ where R is sufficiently large

(III). Minimize E on the subset $W_R \cap W_0^{1,p}(B_R)$ of $W_0^{1,p}(B_R)$ where R is sufficiently large and $1 \leq p \leq \infty$.

Moreover if ρ_0 is the solution of any one of these problems, it is also a solution of any of the other problems.

To prove results about the numerical approximations to solutions of this problem one needs the following definitions.

$C(B_R)$ is the space of all continuous real-valued functions on B_R. It is a Banach space with respect to the norm

$$\|u\|_\infty = \sup_{x \in B_R} |u(x)| .$$

If X is a Banach space and V is a subset of X, then a functional $f: V \to \mathbb{R}$ is said to be weakly (strongly) continuous at a point x in V if for any sequence $\{x_n\}$ in V which converges weakly (strongly) to x in X, one has

$$f(x) = \lim_{n\to\infty} f(x_n) .$$

The functional is weakly (strongly) continuous on a subset W of X provided it is weakly (strongly) continuous at each point in W.

<u>Theorem 2.4.</u> Let $C = W_R \cap C(B_R)$. Then the functional E given by (1.5) and (2.2) is strongly continuous on C, when the topology on C is induced by $C(B_R)$ and L is Lipshitz continuous. Moreover if $\{u_n\}$ is a sequence in C which converges strongly to u in C then there exists a constant K such that

$$|E(u) - E(u_n)| \leq K\|u - u_n\|_\infty . \tag{2.5}$$

<u>Proof.</u> Suppose $\{u_n\}$ converges strongly to u in C. Then there is a constant c such that $\|u_n\|_\infty \leq c$ for all n and $\|u\|_\infty \leq c$.

Now

$$\int_{B_R} [u(x)^{\gamma} - u_n(x)^{\gamma}]dx \le \gamma c^{\gamma-1} \int_{B_R} |u(x) - u_n(x)| dx \le c\|u - u_n\|_\infty \tag{2.6}$$

Here, and henceforth, c is a positive constant, not always the same.

Let $m_n(r) = m_{u_n}(r)$. Then $m_n(r)$ converges to $m(r)$ uniformly for $0 \le r \le R$ and

$$m_n(r) = \frac{4\pi}{M} \int_0^r \int_0^R s\,\rho(s,z)\,ds\,dz \le c\,r^2$$

Similarly $m_u(r) \le c\,r^2$.

Also $$|L(m_n(r)) - L(m_u(r))| \le c|m_u(r) - m_n(r)| \le c\|u - u_n\|_\infty r^2$$

using Lipshitz continuity and Hölder's inequality. Therefore

$$\left| \int_{B_R} u \frac{L(m_u(r))}{r^2} dx - \int_{B_R} u_n \frac{L(m_n(r))}{r^2} dx \right| =$$

$$= \left| \int_{B_R} u \left[\frac{L(m_u(r))}{r^2} - \frac{L(m_n(r))}{r^2} \right] dx + \int_{B_R} (u - u_n) \frac{L(m_n(r))}{r^2} dx \right|$$

$$\le c\|u - u_n\|_\infty .$$

Finally $B: C(B_R) \to C(B_R)$ is a compact operator and

$$\left|\int_{B_R} u(x)Bu(x)dx - \int u_n(x)Bu_n(x)dx\right| =$$

$$= \left|\int_{B_R} Bu(x)[u(x) - u_n(x)]dx + \int_{B_R} u_n(x)[Bu(x) - Bu_n(x)]dx\right| \tag{2.8}$$

$$\leq c\|u - u_n\|_\infty .$$

Adding (2.6), (2.7) and (2.8) one gets (2.5) so E is strongly continuous on C.

<u>Corollary.</u> Let $C^p = W_R \cap W_0^{1,p}(B_R)$. If E, L as in Theorem 2.4 and $p > 3$, then E is a weakly continuous functional on C^p.

<u>Proof.</u> From the Sobolov imbedding theorem, the map $i: W_0^{1,p}(B_R) \to C(B_R)$ in compact if $p > 3$.

Thus if $\{u_n\}$ converges weakly to u in C^p, it converges strongly to u in C and so $E(u_n) \to E(u)$ as $n \to \infty$. Thus the result.

Finally, in actual computations, we use the fact that is u is in W, then so also is

$$u_\epsilon(x) = \epsilon^{-3}u(x/\epsilon) \tag{2.9}$$

A simple change of variables gives

$$E(u_\epsilon) = \epsilon^{3-3\gamma}E_1 + \epsilon^{-2}E_2 - \epsilon^{-1}E_3 \tag{2.10}$$

here $E_1 = \int u^\gamma(x)\,dx$, $E_2 = \frac{1}{2}\int u \frac{L(m_u(r))}{r^2}\,dr$ and $E_3 = \frac{1}{2}\int u\,Bu\,dx$.

Thus for any given u in W with compact support ne may find the value of ϵ for which $E(u_\epsilon)$ is minimized. his requires only a simple algebraic computation but it ives quite good approximations to the radius of the star.

. Discretization of the Problem.

To approximate the solutions of this variational roblem, one would like to have finite dimensional analogues f the problem. In this section we shall establish, and ndicate how to solve, an analog of the problem of minimizing in W_R. This analog is obtained by using the values of ie unknown solution only at points of a grid on B_R and by eplacing the integrals in expression (1.5) for E by pproximate integration formulae.

First one notes that the problem is essentially o dimensional as we are only interested in axially symmetric solutions. Because the functions are centered we shall also assume

$$u(r,z) = u(r,-z) \qquad \text{for all } z$$

Thus instead of studying functions defined on the ball B_R of radius R in $\mathbb{R}^3$ it suffices to consider functio defined on the square $[0,R] \times [0,R]$ in $\mathbb{R}^2$.

Let $\Gamma = \{(r_i, z_j): 0 \le r_i \le R\ ,\ 0 \le z_j \le R\ ,\ 0 \le i \le M\ ,\ 0 \le j \le N\}$ be a rectangular grid on $[0,R] \times [0,R]$ and assume $r_0 = z_0 = 0$ $r_M = z_N = R$.

Associated with this grid is an integration formula

$$I_\Gamma(u) = \sum_{i=0}^{M} \sum_{j=0}^{N} \omega_{ij} u(r_i, z_j) \tag{3.1}$$

which is an approximation to the integral

$$I(u) = \int_0^R \int_0^R r\, u(r,z)\, dr\, dz \tag{3.2}$$

The mesh of the grid is

$$|\Gamma| = \max\{\tfrac{1}{2}(r_i - r_{i-1})^2 (z_j - z_{j-1}) : 0 \le i \le M\ ,\ 0 \le j \le N\}\ .$$

Let $k = (M+1)(N+1)$ and v be a vector in $\mathbb{R}^k$

where $$v = (v_{00}, v_{01}, v_{02}, \dots, v_{0N}, v_{10}, \dots, v_{MN}) \tag{3.3}$$

and consider the functional $E_\Gamma : \mathbb{R}^k \to \mathbb{R}$ defined by

$$E_\Gamma(v) = \sum_{i=0}^{M} \sum_{j=0}^{N} \omega_{ij} A(v_{ij}) - \tfrac{1}{2} \sum_{i=0}^{M} \sum_{j=0}^{N} \sum_{k=0}^{M} \sum_{\ell=0}^{N} b_{ijk\ell} v_{ij} v_{k\ell} + $$

$$+ \tfrac{1}{2} \sum_{i=1}^{M} \frac{L(m_i)}{r_i^2} \left(\sum_{j=0}^{N} \omega_{ij} v_{ij} \right) \tag{3.4}$$

Here $m_0 = 0$ and $m_i = \frac{4\pi}{M} \sum_{k=0}^{i} \left(\sum_{j=0}^{N} \omega_{kj} v_{kj} \right) \quad 1 \le i \le M$.

The array $b_{ijk\ell}$ is obtained by discretizing the integral

$$\int_{B_R} \int_{B_R} B_h(x,y)\, u(x)\, dx\, dy$$

where $$B_h(x,y) = \begin{cases} |x-y|^{-1} & \text{if } |x-y| \ge h \\ h^{-1} & \text{if } |x-y| \le h \end{cases}$$

and h is the mesh of Γ. When $|x-y| > h$, a calculation gives

$$b_{ijk\ell} = \omega_{ij}\, \omega_{k\ell}\, e(x_{ij}, x_{k\ell})$$

where e is a complicated function involving elliptic integrals and where $x_{ij} = (r_i, z_j)$ and $x_{k\ell} = (r_k, z_\ell)$.

Let C_k be the subset of $\mathbb{R}^k$ of all vectors obeying

$$v_{ij} \geq 0 \quad \text{for} \quad 0 \leq i \leq M, \; 0 \leq j \leq N \tag{3.5}$$

and

$$\sum_{i=0}^{M} \sum_{j=0}^{N} \omega_{ij} v_{ij} = \frac{M}{4\pi} \tag{3.6}$$

The discrete variational problem is:

(DP). Minimize E_Γ on the subset C_k of $\mathbb{R}^k$.

The set C_k is a compact, convex subset of $\mathbb{R}^k$ and E_Γ is a continuously differentiable function, so this discrete variational problem always has a solution.

To find the solution we use a gradient method. There is an extensive literature on such methods, but the following iterative scheme has proven very successful for this problem.

Choose $v^{(0)}$ in C_k and assume $v^{(1)}, v^{(2)}, \ldots, v^{(m)}$ have been found. Then let

$$\tilde{w}_{ij}^{(m)} = \begin{cases} 0 & \text{if } v_{ij}^{(m)} = 0 \\ -\dfrac{\partial E_\Gamma}{\partial v_{ij}}(v^{(m)}) & \text{if } v_{ij}^{(m)} > 0 \end{cases}$$

Here, as in the rest of this section $v_{ij}^{(m)}$ or $w_{ij}^{(m)}$ represent the i,j-th component of $v^{(m)}$ or $w^{(m)}$; $0 \le i \le M$, $0 \le j \le N$.

Let $\mu_m = I_\Gamma(\tilde{w}^{(m)}) = \sum_{i=0}^{M} \sum_{j=0}^{N} \omega_{ij} \tilde{w}_{ij}^{(m)}$ and

$$w_{ij}^{(m)} = \begin{cases} 0 & \text{if } v_{ij}^{(m)} = 0 \\ \tilde{w}_{ij}^{(m)} - \mu_m & \text{if } v_{ij}^{(m)} > 0 \end{cases}$$

Fix $0 < \tau < \infty$ depending on the constants of Holder continuity of the derivative of $E_\Gamma(v)$ and write

$$\tilde{v}_{ij}^{(m+1)} = \begin{cases} v_{ij}^{(m)} + \tau w_{ij}^{(m)} & \text{if } v_{ij}^{(m)} + \tau w_{ij}^{(m)} > 0 \\ 0 & \text{otherwise} \end{cases}$$

Let $M_{m+1} = 4\pi I_\Gamma(\tilde{v}^{(m+1)})$, then define

$$v^{(m+1)} = \frac{M}{M_{m+1}} \tilde{v}^{(m+1)}$$

This construction ensures that $v^{(m+1)}$ is in C_k . Moreover with the appropriate choice of τ , one can show (as in Daniel [5] Chapter 4) that the sequence $\{v^{(m)}\}$ converges to a conditional critical point $\hat{v}$ of E_Γ . In other words there exists a real number λ such that

$$\frac{\partial E_\Gamma}{\partial v_{ij}}(\hat{v}) \geq \lambda \omega_{ij} \qquad \text{for all} \quad 0 \leq i \leq M \text{ , } 0 \leq j \leq N$$

and
$$\frac{\partial E_\Gamma}{\partial v_{ij}}(\hat{v}) = \lambda \omega_{ij} \qquad \text{if} \quad \hat{v}_{ij} > 0 \text{ .}$$

This is the discrete version of (2.1) and so $\hat{v}$ is considered to be an approximation to the solution of the variational problem. One can obtain a density distribution for the star by interpolating a function taking the values $\hat{v}_{ij}$ at the grid points (r_i, z_j) .

Originally the author tried to method of steepest descent on this problem. It did not work very well as E_Γ i not very sensitive to changes in v . Often the change in E_Γ along a path of steepest descent was negligible, so it wa not possible to choose with any confidence, a minimizing valu

In actual practice one also has to determine the size R of a side of the region being considered. To do this one uses (2.10). At each iteration, one calculates the discrete analogs of E_1, E_2, and E_3 with the density distribution given by $v^{(m)}$ and with a given $R = R_m$. One minimizes (2.10) as a function of ϵ . If $\epsilon = \epsilon_0$ at th minimum and ϵ_0 is not close to 1 one then adjusts the radius to be R_{m+1} where $R_{m+1} > \epsilon_0 R_m$, and one modifies both the grid points and the density distribution. The new density distribution is given by the discrete analog of (2.9), namely

$$v_{ij}^{(m)} = \epsilon_0^{-3} v_{ij}^{(m)}$$

4. Convergence.

In the last section a method of solving a discretization of the variational problem was described. It will now be shown that if one uses an appropriate sequence of grids and interpolation formulae, the solutions of the discrete problems will converge in a certain sense to a solution of the original problem.

To prove this a new discretization result will be used. In many respects the result is similar to theorem 5 in Daniel [6] but it provides for approximation of one of the constraints. This method also has some similarity to the theory of completely compact families of operators as discussed in Daniel [5] or [6] or Anselone [4]. This

might be expected as if one knows the support S of a soluti
of the variational problem, the from (2.1) one has

$$\frac{k\gamma}{\gamma-1}\rho(x)^{\gamma-1} - \int_{B_R}\frac{\rho(y)}{|x-y|}dy + \int_{r(x)}^{\infty}\frac{L(m_\rho(s))}{s^3}ds = \lambda$$

for x in S

and $$\rho(x) = 0 \qquad \text{for } x \text{ not in}$$

Thus one essentially has a non-linear integral equation for the density ρ.

Let X be a Banach space and $\{f_m : m \in I\}$ be a family of weakly continuous functionals defined on a subset V of X.

Let C be a weakly closed subset of X and $C \subset V$. Then the family $\{f_m\}$ is said to be weakly sequentially equicontinuous on C, if for each weakly convergent sequence $\{x_n\}$ in C with weak limit x and for each $\epsilon > 0$, there is an N such that $n > N$ implies

$$|f_m(x_n) - f_m(x)| < \epsilon \qquad \text{for all } m \text{ in } I.$$

Let $\{g_m\}$ be a family of continuous linear function on X such that

$$\lim_{m\to\infty} g_m(u) = g_0(u) \qquad \text{for all } u \text{ in } X \tag{4.1}$$

where g_0 is also continuous.

Let $D = \{u \in C: g_0(u) = \alpha\}$ and

$$D_m = \{u \in C: g_m(u) = \alpha\} \text{ where } \alpha \neq 0 . \tag{4.2}$$

Define $r_m: D \to D_m$ and $p_m: D_m \to D$ by the formulae

$$r_m(u) = [\alpha/g_m(u)]u \tag{4.3}$$

$$p_m(u) = [\alpha/g(u)]u \tag{4.4}$$

Consider the following problems:

(P_0). Minimize f_0 on D .

(P_m). Minimize f_m on D_m .

Assume that f_0 is bounded below on D and let

$$\mu_o = \inf_{u \in D} f_0(u)$$

$$\Sigma = \{u \in D: f_0(u) = \mu_0\} .$$

Similarly for each $m \geq 1$, assume f_m is bounded below on D_m and let

$$\mu_m = \inf_{u \in D_m} f_m(u)$$

$$\Sigma_m = \{u \in D_m : f_m(u) = \mu_m\} .$$

Theorem 4.1. Let C be a weakly compact subset of the reflexive Banach space X. Let $\{f_0\} \cup \{f_m : m \geq 1\}$ be weakly sequentially equicontinuous on C and assume

(i) Σ_0 and Σ_m are non-empty for all m

(ii) $\lim_{m\to\infty} f_m(u) = f_0(u)$ for all u in C.

Suppose $\{g_0\} \cup \{g_m : m \geq 1\}$ are a family of continuous linear functionals obeying (4.1) and D_0, D_m are given by (4.2). Choose x_m in D_m so that

$$f_m(x_m) \leq \mu_m + \epsilon_m \tag{4.5}$$

where $\epsilon_m > 0$ and $\lim_{m\to\infty} \epsilon_m = 0$. Then each weak limit point $\tilde{x}$ of $\{x_m\}$ minimizes f_0 on D.

Proof. Let p_m and r_m be defined by (4.3) and (4.4), and suppose $\{x_{m_j}\}$ is the subsequence of $\{x_m\}$ with $x_{m_j} \to \tilde{x}$ weakly. Then $\tilde{x}$ is in D.

If x_0 is in Σ_0 then

$$f_0(x_0) = \lim_{n\to\infty} f_n(x_0) = \lim_{n\to\infty} f_n(r_n x_0)$$

using (4.1), (ii) and the weak sequential equicontinuity.

But $\lim_{n\to\infty} f_{n_j}(r_{n_j} x_0) \geq \limsup_{n\to\infty}(f_{n_j}(x_{n_j}) - \epsilon_{n_j} \geq$ $\liminf_{n\to\infty} f_{n_j}(x_{n_j})$. Using $f_{n_j}(x_{n_j}) = f_0(\tilde{x}) + [f_{n_j}(\tilde{x}) - f_0(\tilde{x})] +$ $+ [f_{n_j}(x_{n_j}) - f_{n_j}(\tilde{x})]$, one gets $f_0(x_0) \geq f_0(\tilde{x})$ Therefore $\tilde{x}$ minimizes f_0 on D .

To apply this result to our problem one takes $X = W_0^{1,p}(B_R)$ with $p > 3$, f_0 is the energy functional E given by (1.5) and C is the set of functions in X which are centered, nonnegative, axially symmetric and obey $\|u\|_{1,p} \leq C'$ where C' is sufficiently large. Also $\alpha = M$ and $g_0(u) = \int_{B_R} u\,dx$.

The functionals f_m and g_m will be discretizations of E and g_0 . To define these discretizations one takes a family $\{\Gamma_m\}$ of square grids on $[0,R]\times[0,R]$.

Let $\Gamma_m = \{(r_i, z_j) : 0 \leq i \leq m\ ,\ 0 \leq j \leq m\}$ with $r_0 = z_0 = 0$, $r_m = z_m = R$ and let

$$I_m(u) = I_{\Gamma_m}(u) = \sum_{i=0}^{m} \sum_{j=0}^{m} \omega_{ij}^{(m)} u(r_i, z_j) \tag{4.5}$$

be the associated integration formula.

Then the functional f_m should be given by E_{Γ_m} and the functional g_m by $4\pi I_m$. E_{Γ_m} is given by (3.4) where the grid is Γ_m and the coefficients ω_{ij} are replaced by $\omega_{ij}^{(m)}$ from (4.5).

$$E_{\Gamma_m}(u) =$$

$$= \sum_{i=0}^{m} \sum_{j=0}^{m} \omega_{ij}^{(m)} A(u(r_i, z_j)) - \tfrac{1}{2} \sum_{i=0}^{m} \sum_{j=0}^{m} \sum_{k=0}^{m} \sum_{\ell=0}^{m} b_{ijk\ell}^{(m)} u(r_i, z_j) u(r_k, z$$

$$+ \tfrac{1}{2} \sum_{i=1}^{m} \frac{L(m_i)}{r_i^2} \sum_{j=0}^{m} \omega_{ij} u(r_i, z_j) \qquad (4.7)$$

The integration formulae (4.5) are said to be consistent on $C(B_R)$ provided $\lim_{m\to\infty} I_m(u) = \int_{B_R} u\,dx$ for each u in $C(B_R)$. (4.8)

A method for constructing consistent multiple integration formulae for use in this problem is given in the appendix. The following results show that if one uses a sequence of consistent integration formulae to discretize our problem, then the solutions of the discrete problems converge to a solution of our original problem.

Lemma 4.2. Suppose $E_{\Gamma_m} : C(B_R) \to \mathbb{R}$ is defined by (4.7) and the $\omega_{ij}^{(m)}$ are given by a consistent family of integration formulae. If $\{u_n\}$ converges strongly to u in $C(B_R)$, then there exists L, independent of m, such that

$|E_{\Gamma_m}(u_n) - E_{\Gamma_m}(u)| \leq L\|u - u_n\|_\infty$.

Proof. One has from the mean value theorem (writing E_m instead of E_{Γ_m})

$$|E_m(u) - E_m(v)| = \sum_{i=0}^{m} \sum_{j=0}^{m} \frac{\partial E_m}{\partial u_{ij}}((1-t)u + tv)[u_{ij} - v_{ij}]$$

$$0 < t < 1$$

$$\leq \|u - v\|_\infty \left| \sum_{i=0}^{m} \sum_{j=0}^{m} \frac{\partial E_m}{\partial u_{ij}} ((1-t)u + tv)\right| \tag{4.9}$$

But

$$\frac{\partial E_m}{\partial u_{ij}}(u) = \omega_{ij}A'(u_{ij}) - \tfrac{1}{2}\sum_{k=0}^{m} \sum_{\ell=0}^{m} b_{ijk\ell}u_{k\ell} + \tfrac{1}{2}\omega_{ij}\frac{L(m_i)}{r_i^2} +$$

$$+ \tfrac{1}{2}\sum_{k=i}^{m} \frac{L'(m_u)}{r_k^2}\, \omega_{ij}\Big(\sum_{\ell=0}^{m} \omega_{k\ell}u_{k\ell}\Big)$$

Here the primes represent differentiation.

$$\sum_{i=0}^{m}\sum_{j=0}^{m}\omega_{ij}A'(u_{ij}) \quad \text{converges to} \quad \int_{B_R} A'(u)dx = \frac{k\gamma}{\gamma-1}\int_{B_R} u^{\gamma-1}dx$$

and $\sum_{i=0}^{m}\sum_{j=0}^{m}\sum_{k=0}^{m}\sum_{\ell=0}^{m} b_{ijk\ell}u_{k\ell}$ converges to $\int_{B_R} Bu(x)dx$ as m goes to infinity because the integration formulae are consistent.

Similarly $\sum_{i=1}^{m}\sum_{j=0}^{m}\omega_{ij}\frac{L(m_i)}{r_i^2}$ converges to $\int_{B_R}\frac{L(m(r))}{r^2}dx$ and the last term in the expression converges to

$\int_{B_R}\left[\int_r^{\infty}\frac{L'(m(s))}{s^2}\frac{dm(s)}{ds}ds\right]dx$. Thus $\left|\sum_{i=0}^{m}\sum_{j=0}^{m}\frac{\partial E_m}{\partial u_{ij}}(u)\right|$ converges to

$$\left|\frac{K\gamma}{\gamma-1}\int_{B_R} u^{\gamma-1}(x)\,dx - \tfrac{1}{2}\int_{B_R} Bu(x)dx + \int_{B_R}\left(\int_r^{\infty}\frac{L(m(s))}{s^3}ds\right)dx\right| \qquad (4.$$

The functional in (4.10) is a bounded functional on $C(B_R)$. That is there exists a function $\phi\colon [0,\infty)\to[0,\infty)$ such that if $\|u\|_\infty \le c$, then expression (4.10) is less than $\phi(c)$.

Take $c = \max_{0\le t\le 1}\|(1-t)u+tv\|_\infty$ and $L > \phi(c)$ and one has the result.

eorem 4.3. If the integration formulae $\{I_m\}$ given by .5) are consistent on $C(B_R)$, then the family of functionals $\} \cup \{E_{\Gamma_m}\}$ are weakly sequentially equicontinuous on C and

$$\lim_{m\to\infty} E_{\Gamma_m}(u) = E(u) \qquad \text{for all} \quad u \in C \qquad (4.11)$$

oof. Suppose $\{u_n\}$ converges weakly to u_0 in C. en from the Sobolev imbedding theorem, $\{u_n\}$ converges rongly to u_0 in $C(S_R)$.

Lemma 4.2 then implies that the sequence $\{E_{\Gamma_m}\}$ s weakly sequentially equicontinuous on C.

(4.11) follows as if u is in C, then u is in (B_R) from the Sobolev imbedding theorem so $A(u(x))$ is in (B_R) and $u(x)Bu(x)$ is in $C(B_R)$. Also

$(x)\dfrac{L(m_u(r))}{r^2}$ is in $C(B_R)$ as L is differentiable at 0.

Thus since the integration formulae are consistent n $C(B_R)$ one has $E(u) = \lim_{m\to\infty} E_m(u)$.

Having proved this theorem, one can easily verify hat all the other conditions of Theorem 4.1 hold for this amily of discretizations of the variational problem. Thus if $\{u_m\}$ is a sequence of approximate solutions (in the sens of (4.5)) of the discrete variational problem, then any weak limit point $\tilde{u}$ of this sequence minimizes E on $W_R \cap W_0^{1,p}(B$

However if $\{u_{m_j}\}$ converges weakly to $\tilde{u}$ in $W_0^{1,p}(B_R)$ and $p > 3$ then $\{u_{m_j}\}$ converges strongly to $\tilde{u}$ in $C(B_R)$, so one sees that one actually has convergence in the sup-norm.

Appendix

Consider the problem of approximating

$$) = \int_{a_1}^{b_1}\int_{a_2}^{b_2}\cdots\int_{a_n}^{b_n} \omega_1(x_1)\omega_2(x_2)\cdots\omega_n(x_n)u(x_1,x_2,\ldots,x_n)dx_1^n \qquad \text{(A1)}$$

u in $C(\Omega)$ where $\Omega = [a_1,b_1]\times[a_2,b_2]\times\cdots\times[a_n,b_n]$ the functions $\omega_1,\omega_2,\ldots,\omega_n$ are weight functions i.e., h ω_i is real valued, continuous on $[a_i,b_i]$, positive (a_i,b_i) and is such that

$$\int_{a_i}^{b_i} \omega_i(t)\,|t|^m dt < \infty \qquad \forall \text{ nonnegative integer } m$$

Suppose $\{\phi_m^{(i)}\}$ is the sequence of orthonormal ynomials associated with the weight function ω_i on $_i,b_i]$. Denote the zeros of $\phi_m^{(i)}$ by $\{\tau_j^{mi}: 1 \le j \le m\}$.

The integral $I(v) = \int_{a_i}^{b_i} \omega_i(t)\,v(t)\,dt$ may be

proximated by the expression

$$I_m(v) = \sum_{j=1}^{m} c_{mj} v(\tau_j^{mi}) \qquad \text{(A2)}$$

where the $\{c_{mj}\}$ are given by Lagrangian interpolation. Then

$$\lim_{m\to\infty} I_m(v) = I(v) \qquad \forall\ v \in C[a_i, b_i]$$

(see Laurent [7] p. 318).

To approximate (A1), we shall use the formula

$$I_m(u) = \sum_{j_1=1}^{m} \sum_{j_2=1}^{m} \cdots \sum_{j_n=1}^{m} c_{m_{j_1}} c_{m_{j_2}} \cdots c_{m_{j_n}}\, u(\tau_{j_1}^{m1}, \tau_{j_2}^{m2}, \ldots, \tau_{j_n}^{mn})$$

where the $\{\tau_j^{mi}\}$ as above and the c_{m_j}'s are given as in (A

Write $\Gamma_m = \{(\tau_{j_1}^{m_1}, \tau_{j_2}^{m_2}, \ldots, \tau_{j_n}^{m_n}) : 1 \le m_i \le m\ ;\ 1 \le i \le n$

<u>Theorem.</u> For each $u \in C(\Omega)$, $\lim_{m\to\infty} I_m(u) = I(u)$.

<u>Proof.</u> From the Banach-Steinhaus theorem, it suffices to show that

(i) there exists $c > 0$ such that $\|I_m\| \le c$ for all m.

and (ii) $I_m(v) \to I(v)$ for all $v \in K$ where K is dense in $C(\Omega)$.

(i) holds as $\|I_m\| = \sum_{j_1=1}^{m} \sum_{j_2=1}^{m} \cdots \sum_{j_n=1}^{m} |c_{m_{j_1}} c_{m_{j_2}} \cdots c_{m_{j_n}}|$.

$$= \left(\sum_{j_1=1}^{m} |c_{m_j}| \right) \left(\sum_{j_2=1}^{m} |c_{m_{j_2}}| \right) \cdots \left(\sum_{j_n=1}^{m} |c_{m_{j_n}}| \right).$$

Each of the 1-dimensional formulae is consistent, so there exists $c_i > 0$ such that

$$\sum_{j_i=1}^{m} |c_{m_{ji}}| < c_i \quad \text{for all } i .$$

Thus (i) follows.

Take K to be the tensor product of the spaces of all polynomials on the intervals $[a_i,b_1],[a_2,b_2],\ldots,[a_n,b_n]$ respectively. Then K is dense in $C(\Omega)$ from the Stone-Weierstrass theorem and (ii) holds for any $v \in K$.

It is worth noting that (A3) is the approximation to (A1) obtained by interpolating the function u at the points of the grid u_m using an expression of the form

$$p_m u(x) = v(x) = \sum_{j_1=1}^{m} \cdots \sum_{j_n=1}^{m} a_{j_1 j_2 \cdots j_n} x_1^{j_1} x_2^{j_2} \cdots x_n^{j_n} \qquad \text{(A4)}$$

References

1. Auchmuty, J. F. G. and Beals, R., Models of Rotating Stars, Astrophysical J. 165, (1971), L79-82.

2. Auchmuty, J. F. G. and Beals, R., Variational Solutions of some Nonlinear Free Boundary Problems, Arch. Rat. Mech. and Anal., 43, (1971) pp. 255-271.

3. Chandrasekhar, S., Introduction to the Study of Stellar Structure, Chicago: Univ. of Chicago Press (1939).

4. Anselone, P. M., Collectively Compact Operator Approximation Theory, Prentice-Hall (1971).

5. Daniel, J. W., The Approximate Minimization of Functionals Prentice Hall (1971).

6. Daniel, J. W., Collectively Compact Sets of Gradient Mappings, Nederl. Akad. Wetensch. Proc. Ser. A, 71(1968) 270-279.

7. Laurent, P. J., Approximation et Optimisation, Hermann Paris (1972).

8. Ostriker, J. P. and Mark, J. W. K., Rapidly Rotating Stars I. The Self-Consistent Field Method, Astrophysical J., 151, (1968) 1075-1088.

9. Papaloizou, J. C. B. and Whelan, J. A. J., The Structure of Rotating Stars: The J^2 method and Results for Uniform Rotation, Mon. Not. R. Astr. Soc., 164, (1973) 1-10.

AUTOMATIC SOLUTION OF DIFFERENTIAL EQUATIONS

Y. F. Chang, Indiana Univ.

ABSTRACT

A compiler program has been developed for the automatic computer solution of differential equations. The inputs to this compiler are FORTRAN statements of the equations (differential and algebraic) and of the initial and/or boundary conditions. The automatic computations are performed under a preset error-limit, which is controlled by maintaining the integration stepsize to be well within the radius of convergence. The automatic solution is shown to be both faster and more accurate than standard techniques. This is achieved by using optimum stepsizes determined from the calculations of the radius of convergence at every step. This compiler program is particularly powerful in the solution of boundary-value problems and of "stiff" differential equations. This compiler program will accept up to 99 coupled 4-th order differential equations with ease.

INTRODUCTION

In this article, the compiler programs called ATSCPL and ATSBVP will be described. The inputs to these programs are simple FORTRAN statements of the equations and conditions.

The output of the programs is the automatic solution of the given problem with assured accuracy. The method of solution is based on the use of long Taylor series that are automatically from the given equations and functions.

The foundation of the automatic Taylor series method is the simplification of the differentiation of analytic functions to simple arithmetic operations. The first proposed use of power series in the solution of differential equations was made by Wilson(1949). Gibbons(1960) discussed some extensions of automatic differentiation to analytic functions. Many others had independently adapted Taylor series for the solutions of diverse problems. For example, see Steffensen (1956), Richtmyer(1959), Rabe(1961), Chang & Tuan(1964), Fehlberg(1964), Deprit & Price(1965), Miller(1966), Moore (1966), Szebehely(1966), Chang(1967), and Hartwell(1967). In a recent paper by Barton, et al(1972), the automatic solution of initial-value problems were discussed. They were able to show that the Taylor series method was both faster and more accurate than other numerical integration techniques.

In a brief announcement, Chang(1972) introduced the completely automatic method to be discussed below. The basis of this method is the relationship discovered between the radius of convergence and the Taylor series. This relationship is apparent only after the analysis of very long Taylor series, some as long as 1000 terms. This developement makes possible the calculation of the radius of convergence at every step of the solution, which leads to excellent control of solution accuracy. It will also be shown that there is an optimum length of Taylor series, optimized for the speed of computation. Finally, the automatic method will be applied to specific examples to illustrate the power of the method and the nearly ideal error control.

THE AUTOMATIC COMPILER PROGRAMS

The compiler programs will accept up to 99 coupled 4th-order differential equations with 99 distinct variables and 99 different constants. The input statements for the given equations follow all the rules of FORTRAN. The operations and functions in the equations can be any of those listed in Table I. These operations and functions can be nested in any combination to any degree of complexity.

Table I. Operators and Functions

Symbol	Operation/Function
+, -, *, /, **	add, subtract, multiply, divide, power
SQRT	square root
EXP, ALOG	exponential, natural logarithm
SIN, COS, TAN	trigonometric functions
SINH, COSH, TANH	hyperbolic functions
ASIN, ACOS, ATAN	inverse trigonometric functions

The symbol DIFF is used to denote differentiation. The format of an equational statement is quite simple. For example, Bessel's equation

$$x^2 y'' + x y' + (x^2-n^2) y = 0$$

is written, for n=2, as the following input statements.

```
DIFF(Y,X,2) = - DY/X - Y + 4*Y/X**2
DY = DIFF(Y,X,1)  '
```

The end of the complete set of equations is indicated by the end-of-data mark (').

The equational statements are read by the compiler and all the terms are stored into appropriate locations according to variables, operators, and constants. Then, the information is rearranged into the well known reverse Polish notation, from which is generated the computer program for automatic solution of the input equations.

In addition to the equational statements, there are some five levels of instructions specifying the conditions of the problem and of the solution. This can best be illustrated by a specific example. Consider the problem of Coddington and Levinson(1952) as given by Professor O'Malley.

$$e\, y'' = - y' - (y')^3 , \qquad (1)$$

where e is a positive constant, and the boundary conditions are $y(0)=a$ and $y(1)=b$. It can be shown that e is a scale factor, and the boundary conditions can be changed to $y(0)=0$

and y(1)=c without loss of generality. For e = 1 and c = 1, the complete input data to the compiler is listed in Table II.

Table II. Input Data for Coddington-Levinson Example

Equations	DIFF(Y,X,2) = - DY - DY*DY*DY DY = DIFF(Y,X,1) '
1st-level	PROGRAM EXAMPLE(INPUT,OUTPUT,TAPE1) '
2nd-level	START = 0. END = 1. BC = 1. GK = 3. GUESS = 5. Y(1,1) = 0. Y(1,2) = GK Y(2,2) = GUESS YA = 1.2 '
3rd-level	'
4th-level	'
5th-level	CALL BVSIMP(Y) '

The first level of instruction is used to enter the name of the problem and such statements as COMMON or TYPE needed in complex problems. The second level of instruction contains all the conditions of the problem and of the solution. START and END are the boundaries of the independent variable, x. BC is the far boundary condition, y(1)=1. GK and GUESS are the initializing values for y' at x=0. Y(1,1) is the solution representation of y(0). YA is an upper limit to contain the solution. The third level of instruction and also the fourth level are used only in complex problems where one may wish to stop and start the solution at controlled points. This will be illustrated in another example later. The fifth level of instruction is used to find the result for boundary-value problems. The subroutine BVSIMP is a library program that will automatically calculate the correct initial y' at x=0 for the given boundary condition, BC=1, without iteration. The computer output for the above example will be discussed later in this paper.

TAYLOR SERIES AND THE RADIUS OF CONVERGENCE

The use of Taylor series for the solution of differential equations has been studied by many during the last twenty or more years. The attraction is due to the inherent power and accuracy of solution methods based on Taylor series. Most of the solutions has been obtained with Taylor series of short lengths, up to 9 terms. This is unfortunate, because one cannot begin to appreciate the true beauty of Taylor series until one is studying a series of at least 20 terms. In difficult cases, even 100-term series is not long enough.

It will be shown heuristically that there is a relation between the behavior of the Taylor series terms and the location of poles and singularities in the complex plane of the solution function. It is then simple to calculate the radius of convergence, the distance between the solution point and the singularity. In the case of a simple pole in the complex plane, the radius of convergence is given by

$$RC = H/\exp(A) \quad , \qquad (2)$$

where RC is the radius of convergence, H is the integration step (or the step used in the series expansion), and exp(A) is an exponential envelope on the series terms. When the terms of the Taylor series are plotted (magnitude vs order on a semi-log plot) as in Fig. 1, the constant A is the slope of the straight line connecting the tops of the curves.

Consider the equation

$$y'' + (y')^2 + y = 0 \,. \tag{3}$$

This equation is interesting because the solution has poles spaced in conjugate pairs parallel to the real axis. The spacing between the poles is dependent on the initial conditions. For $y=0.46$ and $y'=-0.17$ at $x=0$, the nearest poles are located at $x=3.5$ and $ix=\pm 1.1$. The Taylor series expansion of y' with $H=3.5$ is shown in Fig. 1. This is a nearly perfect example of Taylor series behavior for a simple pole close to the real axis. Note that the function plotted is y' and not y. This is very important. The function y' has simple poles of order one. The function y is logarithmic in nature. In the calculation of the radius of convergence from a Taylor series, it is very important to determine the proper function that has the simple poles of order one.

There is a direct approach, applicable in cases such as Eq.(3), for finding the order of the poles. Let p be the order of the poles in the function y. Then, the function y' will have poles of order $p+1$, and y'' will have poles of order

$p+2$. In Eq.(3), the first term has poles of order $p+2$; the second term has poles of order $2(p+1)$; the third term will be ignored, because it is linear. Since the order of the poles must match, we have

$$p + 2 = 2p + 2 .$$

The result is $p=0$, and y' has poles of order one.

Applying this simple approach to the Coddington-Levinson example, Eq.(1), we find the first term with poles of order $p+2$. The second term is linear and will be ignored. The last term has poles of order $3(p+1)$. Matching the order of poles in the two significant terms, we have

$$p + 2 = 3p + 3 .$$

The result is $p=-1/2$. There are branch points in the complex plane for y'. In the neighbourhood of the branch point, $y'=k(ax+b)^{-1/2}$, and $y=c(ax+b)^{1/2}$. This result will be used later in the discussion of the solution of the Coddington-Levinson example.

For a look at Taylor series behavior for poles near the imaginary axis, consider the problem of Eq.(3) with initial conditions $y=-1.3$ and $y'=-0.1$ at $x=0$. The nearest poles are located at $x=0.12$ and $ix=\pm1.25$. The Taylor series expansion of y' with H=1.2 is shown in Fig. 2. The appearance of two sets of curves in Fig. 2 is because the elevation angle of the pole is close to 90° resulting in a beat phenomenon. For a pole with an elevation angle close to 60°, the beat phenomenon produces three sets of curves. However many sets of curves, the tops of the curves do form a straight line when the poles are of order one.

The calculation for the radius of convergence by Eq.(2) depends on the existence of an exponential envelope, exp(A), for the series terms. Although it is possible to prove that all simple poles and branch points lead to Taylor series whose terms are bounded by exp(A), there are practical limitations to what can be included in a computer program. The compiler program under discussion is capable of accurate calculation of the radius of convergence when poles of order one occur in up to the fourth-derivative of the solution function. Of the many problems solved using the radius-of-convergence concept during the past six years, none has had solutions with poles of order one in a derivative higher than the second.

Solution functions that have essential singularities are beyond the above analysis. There is no proof that the Taylor series terms for these functions can be bounded by any exponential envelope. Heuristically however, provided that the series is long enough and its terms can be bounded by exp(A), the estimate for the radius of convergence can be quite good.

For an example of an essential singularity, consider the equation

$$x^2 y'' - y^3 + y = 0 .$$

This equation cannot be analyzed by the simple approach used earlier for Eqs.(1) and (3). There is an essential singularity at $x=0$. With initial conditions $y=0.5901736$ and $y'=0.1656055$ at $x=1.7$, the series expansion of the solution with $H=-1.7$ is shown in Fig. 3. This is a 1000-term Taylor series! The abscissa is logarithmic. The two minima occur at the 12th-term and at the 358th-term. It is notable that the series terms are bounded, although not by an exponential envelope. The radius-of-convergence calculation will yield a good estimate of the true location of the singularity.

OPTIMUM SERIES LENGTH

The control of accuracy in the automatic solution is obtained by calculating the radius of convergence at every step. Then, the integration stepsize is adjusted to be the maximum allowable for a given preset error limit. The stepsize is varied so as to make the factor exp(A) proper for the truncation error to be equal to the preset error limit. The truncation error is given by

$$Er = \exp\,(n-1)\cdot A \quad , \tag{4}$$

where n is the length of the series. This error is an upper bound on the error. When the terms of the series are not of the same sign, the truncation error will be much less than that given by Eq.(4). However, in the solution of equations, the point of solution moves along the real axis and passes poles in all possible positions. The result is that the terms of the Taylor series are at times alternating in sign only about half the time. Therefore, it is best to use Eq.(4)

without the addition of any refinements.

From the standpoint of accuracy, there is no limitation on the length of the Taylor series. Provided that the radius of convergence can be calculated, the error can be controlled by decreasing the integration stepsize. It would be desirable to have long series since the radius of convergence is then more accurate. From the standpoint of computation speed, it is very important to have an optimum series length.

There are four basic parts to the automatic solution. These four parts are separated by three levels of nested DO loops. The first part, outside of all DO loops, is independent of the number of steps and the series length. It will be ignored. The second part is repeated for every integration step, but is independent of the series length. The third part is repeated for every step and once for each series term. The fourth part is within the third part and is repeated for each series term. Thus, the total computation time is given by

$$\text{Time} = (A + B\ n + C\ n^2)(\text{number of steps})\ , \qquad (5)$$

where n is the series length and A, B, and C are the computing times for parts two, three, and four, respectively. The number of steps is equal to the total range of the solution divided by the average maximum stepsize. The average maximum stepsize is given by the average radius of convergence multiplied by $(EL)^{1/(n-1)}$, where EL is the error limit. Eq. (5) becomes

$$\text{Time} = (A + B\ n + C\ n^2)(EL)^{-1/(n-1)}\ , \qquad (6)$$

when all the other constant are absorbed into A, B, and C.

The validity of Eq.(6) is tested in the solution of the problem of Eq.(3) using series lengths from 7 to 100. The initial conditions used are y=0.499 and y'=0 at x=0. Under these conditions, there are sets of conjugate poles close to the real axis at x=-4.5, 4.5, 13.5, 22.5, 31.5, etc. These poles are only ±0.8 away from the real axis. The experiment was performed in the range from x=0 to x=20. The results of this experiment are plotted in Fig. 4. The little crosses in Fig. 4 are the experimental computation times obtained on a CDC-6600 computer. The solid line is the curve for Eq.(6) with the parameters

$$\begin{aligned} A &= 0.01015793 \\ B &= 0.00118513 \\ C &= 0.00004326 \\ EL &= 10^{-8} \, . \end{aligned}$$

The optimum length for the Taylor series in the automatic solution is about 20 terms.

Also shown in Fig. 4 are the computation times for the Runga-Kutta method and Hamming's predictor-corrector method. Both the Runga-Kutta(RKGS) and Hamming's(HPCG) methods were modified from the IBM scientific subroutine package with appropriate adjustment for the 15-figure accuracy of the CDC computer. The lines marked "package" are the computation times for the integration methods written as standard subroutines. The lines marked "adapted" are for the methods written specially for the particular equation. As evident in Fig. 4, the automatic solution is superior to both the standard techniques.

AUTOMATIC SOLUTION OF EXAMPLES

The input data and computer output for specific examples will be discussed in this section. The first example is the calculation for the zeros of the Neumann function, $N_1(x)$, for values of x between 1 and 20. The second example deals with the solution of the Coddington-Levinson problem, and gives some indication of the effectiveness of error control. The final example is the round-trip trajectory of a space ship between Earth and Jupiter. This example will be used in an experiment to test the validity of the preset error limit.

The Zeros of $N_1(x)$

Consider the problem of calculating the zeros of the Neumann function, $N_1(x)$, starting with values of N and N' at x=1. The Bessel's equation for $N_1(x)$ is

$$x^2 z'' + x z' + (x^2-1) z = 0 .$$

The initial conditions are z=-0.78121282 and z'=0.869469784 at x=1. The function z is used here to represent $N_1(x)$. In this example, the printout will be controlled so as to print at every integer value of x. The input data for this problem is listed in Table III.

In this solution, ZOLD is used to detect when the value

of Z has changed sign. DXPT is a variable internal to the program; it stands for <u>delta-x for print</u>. Since printing is desired at every integer value of x, DXPT is set equal to 1. ZEROT is a library program for calculating roots of the function. The arguements for ZEROT are Y, ROOT, SIZE, ORDER. Y is the variable whose roots is being sought; here it is Z. ROOT is the value of the root; here it is zero. SIZE is used to control the error; the root is calculated with an error less than SIZEx10^{-6}. ORDER is the order of the differential equation being solved; here it is 2. The calculations for the root performed within ZEROT are straightforward and do not involve iteration.

Table III. Input Data for Zeros of $N_1(x)$

```
Equations      DIFF(Z,X,2) = - DZ/x - Z + Z/X**2
               DZ = DIFF(Z,X,1)  '

1st-level      PROGRAM NEUMANN(INPUT,OUTPUT,TAPE1)  '

2nd-level      START = 0.
               END = 20.
               DXPT = 1.
               Z(1) = -0.78121282
               Z(2) = 0.869469784
               ZOLD = -1.
               GO TO 107
            11 CALL ZEROT(Z,0.,1.,2)
               ZOLD = - ZOLD
               GO TO 109  '
3rd-level      '
4th-level      IF(Z(1)*ZOLD.LT.(-1.E-5)) GO TO 11  '
```

Note that this solution makes used of the 4th-level of instructions. It is used to test the solution for change of sign, upon which the solution is stopped for the calculation of the root by the subroutine ZEROT.

The computer printout for this example is given in Appendix A. Note that the program automatically determined that the order is 1. This means that simple poles are to be found in the first derivative of the function. Once the correct order has been found, the program performs radius-of-convergence calculations at every step and controls the error automatically.

Coddington-Levinson Example

This example has been discussed earlier in this paper. It is the problem of Eq.(1). The input data to the automatic program was listed in Table II. Aside from the pragmatic results given by the computer, this problem has some very interesting features.

The solution of this example can be understood better when the upper boundary is extended to positive infinity. We also can extend the discussion to include

$$e\, y'' = -\, y' - (y')^c ,$$

where c is a positive constant greater than 2. Analysis of this equation can be broken into two ranges; one where y' is less than one, and one where y' is greater than one. For y' much less than one, the solution is

$$y = k - \exp(-x/e) ,$$

where k is an arbitrary constant. For y' much greater than

one, the solution is

$$y = (e/q)^{1/p} (1/r)^{r} (x+b)^{r} ,$$

where b is an arbitrary constant, $p = c-1$, $q = c-2$, and $r = q/p$. For $c=3$, this solution becomes

$$y = (2e)^{1/2} (x+b)^{1/2} .$$

Numerical solution of this problem is needed only in the middle range where y' is about one.

The computer output for the solution of the Coddington-Levinson example with the input data of Table II is listed in Appendix B. In all boundary-value solutions, the initializing value for y' at x=START is the sum GK + GUESS. In this solution, the value of GUESS was reduced from 2.0 to 1.88353 by the automatic program in order to control the error in the solution. The automatic program will reduce the value of GUESS rather than the stepsize. This is to allow for faster solutions. If the solution had been unsuccessful with this value for GUESS, one could force a solution with higher value of GK + GUESS by increasing the value of GK. As it was, the automatic solution was successful.

The solution of the boundary-value problem is the single result for the correct initial condition, y'=4.880296. The validity ratio given is the ratio of the radius of convergence divided by the stepsize. The solution is valid whenever this ratio is larger than about 1.5. In this solution, this ratio was a very high 3.80. To test the accuracy of the result, this example was solved again as an initial-value problem with y=0 and y'=4.880296. The result for y at x=1 was found to be 1.000000795. This is an indication of the accuracy and error control obtainable from the automatic program.

VALIDITY OF PRESET ERROR CONTROL

In Fig. 4, the automatic Taylor series method was compared with the Runga-Kutta and Hamming's methods. It was shown that the automatic method was both faster and more accurate than these standard methods. Now, the automatic method will be compared against these standard methods with a complex example that requires appreciable computation time. We shall also test the validity of the preset error limit.

The example is a 4-body problem in celestial mechanics. It is the trajectory of a space ship on a round-trip travel from Earth to Jupiter and return. The four bodies are the Sun, the Earth, Jupiter, and the space ship. Each of these four bodies have six degrees of freedom; thus, the problem is a set of 12 second-order differential equations. Each of the equation is similar to

$$d^2Xv/dt^2 = - AMe \cdot Xve \cdot Rve - AMj \cdot Xvj \cdot Rvj - AMs \cdot Xvs \cdot Rvs \ ,$$

where Xv is the x-position of the vehicle; AMe, AMj, AMs are the masses of the Earth, Jupiter, and Sun, respectively; Xve, Xvj, Xvs are the x-distance between the vehicle and the other bodies; Rve, Rvj, Rvs are the reciprocal cubes of the direct distances between the respective bodies. For example,

$$Xve = Xv - Xe \quad ,$$

and
$$Rve = (Xve^2 + Yve^2 + Zve^2)^{-3/2} \; .$$

The input data for the Jupiter round-trip example is listed in Appendix C.

The automatic program is written as a subroutine so that the total computation time can be calculated in a calling main program for many different value of the preset error limit. The input to the subroutine is the error limit, ELIN. The output of the automatic subroutine is the distance between vehicle and Earth after 7.355×10^7 seconds of space travel. The actual relative error of the result is calculated from comparison with a double-precision result. A plot of the trajectory is given in Fig.5. This problem was solved on the CDC-6600 computer.

The solution of the Jupiter example was completed for 119 different values of the error limit between 10^{-2} and 10^{-12}. The 119 actual relative errors are plotted against the error limit in Fig. 6. The automatic solution results are plotted as circles. The Hamming's method was used under the same conditions of error limit. The Hamming's results are plotted as crosses in Fig. 6.

The 45° solid line in Fig. 6 is the ideal error control, where the actual relative error is exactly equal to the preset error limit. Note that the automatic solutions have a nearly ideal error control. The circles follow a path parallel to the 45° line. This is experimental indication that the error control in the automatic solution is a very good one. For Hamming's method, the crosses are scattered and far away from the ideal-control line. Incidentally, the Runga-Kutta method did not yield usable results. For an error limit of 10^{-6}, a solution by the Runga-Kutta method required a computer time of 4,790 seconds!

In Fig. 7, the computation times for the automatic method and Hamming's method are compared. Hamming's method is faster than the automatic solution when the errors are larger. Since the automatic method requires almost no programming, it can be considered to be superior overall. The total merit of this new approach to the solution of differential equations is the sum of ease of programming, speed of computation, and the control of accuracy.

The automatic compilers are written in FORTRAN and have been adapted to the CDC-6600 computer as well as the IBM-370 computers. It is a relatively simple matter to adapt these compiler programs for any other large-scale digital computer.

APPENDIX A

COMPUTER OUTPUT FOR THE ZEROS OF $N_1(X)$

Listed below is the computer output for the solution of the zeros of the Neumann function, $N_1(x)$, between x=0 and x=20. The error limit used was 10^{-6}.

```
X= 1.000000E+00
Y(1),(2)= -7.812128E-1  8.694698E-1
                          BY 4-PT ANALYSIS, THE ORDER IS  1
                                   RC= 1.48E+00 STEP 5.20E-1
X= 2.000000E+00
Y(1),(2)= -1.070291E-1  5.638968E-1
                                   RC= 2.20E+00 STEP 8.03E-1
THERE IS A ROOT AT POINT = 2.197133E+00

X= 3.000000E+00
Y(1),(2)=  3.246801E-1  2.686249E-1
                                   RC= 2.62E+00 STEP 1.00E+0
X= 4.000000E+00
Y(1),(2)=  3.979290E-1 -1.164262E-1
                                   RC= 3.90E+00 STEP 1.00E+0
X= 5.000000E+00
Y(1),(2)=  1.478621E-1 -3.380943E-1
                                   RC= 5.55E+00 STEP 1.00E+0
X= 6.000000E+00
Y(1),(2)= -1.750141E-1 -2.590273E-1
                                   RC= 6.30E+00 STEP 1.00E+0
THERE IS A ROOT AT POINT = 5.429673E+00

X=7.000000E+00
Y(1),(2)= -3.026702E-1  1.729089E-2
                                   RC= 9.74E+00 STEP 1.00E+0
X= 8.000000E+00
Y(1),(2)= -1.580602E-1  2.432825E-1
                                   RC= 1.06E+01 STEP 1.00E+0
X= 9.000000E+00
Y(1),(2)=  1.043174E-1  2.383476E-1
                                   RC= 1.07E+01 STEP 1.00E+0
THERE IS A ROOT AT POINT = 8.595998E+00

X= 1.000000E+01
Y(1),(2)=  2.490181E-1  3.076801E-2
                                   RC= 9.72E+00 STEP 1.00E+0
X= 1.100000E+01
Y(1),(2)=  1.637058E-1 -1.837326E-1
```

Appendix A (continued)

```
                                    RC= 1.05E+01 STEP 1.00E+0
X= 1.200000E+01
Y(1),(2)= -5.710146E-2 -2.204806E-1
                                    RC= 1.07E+01 STEP 1.00E+0
THERE IS A ROOT AT POINT = 1.174915E+01

X= 1.300000E+01
Y(1),(2)= -2.100839E-1 -6.204673E-2
                                    RC= 9.62E+00 STEP 1.00E+0
X= 1.400000E+01
Y(1),(2)= -1.666454E-1  1.390983E-1
                                    RC= 1.04E+01 STEP 1.00E+0
X= 1.500000E+01
Y(1),(2)=  2.107539E-2  2.040611E-1
                                    RC= 1.06E+01 STEP 1.00E+0
THERE IS A ROOT AT POINT = 1.489743E+01

X= 1.600000E+01
Y(1),(2)=  1.779776E-1  8.468698E-2
                                    RC= 9.56E+00 STEP 1.00E+0
X= 1.700000E+01
Y(1),(2)=  1.672059E-1 -1.024750E-1
                                    RC= 1.03E+01 STEP 1.00E+0
X= 1.800000E+01
Y(1),(2)=  8.153780E-3 -1.880070E-1
                                    RC= 1.05E+01 STEP 1.00E+0
X= 1.900000E+01
Y(1),(2)= -1.495623E-1 -1.016479E-1
                                    RC= 1.06E+01 STEP 1.00E+0
THERE IS A ROOT AT POINT = 1.804339E+01

X= 2.000000E+01
Y(1),(2)= -1.655127E-1  7.091810E-2
```

APPENDIX B

THE CODDINGTON-LEVINSON EXAMPLE

Listed below is the computer output for the solution of the Coddington-Levinson example of a nonlinear boundary-value problem. The error limit used was 10^{-6}.

```
 FOR GUESS = 2.000000E+00
X= 0.
Y(1),(2)=  0.                 6.000000E+00
                                         TOP-LINE ANALYSIS, the order is 2
 FOR GUESS = 1.883529E+00
X= 5.600000E-03
Y(1),(2)=  3.018086E-02  4.970466E+00
 FOR GUESS = 1.883529E+00
X= 1.370000E-02
Y(1),(2)=  6.693115E-02  4.171385E+00
 FOR GUESS = 1.883529E+00
X= 2.500000E-02
Y(1),(2)=  1.099555E-01  3.499828E+00
 FOR GUESS = 1.883529E+00
X= 4.100000E-02
Y(1),(2)=  1.609752E-01  2.925867E+00
 FOR GUESS = 1.883529E+00
X= 6.300000E-02
Y(1),(2)=  2.196415E-01  2.446679E+00
 FOR GUESS = 1.883529E+00
X=  9.40000E-02
Y(1),(2)=  2.885696E-01  2.034063E+00
 FOR GUESS = 1.883529E+00
X=  1.37000E-01
Y(1),(2)=  3.678711E-01  1.682612E+00
 FOR GUESS = 1.883529E+00
X= 1.980000E-01
Y(1),(2)=  4.603799E-01  1.375171E+00
 FOR GUESS = 1.883529E+00
X= 2.760000E-01
Y(1),(2)=  5.572686E-01  1.127306E+00
 FOR GUESS = 1.883529E+00
X= 3.900000E-01
Y(1),(2)=  6.716141E-01  8.964045E-01
 FOR GUESS = 1.883529E+00
X= 5.500000E-01
Y(1),(2)=  7.974022E-01  6.915570E-01
```

Appendix B (continued)

```
 FOR GUESS = 1.883529E+00
X= 7.800000E-01
Y(1),(2)=  9.335173E-01  5.066103E-01
 FOR GUESS = 1.883529E+00
X= 1.000000E+00
Y(1),(2)=  1.031299E+00  3.891753E-01

THE CORRECT INITIAL CONDITION IS 4.880296E+00
   FOR FAR-BOUNDARY CONDITION OF 1.00E+00
        WITH A VALIDITY RATIO OF   3.80
```

APPENDIX C

INPUT DATA FOR JUPITER ROUND-TRIP

Listed below are the input data used to solve the problem of sending a space ship on a round-trip ballistic flight from Earth to Jupiter and return.

```
Equations   DIFF(XV,T,2)=-AME*XVE*RVE-AMJ*XVJ*RVJ-AMS*XVS*RVS
            DIFF(YV,T,2)=-AME*YVE*RVE-AMJ*YVJ*RVJ-AMS*YVS*RVS
            DIFF(ZV,T,2)=-AME*ZVE*RVE-AMJ*ZVJ*RVJ-AMS*ZVS*RVS
            DIFF(XE,T,2)=-AMJ*XEJ*REJ-AMS*XES*RES
            DIFF(YE,T,2)=-AMJ*YEJ*REJ-AMS*YES*RES
            DIFF(ZE,T,2)=-AMJ*ZEJ*REJ-AMS*ZES*RES
            DIFF(XJ,T,2)=+AME*XEJ*REJ-AMS*XJS*RJS
            DIFF(YJ,T,2)=+AME*YEJ*REJ-AMS*YJS*RJS
            DIFF(ZJ,T,2)=+AME*ZEJ*REJ-AMS*ZJS*RJS
            DIFF(XS,T,2)=+AME*XES*RES+AMJ*XJS*RJS
            DIFF(YS,T,2)=+AME*YES*RES+AMJ*YJS*RJS
            DIFF(ZS,T,2)=+AME*ZES*RES+AMJ*ZJS*RJS
            RVE = (XVE**2 + YVE**2 +ZVE**2)**(-1.5)
            RVJ = (XVJ**2 + YVJ**2 +ZVJ**2)**(-1.5)
            RVS = (XVS**2 + YVS**2 +ZVS**2)**(-1.5)
            REJ = (XEJ**2 + YEJ**2 +ZEJ**2)**(-1.5)
            RES = (XES**2 + YES**2 +ZES**2)**(-1.5)
            RJS = (XJS**2 + YJS**2 +ZJS**2)**(-1.5)
            XVE = XV - XE
            YVE = YV - YE
            ZVE = ZV - ZE
            XVJ = XV - XJ
            YVJ = YV - YJ
            ZVJ = ZV - ZJ
            XVS = XV - XS
            YVS = YV - YS
            ZVS = ZV - ZS
            XEJ = XE - XJ
            YEJ = YE - YJ
            ZEJ = ZE - ZJ
            XES = XE - XS
            YES = YE - YS
            ZES = ZE - ZS
            XJS = XJ - XS
            YJS = YJ - YS
            ZJS = ZJ - ZS
```

Appendix C (continued)

```
          AME = AMS/332851.3
          AMJ = AMS/1047.39
          AMS = 1.32452139E20  '  .

1st-level SUBROUTINE JUPITER(ELIN,DVE)  '

2nd-level EL = ELIN
          M = 7
          START = 0.
          END = 7.355E7
          H = 1000.
          XV(1) = -1.41969280135E11
          XV(2) = -1.54057864559E4
          YV(1) =  4.3083952693E10
          YV(2) = -4.370268227E4
          ZV(1) =  4.48894332309E4
          ZV(2) = -5.6378802E2
          XE(1) = -1.41963187423E11
          XE(2) = -9.13745555591E3
          YE(1) =  4.30815180477E10
          YE(2) = -2.86096660851E4
          ZE(1) =  0.
          ZE(2) =  0.
          XJ(1) = -8.49631455977E10
          XJ(2) =  1.28549804288E4
          YJ(1) = -7.84711554438E11
          YJ(2) = -7.89887027699E2
          ZJ(1) =  4.98899147707E9
          ZJ(2) = -2.85975487774E2
          XS(1) = 0.
          XS(2) = 0.
          YS(1) = 0.
          YS(2) = 0.
          ZS(1) = 0.
          ZS(2) = 0.
          XPRT = END  '

3rd-level IF(POINT.LT.(7.34999E7))GO TO 109
          DVE=SQRT((XV(1)-XE(1))**2+(YV(1)-YE(1))**2+
         *   (ZV(1)-ZE(1))**2)
          return '
4th-level   '
```

REFERENCES

Barton, D., Willers, I.M., and Zahar, R.V.M. (1972). The Automatic Solution of Systems of Ordinary Differential Equations by the Method of Taylor Series, Comp. J., Vol. 14, p. 243.

Chang, Y.F. (1967). The conduction-Diffusion Theory of Semiconductor Junctions, Jour. Appl. Phys., Vol. 38, p. 534.

Chang, Y.F. (1972). Compiler Program Solves Equations, Computer Decisions, June 1972, p. 59.

Chang, Y.F. and Tuan, S.F. (1964). Possible Experimental Consequences of Triangle Singularities in Single-Particle Production Processes, Phys. Rev., Vol. 136, p. B741.

Coddington, E.A. and Levinson, N. (1952). A Boundary Value Problem, Proc. Am. Math. Soc., Feb. 1952, p. 81.

Fehlberg, E. (1964). Numerical Integration of Differential Equations by Power Series Expansions, NASA Report 1964, TN D-2356.

Gibbons, A.(1960). A Program for the Automatic Integration of Differential Equations Using the Method of Taylor Series, Comp. J., Vol. 3, p. 108.

Hartwell, J.G. (1967). Simultaneous Integration of N-Bodies by Analytic Continuation with Recursively Formed Derivatives, J. Astro. Sci., Vol. 14, p. 173.

Moore, R.A. (1966). Interval Analysis, Prentice-Hall, Englewood Cliffs, New Jersey.

References (continued)

Miller, J.C.P. (1966). The Numerical Solution of Ordinary Differential Equations. Numerical Analysis, An Introduction J. Walsh, ed., Chapter 4, Academic Press, London.

Rabe, E. (1961). Determination and Survey of Periodic Trojan Orbits in the Restricted Problem of Three Bodies, Astron. J., Vol. 66, p. 500.

Richtmyer, R.D. (1959). Detached-Shock Calculations by Power Series, AEC R&D Report, NYU-7972, New York University.

Steffensen, J.F. (1956). On the Restricted Problem of Three Bodies, Kgl. Danske Videnskab. Selskab, Mat. Fys. Medd., Vol. 30, p. 1.

Szebehely, V. (1966). Solution of the Restricted Problem of Three Bodies by Power Series, Astron. J., Vol. 71, p. 968.

Wilson, E.M. (1949). A Note on the Numerical Integration of Differential Equations, Quar. J. Mech. & Appl. Math., Vol. 2, p. 208.

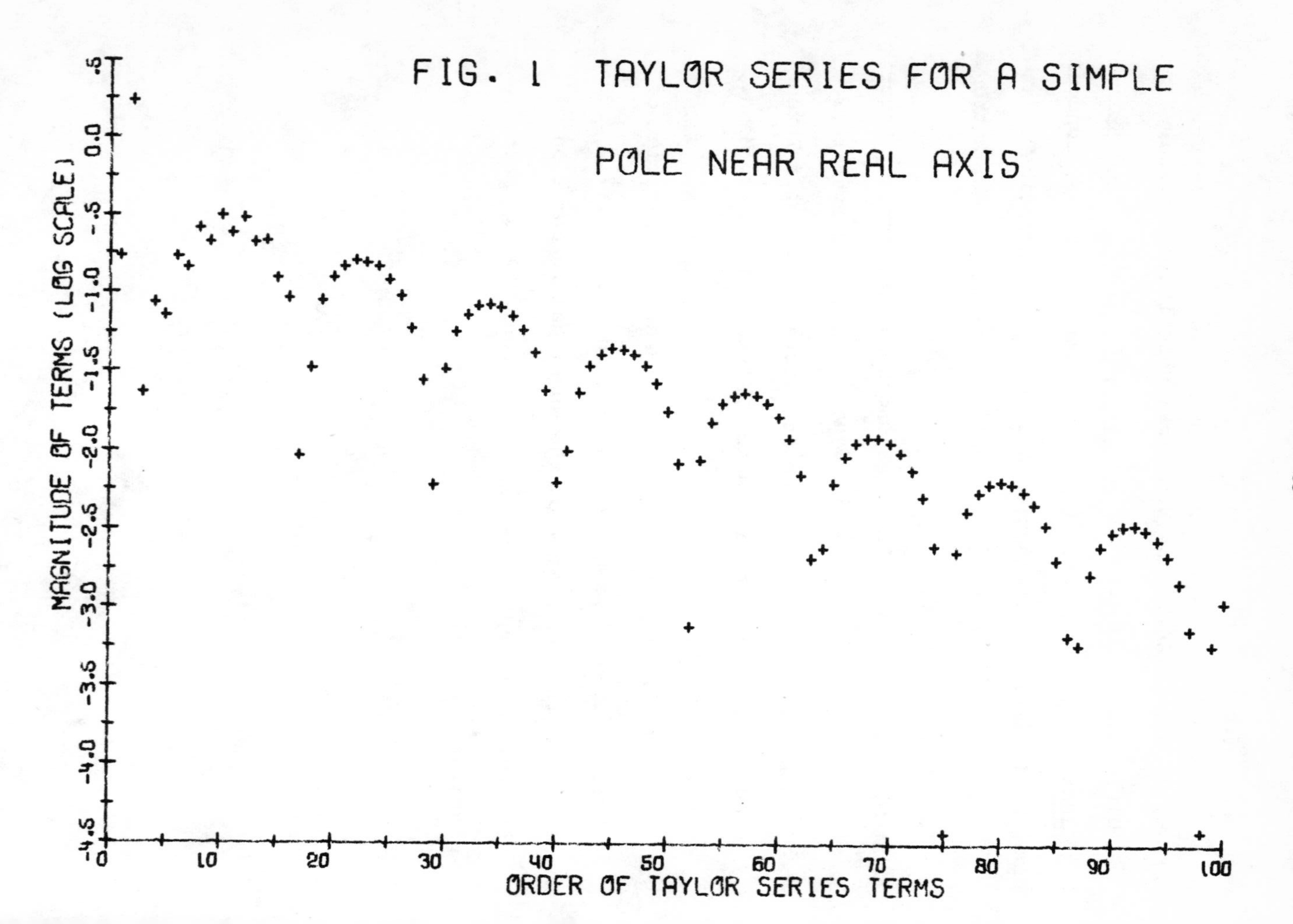

FIG. 1 TAYLOR SERIES FOR A SIMPLE POLE NEAR REAL AXIS

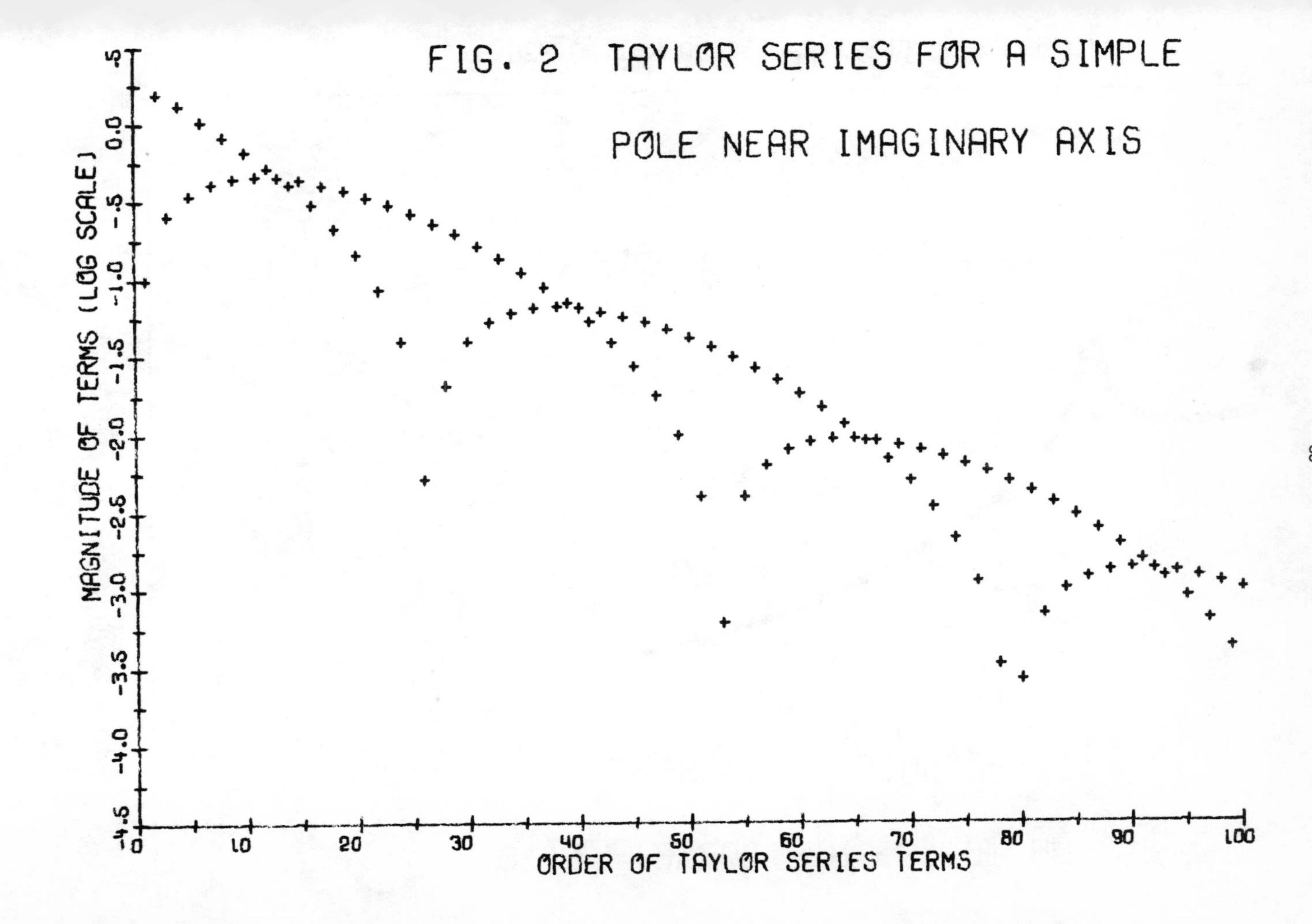
FIG. 2 TAYLOR SERIES FOR A SIMPLE
POLE NEAR IMAGINARY AXIS
MAGNITUDE OF TERMS (LOG SCALE)
.5
0.0
-.5
-1.0
-1.5
-2.0
-2.5
-3.0
-3.5
-4.0
-4.5
0
10
20
30
40
50
60
70
80
90
100
ORDER OF TAYLOR SERIES TERMS

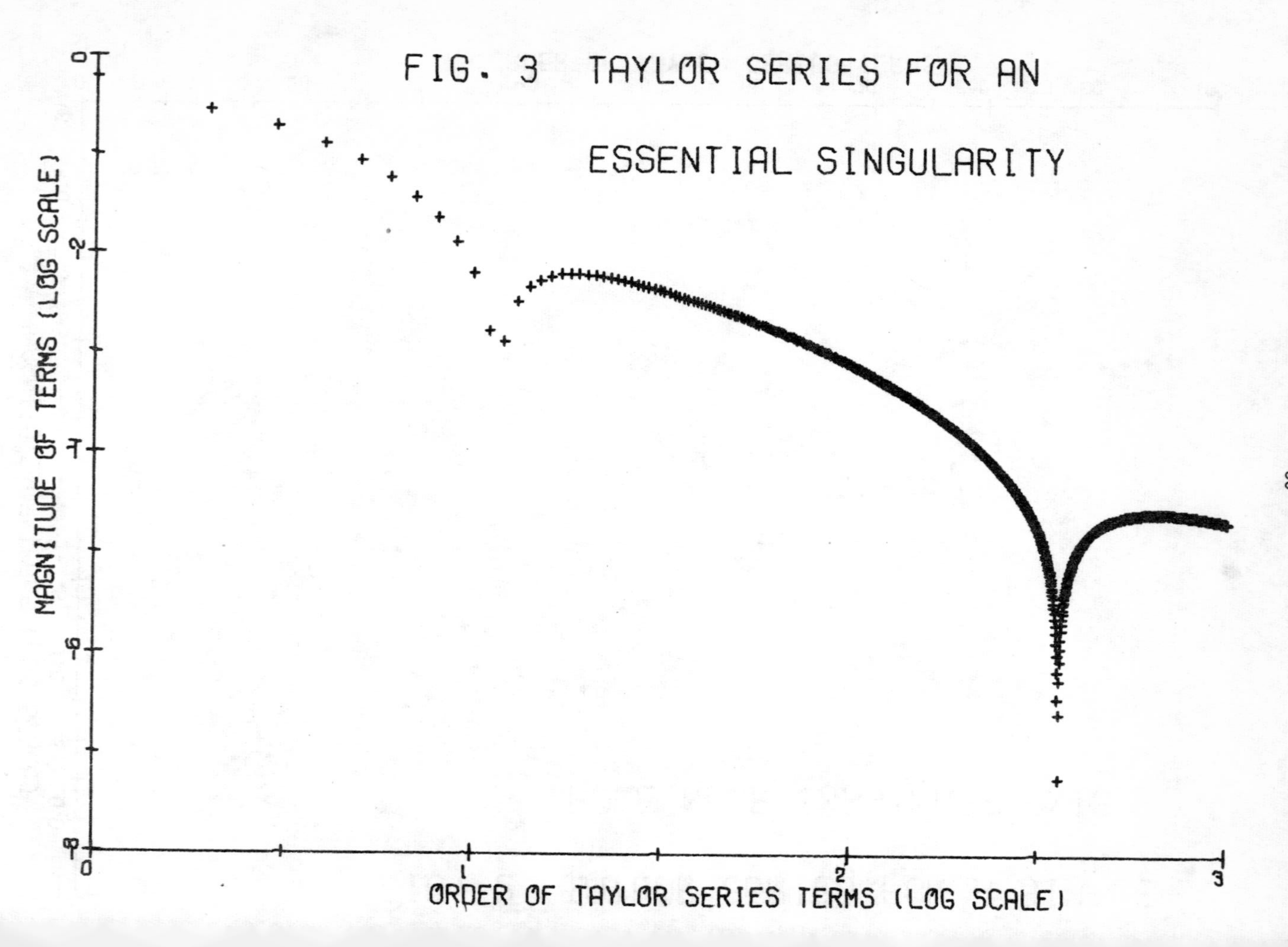
FIG. 3 TAYLOR SERIES FOR AN
ESSENTIAL SINGULARITY
MAGNITUDE OF TERMS (LOG SCALE)
ORDER OF TAYLOR SERIES TERMS (LOG SCALE)
0
-2
-4
-6
-8
0
1
2
3

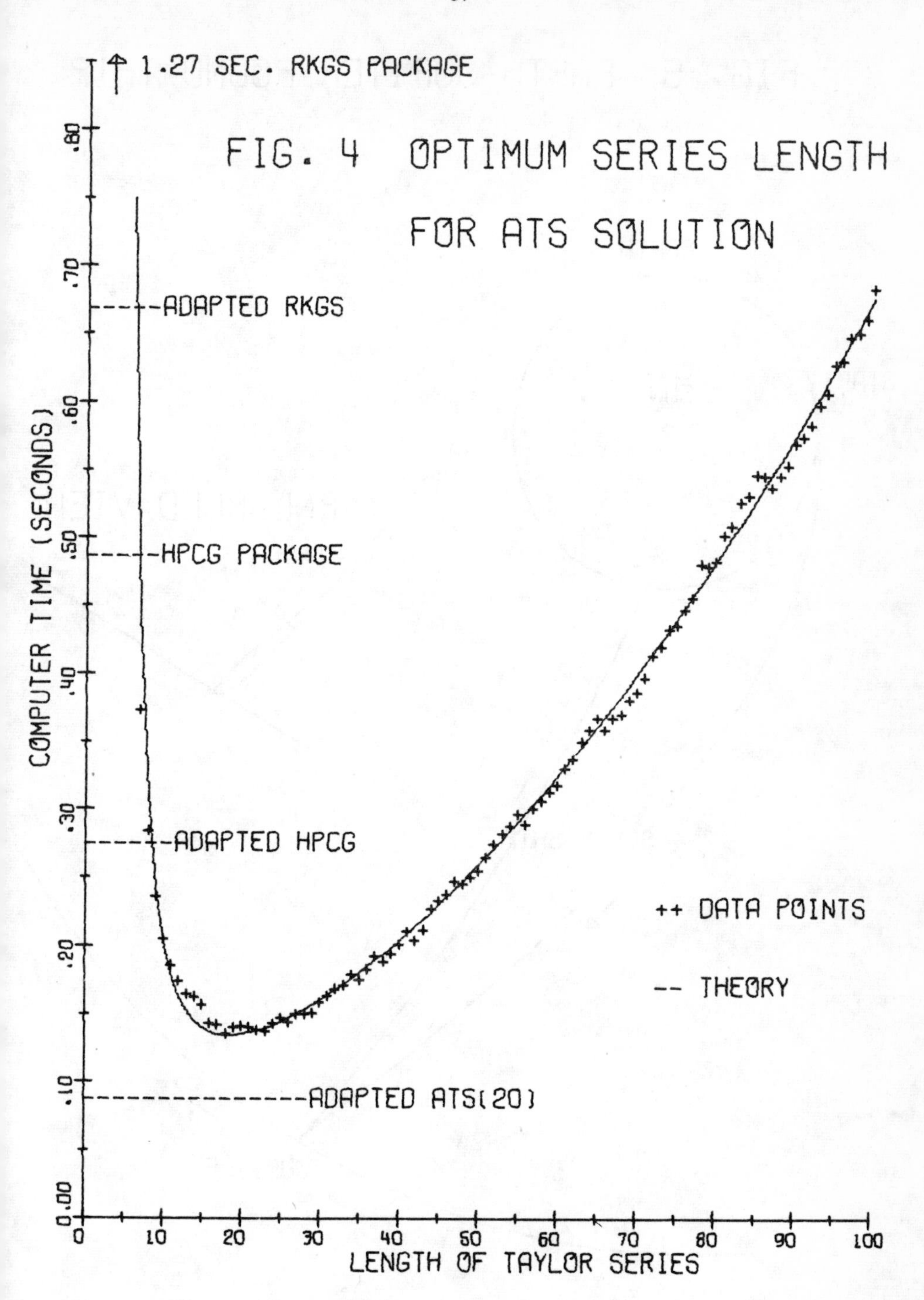

1.27 SEC. RKGS PACKAGE
FIG. 4 OPTIMUM SERIES LENGTH
FOR ATS SOLUTION
ADAPTED RKGS
HPCG PACKAGE
ADAPTED HPCG
ADAPTED ATS(20)
++ DATA POINTS
-- THEORY
COMPUTER TIME (SECONDS)
LENGTH OF TAYLOR SERIES
0.00
.10
.20
.30
.40
.50
.60
.70
.80
0
10
20
30
40
50
60
70
80
90
100

FIG. 5 EARTH-JUPITER ROUND TRIP

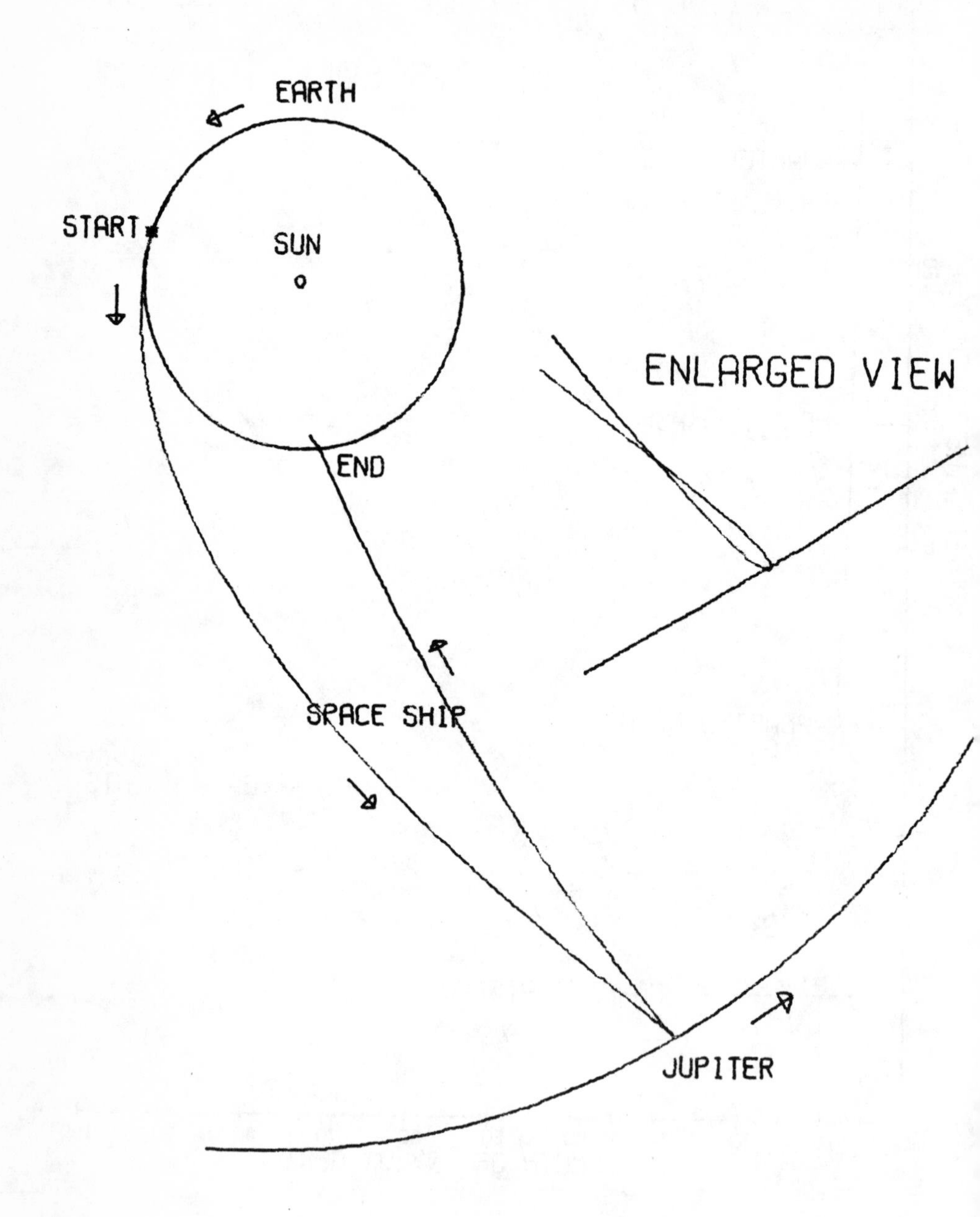

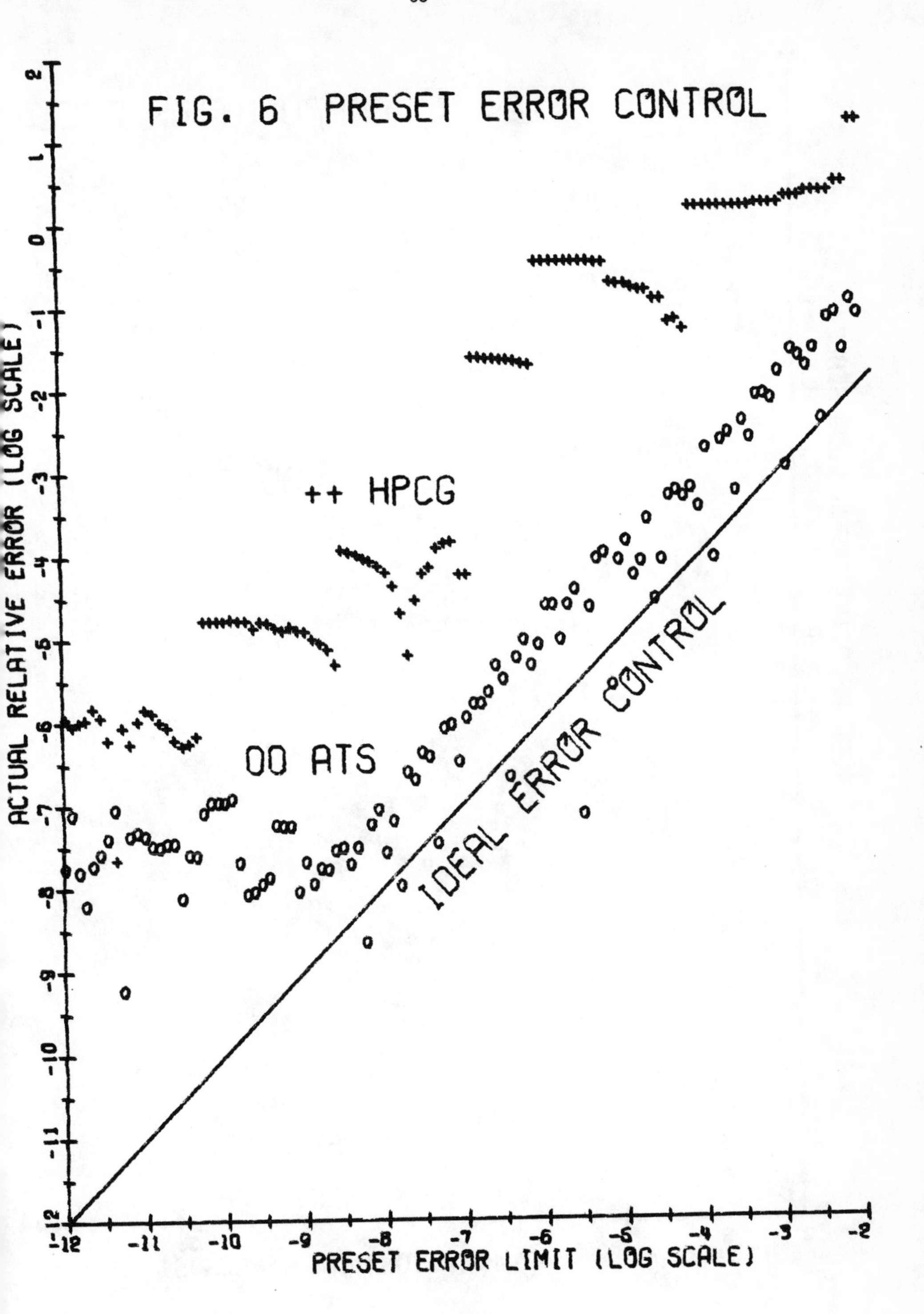
FIG. 6 PRESET ERROR CONTROL
++ HPCG
OO ATS
IDEAL ERROR CONTROL
ACTUAL RELATIVE ERROR (LOG SCALE)
PRESET ERROR LIMIT (LOG SCALE)
2
1
0
-1
-2
-3
-4
-5
-6
-7
-8
-9
-10
-11
-12
-11
-10
-9
-8
-7
-6
-5
-4
-3
-2

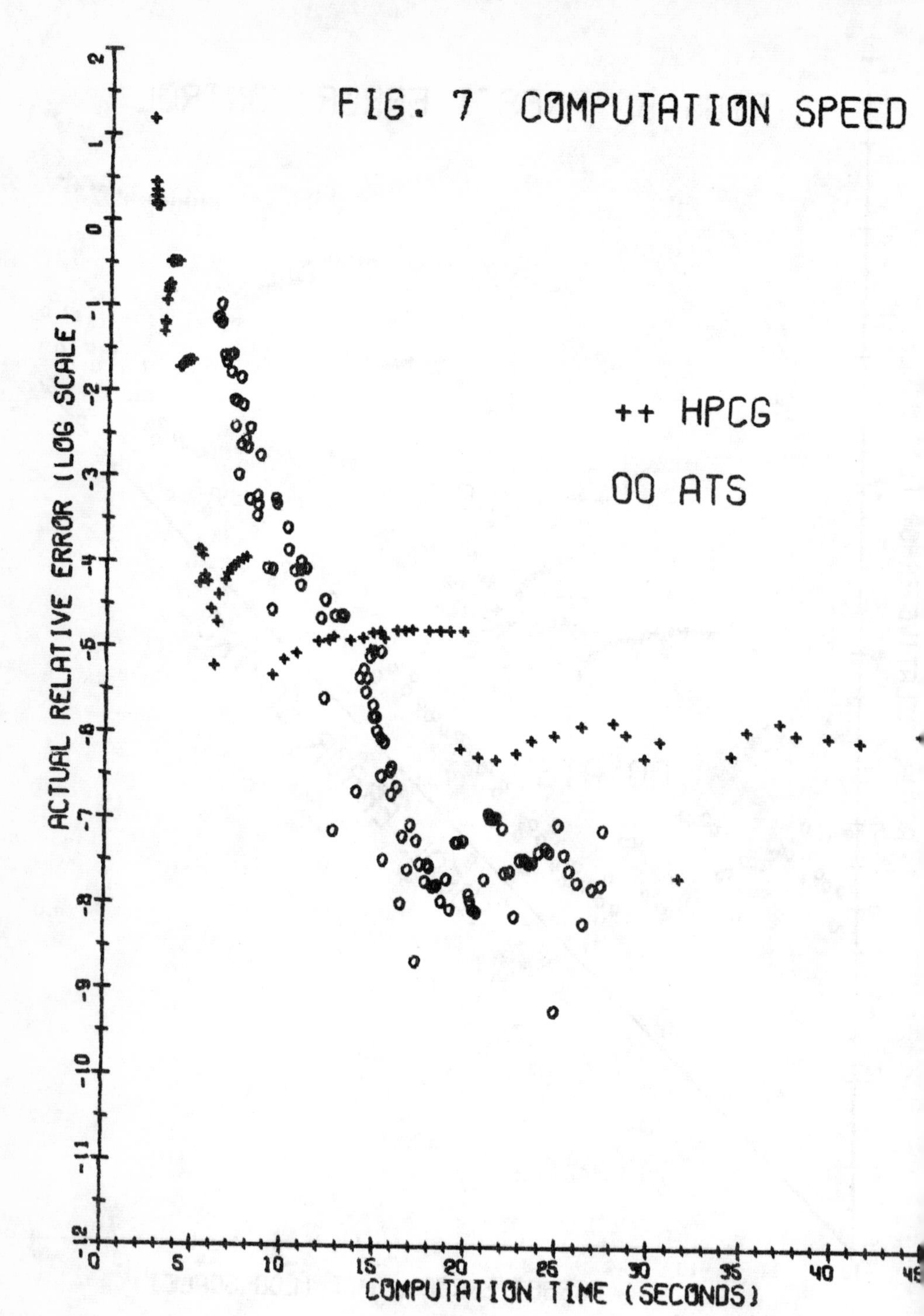
FIG. 7 COMPUTATION SPEED
++ HPCG
OO ATS
ACTUAL RELATIVE ERROR (LOG SCALE)
2
1
0
-1
-2
-3
-4
-5
-6
-7
-8
-9
-10
-11
-12
0
5
10
15
20
25
30
35
40
45
COMPUTATION TIME (SECONDS)

NTEGRAL OPERATORS FOR PARABOLIC EQUATIONS AND THEIR APPLICATION*

by

David Colton

. Introduction.

The theory of integral operators for elliptic partial diff-
rential equations was initiated by S. Bergman ([1]) and I. N.
ekua ([20]) some forty years ago and since that time has blossomed
nto an important area of research in both pure and applied mathem-
tics. The method is based on the construction of an integral
perator which maps analytic functions of a complex variable onto
solutions of the elliptic equation being investigated, and, among
the many applications of these operators, three have attracted
the most attention:

1) The analytic and numerical approximation of solutions to boundary value problems for elliptic equations ([4],[12], [20]);
2) Inverse methods for solving boundary value problems in fluid mechanics ([2],[10],[17]);
3) The analytic continuation of solutions to elliptic equations ([1],[11],[15]).

---

* This research was supported in part by AFOSR Grant 74-2592

In view of the success of the method of integral operators in the investigation of elliptic equations, it is somewhat surprising that the development of a corresponding theory for parabolic equations has been delayed until very recently (however see [3],[13],[14]). It is the purpose of this talk to outline some recent progress we have made in the construction of integral operators for parabolic equations in one space variable and to indicate some of the possible applications of these operators. Considerable progress has also been made in the area of parabolic equations in two space variables, and we hope to provide a survey of these results at a future date.

II. Integral Operators for Parabolic Equations.

We consider the general linear second order parabolic equation in one space variable written in normal form

$$u_{xx} + a(x,t)u_x + b(x,t)u = c(x,t)u_t \quad . \tag{2.1}$$

The substitution

$$u(x,t) = v(x,t) \exp\{-\tfrac{1}{2}\int_0^x a(\xi,t)d\xi\} \tag{2.2}$$

reduces (2.1) to an equation of the same form for $v(x,t)$, but with $a(x,t) \equiv 0$. Hence without loss of generality we can consider the equation

$$u_{xx} + q(x,t)u = c(x,t)\, u_t \quad . \tag{2.3}$$

In this talk we will assume for the sake of simplicity that, unless otherwise stated, $q(x,t)$ and $c(x,t)$ are analytic functions of their independent (complex) variables for $|x| < \infty$ and $|t| \le t_0$, where t_0 is some positive constant. In the theory we are about to describe two distinct situations present themselves: one is the case when the solution $u(x,t)$ of (2.3) (or (2.1)) is an analytic function of its independent variables in some neighborhood of a point $(0,t_1), 0 < t_1 < t_0$, and the other is when we merely assume that $u(x,t)$ is a classical solution of (2.3) (or (2.1)) in a neighborhood of (o,t_1), i.e. $u(x,t)$ is twice continuously differentiable in such a neighborhood. In the latter case we can immediately conclude from the known regularity properties of solutions to parabolic equations ([9]) that $u(x,t)$ is in fact analytic in x and infinitely differentiable (but not necessarily analytic) with respect to t. These two different situations lead us to the construction of two types of integral operators.

In the case when $u(x,t)$ is an analytic solution of (2.3) in some neighborhood of the point $(0,t_1)$ we define the operator $\underset{\sim}{P}_1$ which maps ordered pairs of analytic functions onto analytic solutions of (2.3) by

$$
\begin{aligned}
u(x,t) = \underset{\sim}{P}_1\{f,g\} = & -\frac{1}{2\pi i}\int_{|t-\tau|=\delta} E^{(1)}(x,t,\tau)f(\tau)d\tau \\
& -\frac{1}{2\pi i}\int_{|t-\tau|=\delta} E^{(2)}(x,t,\tau)g(\tau)d\tau
\end{aligned}
\tag{2.4}
$$

where $f(t) = u(0,t)$, $g(t) = u_x(0,t)$, δ is a sufficiently small positive number, and $E^{(1)}(x,t,\tau)$, $E^{(2)}(x,t,\tau)$ are defined by

$$E^{(1)}(x,t,\tau) = \frac{1}{t-\tau} + \sum_{n=2}^{\infty} x^n p^{(1,n)}(x,t,\tau) \tag{2.5a}$$

$$E^{(2)}(x,t,\tau) = \frac{x}{t-\tau} + \sum_{n=3}^{\infty} x^n p^{(2,n)}(x,t,\tau) \tag{2.5b}$$

where

$$p^{(1,1)} = 0$$

$$p^{(1,2)} = -\frac{c}{2(t-\tau)^2} - \frac{q}{2(t-\tau)} \tag{2.6a}$$

$$p^{(1,k+2)} = -\frac{2}{k+2} p_x^{(1,k+1)} - \frac{1}{(k+2)(k+1)}[p_{xx}^{(1,k)} + qp^{(1,k)} - cp_t^{(1,k)}];$$

$k \geq 1$

and

$$p^{(2,2)} = 0$$

$$p^{(2,3)} = -\frac{c}{6(t-\tau)^2} - \frac{q}{6(t-\tau)} \tag{2.6b}$$

$$p^{(2,k+2)} = -\frac{2}{k+2} p_x^{(2,k+1)} - \frac{1}{(k+2)(k+1)}[p_{xx}^{(2,k)} + qp^{(2,k)} - cp_t^{(2,k)}];$$

$k \geq 2$.

It is shown in [7] that the series (2.5a) and (2.5b) converge absolutely and uniformly on compact subsets of $|x| < \infty$, $|t| < t_0$, $|\tau| < t_0$, $\tau \neq t$, and estimate are obtained on the precise rates of convergence. From the Cauchy-Kowalewski theorem we can conclude that every analytic solution of (2.3) in a neighborhood of $(0,t_1)$ can be represented in the form (2.4).

We now consider the case when $u(x,t)$ is a classical (but not necessarily analytic) solution of (2.3) in the rectangle $D = \{(x,t): -x_0 < x < x_0,\ 0 < t < t_0\}$ where x_0 and t_0 are positive constants. We assume that $c(x,t) > 0$ for $(x,t) \in D$ and make the change of variables

$$\xi = \int_0^x \sqrt{c(s,t)}\, ds \qquad\qquad (2.7)$$
$$\tau = t$$

which (after a further change of variables of the form (2.2)) reduces (2.3) to an equation of the same form as (2.3), but with $c(x,t) = 1$. Hence we now assume $c(x,t) = 1$ to begin with and define the operator $\underset{\sim}{P}_2$ which maps solutions of the heat equation

$$h_{xx} = h_t \qquad\qquad (2.8)$$

onto (classical) solutions of (2.3) (with $c(x,t) = 1$) by

$$u(x,t) = \underset{\sim}{P}_2\{h\} = h(x,t) + \tfrac{1}{2}\int_{-x}^{x} [K(s,x,t) + M(s,x,t)]h(s,t)\,ds \qquad\qquad (2.9)$$

where $K(s,x,t)$ and $M(s,x,t)$ are solutions of the Goursat problems

$$K_{xx} - K_{ss} + q(x,t)\,K = K_t \qquad\qquad (2.10a)$$

$$K(x,x,t) = -\tfrac{1}{2}\int_0^x q(s,t)ds, \; K(0,x,t) = 0 \tag{2.10b}$$

and

$$M_{xx} - M_{ss} + q(x,t)\, M = M_t \tag{2.11a}$$

$$M(x,x,t) = -\tfrac{1}{2}\int_0^x q(s,t)\, ds, \; M_s(0,x,t) = 0 \tag{2.11b}$$

respectively. The existence of the functions $K(s,x,t)$ and $M(s,x,t)$ and their analyticity for $(x,t) \in D$, $-x_0 < s < x_0$, is established in [5] by the method of successive approximations, and estimates are obtained on the rates of convergence of these approximations. From the invertibility of Volterra integral operator of the second kind it can be shown that every classical solution of (2.3) defined in D can be represented in the form (2.9) where $h(x,t)$ is a solution of (2.8) in D. In the case when $u(x,t)$ is defined in $D^+ = \{(x,t) = 0 < x < x_0,\ 0 < t < t_0\}$, is continuously differentiable for $0 \le x < x_0$, $0 < t < t_0$, and satisfies $u(0,t) = 0$ it can be shown that $u(x,t)$ can be represented in the form

$$u(x,t) = h(x,t) + \int_0^x K(s,x,t)\, h(s,t)\, ds \tag{2.12}$$

where $h(x,t)$ is a solution of (2.8) in D^+ satisfying $h(0,t) = 0$. Similarly, when $u(x,t)$ is defined in D^+, is continuously differentiable for $0 \le x < x_0$, $0 < t < t_0$, and satisfies $u_x(0,t) =$ we have that $u(x,t)$ can be represented in the form

$$u(x,t) = h(x,t) + \int_0^x M(s,x,t)\, h(s,t)\, ds \tag{2.13}$$

where $h(x,t)$ is a solution of (2.8) in D^+ satisfying $h_x(0,t) = 0$.

III. The First Initial-Boundary Value Problem in a Rectangle.

In this section we will show how the operator $\underset{\sim}{P}_2$ can be used to construct approximate solutions to the first initial-boundary value problem for the reduced equation

$$u_{xx} + q(x,t)u = u_t \tag{3.1}$$

defined in the rectangle $D = \{(x,t): -x_0 < x < x_0,\ 0 < t < t_0\}$. We consider a solution of (3.1) defined in D which continuously assumes the initial-boundary data

$$\begin{aligned} u(-x_0,t) &= f(t),\ u(x_0,t) = g(t); \quad 0 \le t \le t_0 \\ u(x,0) &= h(x); \quad -x_0 \le x \le x_0 . \end{aligned} \tag{3.2}$$

Our aim is to construct a solution of (3.1) which approximates $u(x,t)$ arbitrarily closely in the maximum norm on compact subsets of D. To this end we first introduce the heat polynomials $h_n(x,t)$ defined by Rosenbloom and Widder in [18]. These polynomials are solutions of the heat equation (2.8) defined by

$$\begin{aligned} h_n(x,t) &= n! \sum_{k=0}^{\left[\frac{n}{2}\right]} \frac{x^{n-2k}\, t^k}{(n-2k)!k!} \\ &= (-t)^{n/2} H_n \left(\frac{x}{(-4t)^{\frac{1}{2}}}\right) \end{aligned} \tag{3.3}$$

where $H_n(z)$ denotes the Hermite polynomial of degree n. It was shown in [8] that the set $\{h_n(x,t)\}$ is complete with respect to the maximum norm in the space of solutions to the heat equation which are defined in D and continuous in the closure $\bar{D}$ of D, i.e. if $h(x,t)$ is a classical solution of (2.8) in D and continuous in $\bar{D}$, then for any $\epsilon > 0$ there exist constants $a_1, \ldots, a_n$ such that for N sufficiently large

$$\max_{(x,t)\in\bar{D}} \left|h(x,t) - \sum_{n=0}^{N} a_n h_n(x,t)\right| < \epsilon . \quad (3.4)$$

We now define $u_n(x,t)$ by

$$u_n(x,t) = \underset{\sim}{P}_2 \{h_n\} . \quad (3.5)$$

A short calculation ([8]) shows that the set $\{u_n(x,t)\}$ is a complete family of solutions with respect to the maximum norm in the space of solutions to (3.1) which are defined in D and continuous in $\bar{D}$. If we now orthonormalize the set $\{u_n(x,t)\}$ over the base and vertical sides of D we are led to the following theorem ([8]):

<u>Theorem</u>: Let $u(x,t)$ be a solution of (3.1) in D which is continuous in $\bar{D}$ and satisfies the initial-boundary data (3.2). Let D_0 be a compact subset of D. For $n = 0,1,\ldots,N$ define a_n by

$$a_n = \int_0^{t_0} f(t)\, u_n(-x_0,t)\,dt + \int_{-x_0}^{x_0} h(x)\, u_n(x,0)\,dx + \int_0^{t_0} g(t) u_n(x_0,t)\,dt$$

Then, given $\epsilon > 0$, there exist an N such that

$$\max_{(x,t)\in D_0} \left| u(x,t) - \sum_{n=0}^{N} a_n u_n(x,t) \right| < \epsilon .$$

The operator $\underset{\sim}{P}_2$ can also be used to construct approximate solutions to the first initial-boundary value problem for (3.1) defined in the quarter space $0 < x < \infty$, $0 < t < \infty$ ([8]).

IV. The Inverse Stefan Problem.

A wide variety of problems associated with the conduction of heat in a medium undergoing a change of phase have their mathematical formulation in terms of a free boundary problem associated with a parabolic equation in one space variable (c.f. [19]). In general the values of the various transfer coefficients and thermodynaimc properties of the material vary with respect to space and time (c.f. [16], chapter 13), i.e. instead of the heat equation (2.8) we are led to a parabolic equation with variable coefficients. After making a change of dependent variables the above considerations lead in the simplest case to the following free boundary problem (the single phase Stefan problem for the heat equation with variable coefficients):

$$u_{xx} + q(x,t)u = c(x,t)u_t, \quad 0 < x < s(t), \quad t > 0 \qquad (4.1a)$$

$$\left.\begin{aligned} &u(s(t),t) = 0, \quad t > 0 \\ &u_x(s(t),t) = -\dot{s}(t), \quad t > 0 \\ &u(0,t) = \varphi(t), \quad t > 0 \\ &s(0) = 0 \quad . \end{aligned}\right\} \qquad (4.1b)$$

The problem is to find $s(t)$ and $u(x,t)$ from (4.1a) and (4.1b) under the assumption that $\varphi(t)$ is known, and difficulties arise due to the fact that this problem is nonlinear with respect to $s(t)$. An alternate approach to the above problem is to consider the inverse problem (c.f. [19], pp. 71-80), i.e. given $s(t)$ to find $u(x,t)$ (and $\varphi(t) = u(0,t)$). This inverse approach has the advantage of being linear; however the problem now becomes improperly posed in the real domain (c.f. [13]). The inverse problem is well posed in the complex domain and hence we are led to the assumption that $s(t)$ be analytic for $0 < t < t_0$. Under this assumption the inverse Stefan problem can be easily solved through the use of the operator $\underset{\sim}{P}_1$ (We note that use of the Cauchy-Kowalewski theorem in this case is unsuitable for our purposes since the calculations involved are far too tedious for practical application and, more seriously, the power series solution construct via such a method may not converge in the full region where the solution is needed, i.e. in a region containing a segment of the t axis). We first place the cycle $|t - \tau| = \delta$ on the two dimensional manifold $x = s(t)$ in the space of two complex variables and note that since $u(s(t),t) = 0$ the integral in (2.4) which contains $E^{(1)}(x,t,\tau)$ vanishes. We are thus led to the following representation of the solution to the inverse Stefan problem:

$$u(x,t) = \frac{1}{2\pi i} \int_{|t-\tau|=\delta} E^{(2)}(x-s(\tau),t,\tau)\, \dot{s}(\tau)\, d\tau . \qquad (4.2)$$

A series representation for the solution can now be obtained by computing the residue in (4.2). This calculation is particularly simple in the case when (4.1a) is the heat equation (2.8). In this case we have from (2.5b) and (2.6b) that

$$E^{(2)}(x,t,\tau) = \frac{x}{t-\tau} + \sum_{j=1}^{\infty} \frac{x^{2j+1}(-1)^j j!}{(2j+1)!(t-\tau)^{j+1}} \tag{4.3}$$

and computing the residue in (4.2) leads to

$$u(x,t) = \sum_{n=1}^{\infty} \frac{1}{(2n)!} \frac{\partial^n}{\partial t^n} [x-s(t)]^{2n}, \tag{4.4}$$

a result which is in agreement with that of C. D. Hill ([13]).

V. Reflection Principles.

Let $u(x,t)$ be a classical solution of

$$u_{xx} + a(x,t)\, u_x + b(x,t)\, u = u_t \tag{5.1}$$

defined in $D^+ = \{(x,t): 0 < x < x_0,\ 0 < t < t_0\}$ and let $u(x,t)$ be continuously differentiable in $0 \le x < x_0,\ 0 < t < t_0$. We will also assume that $a(x,t)$ and $b(x,t)$ are analytic in the rectangle $D = \{(x,t): -x_0 < x < x_0,\ 0 < t < t_0\}$. We suppose at $x = 0$ that $u(x,t)$ satisfies a boundary condition of the form

$$\alpha(t)\, u(0,t) + \beta(t)\, u_x(0,t) = 0 \tag{5.2}$$

where $\alpha(t)$ and $\beta(t)$ are analytic for $0 < t < t_0$. Our aim is to determine under what conditions $u(x,t)$ can be uniquely continued as a classical solution of (5.1) into all of D. We first make the change of variables (2.2) and transform (5.1) and (5.2) int

$$u_{xx} + q(x,t)u = u_t \tag{5.3}$$

$$(\alpha(t) - \tfrac{1}{2}\,\beta(t)\,a(0,t))\,u(0,t) + \beta(t)\,u_x(0,t) = 0 \tag{5.4}$$

respectively, where $q(x,t)$ is analytic in D.

In the special case when $\beta(t) \equiv 0$ and $\alpha(t) = 1$ (i.e. Dirichlet boundary conditions) (5.4) reduces to

$$u(0,t) = 0 \tag{5.5}$$

and hence from section two of this paper we can represent $u(x,t)$ in the form

$$u(x,t) = h(x,t) + \int_0^x K(s,x,t)\,h(s,t)\,ds \tag{5.6}$$

where $h(x,t)$ is a solution of the heat equation (2.8) in D^+ satisfying $h(0,t) = 0$. From the reflection principle for the heat equation we have that $h(x,t)$ can be uniquely continued into all of D by the rule

$$h(x,t) = -h(-x,t) \tag{5.7}$$

and hence (5.6) provides the continuation of $u(x,t)$ into all of D([5]). In the case when $2\alpha(t) - \beta(t)\,a(0,t) \equiv 0$ for for $0 < t < t_0$ (5.4) reduces to

$$u_x(0,t) = 0 \tag{5.8}$$

and from section two of this paper we can represent $u(x,t)$ in the form

$$u(x,t) = h(x,t) + \int_0^x M(s,x,t)\, h(s,t)\, ds \tag{5.9}$$

where $h(x,t)$ is a solution of the heat equation (2.8) in D^+ satisfying $h_x(0,t) = 0$. From the reflection principle for the heat equation we can uniquely continue $h(x,t)$ into all of D by the rule

$$h(x,t) = h(-x,t) \tag{5.10}$$

and (5.9) now provides the continuation of $u(x,t)$ into all of D ([5]).

We now consider the general case when $2\alpha(t) - \beta(t)a(0,t) \neq 0$ for $0 < t < t_0$ and rewrite (5.4) in the form

$$u(0,t) + a(t)\, u_x(0,t) = 0 \tag{5.11}$$

where $a(t) = \beta(t)\, [\alpha(t) - \tfrac{1}{2}\beta(t)\, a(0,t)]^{-1}$. We will assume that $\beta(t) \neq 0$ for $0 < t < t_0$, i.e. $a(t) \neq 0$ for $0 < t < t_0$. Let $h^{(1)}(x,t)$ be a solution of the heat equation (2.8) defined in D^+ such that $h^{(1)}(0,t) = 0$ and define $h^{(2)}(x,t)$ by

$$h^{(2)}(x,t) = -a(t)\, h_x^{(1)}(x,t)\ . \tag{5.12}$$

Then it is shown in [6] that $u(x,t)$ can be represented in the form

$$u(x,t) = h^{(1)}(x,t) + h^{(2)}(x,t) + \int_0^x K^{(1)}(s,x,t)\, h^{(1)}(s,t)\, ds + \int_0^x K^{(2)}(s,x,t)\, h^{(2)}(s,t)\, ds \quad (5.$$

where $K^{(1)}(s,x,t)$ and $K^{(2)}(s,x,t)$ are solutions to the Goursat problems

$$K^{(1)}_{xx} - K^{(1)}_{ss} + q(x,t)\, K^{(1)} = K^{(1)}_t \quad (5.14a)$$

$$K^{(1)}(x,x,t) = -\tfrac{1}{2}\int_0^x q(s,t)\, ds, \quad K^{(1)}(0,x,t) = 0 \quad (5.14b)$$

and

$$K^{(2)}_{xx} - K^{(2)}_{ss} + \left(q(x,t) - \frac{\dot{a}(t)}{a(t)}\right) K^{(2)} = K^{(2)}_t \quad (5.15a)$$

$$K^{(2)}(x,x,t) = -\tfrac{1}{2}\int_0^x \left(q(s,t) - \frac{\dot{a}(t)}{a(t)}\right) ds, \quad K^{(2)}_s(0,x,t) = 0 \quad (5.15b$$

respectively. The existence of the functions $K^{(1)}(s,x,t)$ and $K^{(2)}(s,x,t)$ and their analyticity for $(x,t) \in D$, $-x_0 < s < x_0$, follows from the results of [5]. The reflection principle for solutions to (5.1) in D^+ satisfying the boundary condition (5.2) and the assumptions $2\alpha(t) - \beta(t)a(0,t) \neq 0$ and $\beta(t) \neq 0$ for $0 < t < t_0$ now follows from the representation (5.13) and the reflection principle (5.7) for the heat equation. It is of interest to note that this generalized reflection principle is new even for the special case of the heat equation (2.8).

References

1. S. Bergman, *Integral Operators in the Theory of Linear Partial Differential Equations*, Springer-Verlag, Berlin, 1961.

2. S. Bergman, Operator methods in the theory of compressible fluids, *Proceedings of Symposia in Applied Mathematics*, American Mathematical Society, Providence, Rhode Island, Vol. 1, 1949, 19-40.

3. S. Bergman, On singularities of solutions of certain differential equations in three variables, *Trans. Amer. Math. Soc.* 85 (1957), 462-488.

4. S. Bergman and M. Schiffer, *Kernel Functions and Differential Equations in Mathematical Physics*, Academic Press, New York, 1953.

5. D. Colton, Integral Operators and reflection principles for parabolic equations in one space variable, *J. Diff. Eqns.*, to appear.

6. D. Colton, Generalized reflection principles for parabolic equations in one space variable, *Duke Math. J.*, to appear.

7. D. Colton, The non-characteristic Cauchy problem for parabolic equations in one space variable, *SIAM J. Math. Anal.*, to appear.

8. D. Colton, The approximation of solutions to initial-boundary value problems for parabolic equations in one space variable, submitted for publication.

9. A. Friedman, Partial Differential Equations of Parabolic Type, Prentice Hall, Englewood Cliffs, New Jersey, 1964.

10. P. R. Garabedian, An example of axially symmetric flow with a free surface, Studies in Mathematics and Mechanics Presented to Richard von Mises, Academic Press, New York, 1954, 149-159.

11. R. P. Gilbert, Function Theoretic Methods in Partial Differentia Equations, Academic Press, New York, 1969.

12. R. P. Gilbert, Constructive Methods for Elliptic Partial Differential Equations, Springer-Verlag Lecture Note Series, Berlin, to appear.

13. C. D. Hill, A method for the construction of reflection laws for a parabolic equation, Trans. Amer. Math. Soc. 20(1967), 357-372.

14. C. D. Hill, Parabolic equations in one space variable and the non-characteristic Cauchy problem, Comm. Pure. Appl. Math. 20 (1967), 619-633.

15. H. Lewy, On the reflection laws of second order differential equations in two independent variables, Bull. Amer. Math. Soc. 65 (1959), 37-58.

16. A. V. Luikov, Analytical Heat Diffusion Theory, Academic Press, New York, 1968.

17. R. von Mises and M. Schiffer, On Bergman's integration method in two dimensional compressible fluid flow, Advances in Applied Mechanics, Vol. 1, Academic Press, New York, 1948, 249-285.

18. P. C. Rosenbloom and D. V. Widder, Expansions in terms of heat polynomials and associated functions, Trans. Amer. Math. Soc. 92 (1959), 220-266.

19. L. I. Rubinstein, The Stefan Problem. American Mathematical Society, Providence, Rhode Island, 1971.

20. I. N. Vekua, New Methods for Solving Elliptic Equations, John Wiley, New York, 1967.

Department of Mathematics
Indiana University
Bloomington, Indiana 47401

Galerkin Methods for Modeling Gas Pipelines

Todd Dupont, University of Chicago, Chicago, Ill. 60637

1. Introduction. We shall consider a quasi-linear, first-order hyperbolic system which describes the transient behavior of isothermal gas flowing in a pipe. The goal of this paper is to provide *a priori*, asymptotic error estimates for two Galerkin methods for the approximate solution of this system. The methods to be studied were described by RACHFORD and DUPONT in [2]

In section 2, after discussing the example which motivates this work, a slightly more general differential system is describe This more general system is the one that is analyzed; the assumptions that are made about this system are based on the example. In section 3 the two Galerkin methods, continuous-time and linearized Crank-Nicolson, are defined and the convergenc results are stated. Sections 4 and 5 are devoted to proving these results for the continuous-time and discrete-time cases, respectively.

2. <u>The differential problem</u>. Let $G(x,t)$ and $\rho(x,t)$ for $0 \le x \le L$ and $0 \le t$ be the mass rate of flow per unit cross sectional area and the density, respectively, of a gas flowing in a pipe of uniform cross sectional area. The variables x and t denote the distance along the pipe and time, respectively. In some situations of physical interest [2] it is reasonable to assume that, because heat transfer between the gas, the pipe, and its surroudings, the temperature of the gas is constant. In such cases the behavior of the system is described by the equations (in consistent units)

$$
\begin{aligned}
&\rho_t + G_x = 0 , \\
(1) \qquad &G_t + \gamma^2 \rho_x + (G^2/\rho)_x + \frac{fG|G|}{2D\rho} + \rho g h_x = 0,
\end{aligned}
$$

for $0 \le x \le L$, $0 \le t$, where $\gamma = \gamma(\rho)$ is the isothermal speed of sound, D is the internal pipe diameter, $f = f(G,\rho)$ is the "Moody friction factor", g is the acceleration due to gravity, and $h(x)$ is the height above a horizontal surface. The subscripts in (1) denote partial differentiation.

For the cases we wish to study the speed, $|G/\rho|$, of the gas is much less than the speed of sound γ, say 10% or less. In these situations it is reasonable to supply one boundary condition at each end of the pipe; in particular one can specify G at $x=0$ and $x=L$ for $t>0$. In order to determine

ρ and G for $t>0$, it is also necessary to specify their initial values.

If we associate $(\rho, G)^T$ with $u(x,t) = (u_1, u_2)^T$ then the initial-boundary-value problem above is a special case of a system of the form

$$u_t + Au_x + F = 0, \quad 0<x<L, \quad 0<t\leq T,$$

(2) $$u_2(0,t) = g_0(t), \quad u_2(L,t) = g_L(t), \quad 0<t\leq T,$$

$$u(x,0) = u_0(x), \quad 0\leq x\leq L,$$

where

(3) $$A = \begin{pmatrix} 0 & \mathfrak{a}_{12} \\ \mathfrak{a}_{21} & \mathfrak{a}_{22} \end{pmatrix}, \quad F = \begin{pmatrix} f_1 \\ f_2 \end{pmatrix},$$

and A and F are functions of x, t, u_1, u_2. It is systems of this type that we shall consider henceforth.

We shall assume that $u(x,t)$ is a smooth (sufficiently differentiable) solution of (2) and that A and F are twice continuously differentiable functions of their four arguments in a neighborhood, in $[0,L]\times[0,T]\times\mathbb{R}\times\mathbb{R}$, of the set

$$\mathcal{S} = \{(x,t,s_1,s_2): \quad 0\leq x\leq L, \; 0\leq t\leq T, \; s_1=u_1(x_1t), \; s_2=u_2(x,t)\}.$$

In addition the functions $\mathfrak{a}_{12}$ and $\mathfrak{a}_{21}$ will be assumed to be bounded below by a positive constant on $\mathcal{S}$ and, therefore, in some neighborho of $\mathcal{S}$. Since $\mathfrak{a}_{12} = 1$ and $\mathfrak{a}_{21} = \gamma^2-(G/\rho)^2$ in the example, this is

reasonable assumption in that case. The approximate solutions
ill be seen to converge uniformly to the solution u. Thus, there
s no loss in generality (and there is some gain in simplicity)
n assuming that these conditions hold everywhere. Hence, A and
 are assumed to be bounded, twice continuously differentiable
unctions on $[0,L]\times[0,T]\times\mathbb{R}\times\mathbb{R}$ with uniformly bounded first
nd second derivatives, and $\mathfrak{a}_{21}$ and $\mathfrak{a}_{12}$ are assumed to be bounded
elow by a positive constant which is independent of their
rguments.

3. Procedures and results. Let r and k be integers satisfying $2 \le r$, $0 \le k < r$; these will be fixed throughout and any constants which arise may depend on k and r. For a positive integer N let $h = L/N$; the behavior of the error as $h \to 0$ will be one of our primary concerns. Define $\mathfrak{m}_k(r,h) = \{V \in C^k([0,1]):$ V is a polynomial of degree $\le r$ on $((j-1)h, jh)$ for $j = 1, \ldots, N\}$ and $\overset{\circ}{\mathfrak{m}}_k(r,h) = \{v \in \mathfrak{m}_k(r,h): v(0) = v(1) = 0\}$. Take

$$(4)\quad \begin{aligned} \mathfrak{m} &= \mathfrak{m}_h = \mathfrak{m}_k(r,h) \times \mathfrak{m}_k(r,h) , \\ \mathfrak{n} &= \mathfrak{n}_h = \mathfrak{m}_k(r,h) \times \overset{\circ}{\mathfrak{m}}_k(r,h) . \end{aligned}$$

The continuous-time Galerkin approximation is a differentiable map $U: [o,T] \to \mathfrak{m}$ such that $U(0) - u_0$ is small and U (considered now as a function of (x,t)) satisfies

$$(5)\quad \begin{aligned} &\left(U_t + A(U)U_x + F(U),\ V\right) = 0,\ V \in \mathfrak{n},\ 0 < t \le T , \\ &U_2(0,t) = g_0(t),\ U_2(L,t) = g_L(t),\ 0 < t \le T , \end{aligned}$$

where $(W,V) = \int_0^L W \cdot V dx = \int_0^L (W_1 V_1 + W_2 V_2) dx$. The functions A and F are evaluated at $(x,t,U_1(x,t), U_2(x,t))$. For definiteness, the initial value $U(0)$ will be chosen so that

$$(6)\quad \begin{aligned} &\left(U(0) - u_0,\ V\right) = 0,\ V \in \mathfrak{n} , \\ &U_2(0,0) = g_0(0) ,\ U_2(L,0) = g_L(0) ; \end{aligned}$$

it is easily seen that (6) defines $U(0) \in \mathfrak{m}$ uniquely. Furthermore, exists and is unique for $0 \le t \le T$ and satisfies the following estimate.

Theorem 1. If u is $(r+1)$-times continuously differentiable on $[0,L] \times [0,T]$, then there is a constant C independent of h such that for $0 \le t \le T$

$$\|(u-U)(\cdot,t)\|_{L^2((0,L))} \le Ch^r. \tag{7}$$

The continuous-time Galerkin approximation is a theoretical modelization, since it is not readily computable, but the analysis is easier in this case and some of the ideas can be more clearly seen. In order to compute approximations we shall discretize time and use a linearized version of the Crank-Nicolson equation in place of (5).

Given a positive integer M the linearized-Crank-Nicolson-Galerkin approximation will be a sequence $\{U^m\}_{m=0}^M$ in $\mathfrak{m}$ such that U^m approximates u at $t = t^m = m\Delta t$, where $\Delta t = T/M$. The initial value U^0 is taken to be the $U(0)$ of (6). For a function S defined at the discrete times t^m, adopt the notations

$$S^m = S(t^m), \quad S^{m+\frac{1}{2}} = \tfrac{1}{2}(S^{m+1} + S^m), \tag{8}$$

$$\partial_t S^m = (S^{m+1} - S^m)/\Delta t .$$

The sequence $\{U^m\}_{m=0}^M$ is required to satisfy

$$(\partial_t U^m + \tilde{A}^m U_x^{m+\frac{1}{2}} + \tilde{F}^m + \tilde{S}^m(U^{m+1} - U^m), v) = 0, \quad v \in \mathfrak{m}, \tag{9}$$

$$U_2^{m+1}(0) = g_0(t^{m+1}) , \quad U_2^{m+1}(L) = g_L(t^{m+1}) ,$$

for $0 \le m < M$, where

$$\text{(10)} \qquad \tilde{A}^m = A(x,t^{m+\frac{1}{2}},U^m) , \quad \tilde{F}^m = F(x,t^{m+\frac{1}{2}},U^m) ,$$
$$\tilde{S}^m = S(x,t^{m+\frac{1}{2}},U^m,U_x^m) .$$

The function $S(x,t,\varphi,\psi)$ is defined by

$$\text{(11a)} \qquad 2S(x,t,\varphi,\psi) = \begin{pmatrix} s_{11} & s_{12} \\ s_{21} & s_{22} \end{pmatrix}$$

where

$$s_{11} = (D_3 a_{12})\psi_2 + D_3 F_1 ,$$

$$\text{(11b)} \qquad s_{12} = (D_4 a_{12})\psi_2 + D_4 F_1 ,$$

$$s_{21} = (D_3 a_{21})\psi_1 + (D_3 a_{22})\psi_2 + D_3 F_2 ,$$

$$s_{22} = (D_4 a_{21})\psi_1 + (D_4 a_{22})\psi_2 + D_4 F_2 ,$$

with D_i denoting partial differentiation with respect to the ith variable; all derivates in (11b) are evaluated at $(x,t,\varphi_1(x),\varphi_2(x))$ and ψ_i denotes $\psi_i(x)$.

The linearized-Crank-Nicolson-Galerkin approximation will be en to satisfy the following error bound.

heorem 2. Suppose that u is $(r+1)$-times continuously dif-
erentiable on $[0,L]\times[0,T]$ and that, for some constant C_1,
t and h tend to zero in such a fachion that $\Delta t \le C_1 h$.
hen, for Δt and h sufficiently small, there exists a unique
equence $\{U^m\}_{m=0}^{M}$ satisfying (10) with $U^0 = U(0)$ of (6).
urther, there is a constnat C depending on u, r, and C_1
uch that for h sufficiently small

$$\sup_{0\le m\le M} \|U^m - u(\cdot,t^m)\|_{L^2((0,L))} \le C[h^r + (\Delta t)^2]\,. \tag{12}$$

4. Continuous-time Galerkin convergence. Before proving Theorem
we shall collect a few results for later use. Adopt the notation

(13) $$\|r\| = \|r\|_{L^2((0,L))} .$$

The following result is easily seen.

Lemma 1. There is a constant C, depending only on r, such that if $b \in \mathfrak{M}_k(r,h)$,

(14) $$h^{\frac{1}{2}}\|b\|_{L^\infty((0,L))} + h\|b'\| + h^{3/2}\|b'\|_{L^\infty((0,L))} \le C\|b\| .$$

In producing error estimates for the Galerkin procedures it would be convenient if we could use a test function V which is the product of a smooth function and an element of $\mathfrak{h}$; the following lemma is used to estimate the error caused by using such a test function.

Lemma 2. There is a constant C such that if b is a continuousl differentiable function on $[0,L]$, $\varphi \in \mathfrak{M}_k(r,h)$, and if $\psi_1, \psi_2 \in \mathfrak{M}_k(r,h)$ are defined by

(15a) $$(b\varphi - \psi_1, V) = 0 , \quad V \in \mathfrak{M}_k(r,h) ,$$

(15b) $$(b\varphi - \psi_2, V) = 0 , \quad V \in \overset{0}{\mathfrak{M}}_k(r,h) , \quad (b\varphi - \psi_2)(x) = 0 ,$$
$$x = 0, L ,$$

then

$$\|b\varphi - \psi_\ell\| \le Ch\|b'\|_{L^\infty((0,L))}\|\varphi\| \ , \ \ell = 1,2 \ . \tag{16}$$

roof. In [1] an approximation χ of $b\varphi$ is constructed by a ocal interpolation process such that

$$\|b\varphi-\chi\| \le Ch\|b'\|_{L^\infty((0,L))}\|\varphi\| \ ;$$

he process is actually carried out in [1] for periodic functions nd is a little easier in this case. The χ can be taken such hat $(b\varphi - \chi)(x) = 0$ at $x = 0$ and $x = L$. But then (16) ollows since ψ_1 is the closest element of $\mathfrak{m}_k(r,h)$ to $b\varphi$, ith distance given by the norm $\|\cdot\|$, and ψ_2 is the closest lement of $\mathfrak{m}_k(r,h)$ which interpolates the end values.

roof of Theorem 1. Theorem 1 will be proved by showing that there s a constant C such that for any $t \in (0,T]$ if

$$\|(U-u)(\cdot,s)\| \le h^{3/2} \ , \quad 0 \le s \le t \ , \tag{17}$$

hen

$$\|(U-u)(\cdot,s)\| \le Ch^r \ , \quad 0 \le s \le t \ . \tag{18}$$

hat this suffices to prove the theorem can be seen as follows. f we take $0 < h \le h_0 = (2C)^{2/(3-2r)}$, then (18) implies that

$$\|(U-u)(\cdot,s)\| \le \tfrac{1}{2}h^{3/2} \ , \quad 0 \le s \le t \ .$$

hus, since $\|(U-u)(\cdot,s)\|$ is a continuous function of s , it ollows that if $t < T$, then there is a $t_1 \in (t,T]$ such that 17) holds for $0 \le s \le t_1$. Hence, if h is so small that $(U-u)(\cdot,0)\| < h^{3/2}$ and $h \le h_0$, then (18) holds for $0 \le s \le T$.

Define a map $W: [0,T] \to \mathfrak{m}$ such that

$$(u - W, V) = 0 \ , \quad V \in \mathfrak{h} \ , \quad 0 \le t \le T \ ,$$

(19)

$$(u_2 - W_2)(x,t) = 0 \ , \quad x = 0 \ , \quad x = L \ , \quad 0 \le t \le T \ .$$

Let $\eta = u - W$. It is well-known that there is a C, independent of h, such that for $0 \le t \le T$

(20) $$\|\eta(\cdot,t)\| + h^{\frac{1}{2}}\|\eta(\cdot,t)\|_{L^\infty((0,L))} + h\|\eta_x(\cdot,t)\| + h^{3/2}\|\eta_x(\cdot,t)\|_{L^\infty((0,L))} \le Ch^{r+1}\left\|\left(\frac{\partial}{\partial x}\right)^{r+1} u(\cdot,t)\right\|$$

We shall show that U is close to u by showing that U is close to W and then using (20).

Let

(21) $$D(x,t) = \begin{pmatrix} 1 & 0 \\ 0 & d(x,t) \end{pmatrix} ,$$

where $d(x,t) = (a_{21}/a_{12})(x,t,u(x,t))$. Note that by assumption D and D^{-1} are uniformly positive definite and continuously differentiable with bounded first derivatives.

Let $\mathfrak{v}(x,t) = (U - W)(x,t)$ and define $\gamma: [0,T] \to \mathfrak{h}$ by

(22) $$(\gamma - D^{-1}\mathfrak{v}, V) = 0 \ , \quad V \in \mathfrak{h} \ , \quad 0 \le t \le T \ .$$

e that, since $(D^{-1}\upsilon)_2(x,t) = 0$ if $x = 0$ or L, Lemma 2
lies that there is a C such that for $0 \le t \le T$

$$\|(\gamma - D^{-1}\upsilon)(\cdot,t)\| \le Ch\|\upsilon(\cdot,t)\| . \tag{}$$

Next note that

$$\tfrac{1}{2}\frac{d}{dt}(\|D^{-\frac{1}{2}}\upsilon\|^2) = (\upsilon_t, D^{-1}\upsilon) + \tfrac{1}{2}(\upsilon, (D^{-1})_t\upsilon)$$

$$= (\upsilon_t, \gamma) + \tfrac{1}{2}(\upsilon, (D^{-1})_t\upsilon) \tag{}$$

$$\le (\upsilon_t, \gamma) + C\|D^{-\frac{1}{2}}\upsilon\|^2 .$$

Assume that for some $t \in (0,T]$ (17) holds. Then using (20)
see that there is a constant C such that

$$\|U_x\|_{L^\infty((0,L))} \le C ; \tag{5}$$

te that C depends only on the constants in (20) and (14).

The first term on the right hand side of (24) can be re-
itten as

$$(\upsilon_t, \gamma) = (U_t - u_t, \gamma)$$

$$= (A(u)u_x + F(u) - A(U)U_x - F(U), \gamma) ,$$

ere (19) and the fact that $\gamma \in h$ were used to replace W_t
u_t . This relation implies that

$$(\upsilon_t, \gamma) \le (A(u)(\eta-\upsilon)_x + (A(u)-A(U))U_x + F(u) - F(U), \gamma)$$

$$\le -(A(u)\upsilon_x, \gamma) + C[\|\eta_x\| + \|\eta\| + \|\upsilon\|]\|\gamma\| , \tag{6}$$

where we used (25) and the assumption that A and F have bounded (first) derivatives. From (23), Lemma 1, and the fact that (D^{-1}) is bounded, it follows that

$$-(A(u)\upsilon_x,\gamma) \le -(A(u)\upsilon_x, D^{-1}\upsilon) + C\|\upsilon_x\|\, h\, \|\upsilon\|$$

$$\le -(A(u)D(D^{-1}\upsilon)_x, D^{-1}\upsilon) + C\|\upsilon\|^2 . \tag{27}$$

Since $A(u)D$ is symmetric,

$$-(A(u)\upsilon_x,\gamma) \le \tfrac{1}{2}(A(u)D(D^{-1}\upsilon))\cdot D^{-1}\upsilon\Big|_{x=0}^{x=L}$$

$$+ \tfrac{1}{2}((A(u)D)_x D^{-1}\upsilon, D^{-1}\upsilon) + C\|\upsilon\|^2 . \tag{28}$$

Because $\upsilon_2(0,t) = \upsilon_2(L,t) = 0$, the boundary terms in (28) vanis From (22) it follows that $\|\gamma\| \le \|D^{-1}\upsilon\|$; hence (26) and (28) imply that

$$(\upsilon_t,\gamma) \le C[\|D^{-\frac{1}{2}}\upsilon\| + \|\eta_x\| + \|\eta\|]\,\|D^{-\frac{1}{2}}\upsilon\| .$$

This relation and (24) combine to give

$$\frac{d}{dt}\|D^{-\frac{1}{2}}\upsilon\| \le C[\|D^{-\frac{1}{2}}\upsilon\| + \|\eta_x\| + \|\eta\|] . \tag{29}$$

Since, by the choices of $W(0)$ and $U(0)$, $\upsilon(0) = 0$, (29) implies that there is a C such that

$$\|\upsilon(\cdot,s)\| \le Ch^r , \quad 0 \le s \le t .$$

nce, by (20), (18) holds, and this proves the theorem.

The constant C can be seen to depend on r, T,

$$\sup_{0\le t\le T} \left\|\left(\frac{\partial}{\partial x}\right)^{\ell} u(\cdot,t)\right\| , \quad \text{and} \quad \int_0^T \left\|\left(\frac{\partial}{\partial x}\right)^{r+1} u(\cdot,t)\right\| dt ,$$

ere $\ell = \max(r,3)$.

Note that Lemma 1 and the bound on $\|\upsilon(\cdot,s)\|$ imply that

$$\|\upsilon(\cdot,s)\|_{L^\infty((0,L))} \le Ch^{r-\frac{1}{2}} , \quad 0 \le s \le T .$$

nce it follows from 20 that

$$\|(U-u)(\cdot,s)\|_{L^\infty((0,L))} \le Ch^{r-\frac{1}{2}} , \quad 0 \le s \le T .$$

us, the claim of section 2 that the approximate solutions con-
rge uniformly to the solution is verified in this case.

5. Proof of Theorem 2. The proof of Theorem 2 is very much like that of Theorem 1, except that it is messier. The proof is carri out by showing that there are constants C and $\tau > 0$ such that if $0 < \Delta t \leq \tau$, $0 < n \leq M$, U^m exists for $0 \leq m < n$, and

(30) $$\|U^m - u^m\| \leq h^{3/2}, \quad 0 \leq m < n,$$

then U^n exists and

(31) $$\|U^m - u^m\| \leq C[h^r + (\Delta t)^2], \quad 0 \leq m \leq n.$$

Since $\Delta t \leq C_1 h$ we see that for h sufficiently small (31) implies that

$$\|U^m - u^m\| \leq h^{3/2}, \quad 0 \leq m \leq n;$$

hence (31) holds for $n = M$.

To see that U^n exists, first note that by picking a basis for $\mathfrak{m}$ the problem of finding U^n can be reduced to the solution of a set of linear equations in which the number of unknowns is equal to the number of equations. Thus, it suffices to show that there exists $\tau > 0$ such that if $0 < \Delta t \leq \tau$ and $U \in \mathfrak{h}$ satisfi

(32) $$\left(\frac{1}{\Delta t}U + \tfrac{1}{2}\tilde{A}^{n-1}U_x + \tilde{S}^{n-1}U, V\right) = 0, \quad V \in \mathfrak{h},$$

then U is zero. Let D be as in (21) with arguments $(x, t^{n-\frac{1}{2}}, U^{n-1})$. Then take $\gamma \in \mathfrak{h}$ such that

$$(\gamma - D^{-1}U, V) = 0, \quad V \in \mathfrak{h},$$

d note that Lemma 2 implies that

$$\|\gamma - D^{-1}U\| \le Ch\|U\|.$$

e $V = \gamma$ in (32), and apply (30) with Lemma 1, and then
ply the argument of (27) and (28) to see that

$$\|D^{-\frac{1}{2}}U\|^2 \le C\Delta t\|U\|^2 .$$

nce $U \equiv 0$ for Δt sufficiently small.

Take $\upsilon^m = U^m - W^m$ and $\eta^m = u^m - W^m$. Take $(\tilde{D})^m$ to be as
(21) with $(x, \tilde{t}^{m+\frac{1}{2}}, u^m)$ as arguments, where $\tilde{t}^{m+\frac{1}{2}} = t^{m+\frac{1}{2}}$,
$\le m < M$, and $\tilde{t}^{M+\frac{1}{2}} = T$. Let $\gamma^m \in h$ be such that

$$(\gamma^m - (\tilde{D}^{-1})^m \upsilon^{m+\frac{1}{2}}, V) = 0 , \quad V \in h .$$

emma 2 then implies that

$$\|\gamma^m - (\tilde{D}^{-1})^m \upsilon^{m+\frac{1}{2}}\| \le Ch\|\upsilon^{m+\frac{1}{2}}\| . \tag{33}$$

Note that

$$\begin{aligned} \tfrac{1}{2}\partial_t(\|\tilde{D}^{-\frac{1}{2}}\upsilon\|^2)^m &= \frac{1}{2\Delta t}[((\tilde{D}^{-1})^{m+1}\upsilon^{m+1}, \upsilon^{m+1}) - ((\tilde{D}^{-1})^m \upsilon^m, \upsilon^m)] \\ &= ((\tilde{D}^{-1})^m \partial_t \upsilon^m, \upsilon^{m+\frac{1}{2}}) \\ &+ \tfrac{1}{2}((\partial_t(\tilde{D}^{-1})^m)\upsilon^{m+1}, \upsilon^{m+1}) \qquad (34) \\ &\le (\partial_t \upsilon^m, \gamma^m) + C\|\upsilon^{m+1}\|^2 = (\partial_t(U-u)^m, \gamma^m) + C\|\upsilon^{m+1}\|^2 . \end{aligned}$$

Using (2), (9), and Taylor's formula, it can be seen that

$$(\partial_t (u-u)^m, \gamma^m)$$

$$= (\rho^m + \tilde{A}^m(u) u_x^{m+\frac{1}{2}} + \tilde{F}^m(u) + S^m(u,u_x)(u^{m+1}-u^m) \tag{35}$$

$$- [\tilde{A}^m(U) U_x^{m+\frac{1}{2}} + \tilde{F}^m(U) + S^m(U,U_x)(U^{m+1}-U^m)], \gamma^m) \quad ,$$

where $\|\rho^m\| \le C(\Delta t)^2$, where C involves bounds for derivatives of u up through third order.

Note that

$$(\tilde{A}^m(u) u_x^{m+\frac{1}{2}} - \tilde{A}^m(U) U_x^{m+\frac{1}{2}}, \gamma^m)$$

$$= (\tilde{A}^m(u)(\eta-\upsilon)_x^{m+\frac{1}{2}}, \gamma^m) + ([\tilde{A}^m(u) - \tilde{A}^m(U)]\upsilon_x^{m+\frac{1}{2}}, \gamma^m)$$

$$+ ([\tilde{A}^m(u) - \tilde{A}^m(U)] W_x^{m+\frac{1}{2}}, \gamma^m)$$

$$\le -(\tilde{A}^m(u)\upsilon_x^{m+\frac{1}{2}}, \gamma) + Ch^r\|\gamma^m\| \tag{36}$$

$$+ C\|(\eta-\upsilon)^m\|_{L^\infty((0,L))}\|\upsilon_x^{m+\frac{1}{2}}\|\|\gamma^m\|$$

$$+ C\|(\eta+\upsilon)^m\|\|\gamma^m\|$$

$$\le -(\tilde{A}^m(u)\upsilon_x^{m+\frac{1}{2}}, \gamma) + C[h^r + \|\upsilon^m\| + \|\upsilon^{m+1}\|]\|\gamma^m\| \quad ,$$

where we used the induction hypothesis (30), Lemma 1, and

relation (20) to see that $\|(\eta-\upsilon)^m\|_{L^\infty((0,L))} \leq Ch$. Also note that

$$(37) \qquad (\rho^m + \tilde{F}^m(u) - \tilde{F}^m(U), \gamma) \leq C[h^r + (\Delta t)^2 + \|\upsilon^m\|]\|\gamma^m\| .$$

The remaining terms on the right hand side of (35) can be estimated as follows

$$(S^m(u,u_x)(u^{m+1}-u^m) - S^m(U,U_x)(U^{m+1}-U^m), \gamma^m)$$

$$= (S^m(u,u_x)\Delta t \partial_t \eta^m, \gamma^m)$$

$$+ ([S^m(u,u_x) - S^m(U,U_x)]\Delta t \partial_t W^m, \gamma^m)$$

$$+ (S^m(U,U_x)(\upsilon^{m+1} - \upsilon^m), \gamma^m)$$

$$(38) \qquad \leq C\Delta t\, h^{r+1}\|\gamma^m\|$$

$$+ C[\|(\eta-\upsilon)^m\| + \|(\eta-\upsilon)^m_x\|]\Delta t\|\partial_t W^m\|_{L^\infty}\|\gamma^m\|$$

$$+ C(\|\upsilon^{m+1}\| + \|\upsilon^m\|)\|\gamma^m\|$$

$$\leq C[h^r + \|\upsilon^m\| + \|\upsilon^{m+1}\|] ,$$

where $\Delta t \leq C_1 h$ and Lemma 1 were used to replace $\Delta t\|\upsilon^m_x\|$ by $\|\upsilon^m\|$.

When (36), (37), and (38) are used in (35) and the result sed in (34) we see that

$$\partial_t(\|\hat{D}^{-\frac{1}{2}}\nu\|^2)^m \le -(\tilde{A}^m(u)\nu_x^{m+\frac{1}{2}},\gamma)$$

$$+ C[h^{2r} + (\Delta t)^4 + \|\nu^m\|^2 + \|\nu^{m+1}\|^2] . \tag{39}$$

After treating the $-(\tilde{A}^m(u)\nu_x^{m+\frac{1}{2}},\gamma)$ term just as in (27) and (28) we see that

$$\partial_t(\|\hat{D}^{-\frac{1}{2}}\nu\|^2)^m \le C[h^{2r} + (\Delta t)^4 + \|\nu^m\|^2 + \|\nu^{m+1}\|^2] . \tag{40}$$

Since $\nu^0 = 0$ the discrete analog of Gronwall's inequality implies that for Δt sufficiently small $\|\nu^m\| \le C[h^r + (\Delta t)^2]$, $0 \le m \le n$. Hence (20) implies that (31) holds and thus proves the theorem.

Note that just as in section 4 we can get an $L^\infty((0,L))$ estimate for the error by using (20) and Lemma 1 applied ν. In particular it follows that

$$\|U^m - u^m\|_{L^\infty((0,L))} \le C[h^{r-\frac{1}{2}} + (\Delta t)^{3/2}] .$$

References

1. T. Dupont and L. Wahlbin, L^2 optimality of weighted-H^1 projections into piecewise polynomial spaces, to appear.

2. H. H. Rachford, Jr., and T. Dupont, A fast, high accuracy model for transient flow in gas pipeline systems by variational methods, to appear in Soc. Petroleum Eng. Journ., also SPE 4005 Fall, 1972.

The Application of Sparse Matrix Methods to the Numerical Solution of Nonlinear Elliptic Partial Differential Equations

S. C. Eisenstat, M. H. Schultz, and A. H. Sherman*

Abstract

We present a new algorithm for solving general semilinear, elliptic partial differential equations. The algorithm is based on Newton's Method but uses an approximate iterative method to solve the linear systems that arise at each step of Newton's Method. We show that the algorithm can maintain the quadratic convergence of Newton's Method and that it may be substantially faster than other available methods for semilinear or nonlinear partial differential equations.

* Department of Computer Science
Yale University

This research was supported in part by the Office of Naval Research, N0014-67-A-0097-0016.

1. The Model Problem

Let D be a bounded region of the plane, and let ∂D denote the boundary of D. We are interested in solving problems of the form

$$-\Delta u = f(u) \qquad \text{on } D \tag{1.1}$$

$$\text{with } u = 0 \qquad \text{on } \partial D \tag{1.1}$$

where f and D are such that

$$f_u(u) \leq \lambda < \Lambda$$

(where Λ is the fundamental eigenvalue of $-\Delta$ on D), so that u(x,y) exists and is unique. Under suitable conditions, our results apply to more general semilinear, self-adjoint elliptic boundary value problems in two and three dimensions and to certain nonlinear problems.

To obtain a numerical solution to (1.1), we must reduce the continuous, infinite dimensional problem to a discrete, finite dimensional one. Several techniques are available to do this, and we consider the use of finite difference and finite element approximations.

For the finite difference approximation, we replace the differential operator of (1.1) by a five-point difference approximation on a square grid D_h with boundary ∂D_h. Here h is the vertical or horizontal distance between two adjacent grid points. Letting U_{ij} be the approximation to u(ih,jh), we see that we may derive a problem given by the equations

$$U_{i-1,j} + U_{i,j-1} + U_{i,j+1} + U_{i+1,j} - 4U_{ij} = h^2 f(U_{ij}) \tag{1.2}$$

for $(ih,jh) \in D_h$

$$\text{and } U_{ij} = 0 \qquad \text{for } (ih,jh) \in \partial D_h \tag{1.2b}$$

hich is equivalent to (1.1). More concisely, we may rewrite (1.2) as an by N system of nonlinear equations

$$A(\underset{\sim}{U}) = \underset{\sim}{F}(\underset{\sim}{U}) \tag{1.3}$$

here

$$\underset{\sim}{U} = \{U_{ij}: (ih,jh) \in D_h\}$$

$$\underset{\sim}{F}(\underset{\sim}{U}) = \{h^2 f(U_{ij}): (ih,jh) \in D_h\}$$

is the number of grid points in D_h, and A is obtained from the finite ifference equations (1.2). The system (1.3) will have a unique soluion $\underset{\sim}{U}$ and U_{ij} will be a close approximation to $u(ih,jh)$ under the same onditions as we imposed on the system (1.1) earlier.

To solve (1.3) for $\underset{\sim}{U}$, we use Newton's Method. Letting $\underset{\sim}{U}^{(0)}$ be a iven initial guess, we obtain a sequence of successive approximations $\underset{\sim}{U}^{(K)}\}$ to $\underset{\sim}{U}$

$$\underset{\sim}{U}^{(K+1)} = \underset{\sim}{U}^{(K)} - J(\underset{\sim}{U}^{(K)})^{-1}[A\underset{\sim}{U}^{(K)} - \underset{\sim}{F}(\underset{\sim}{U}^{(K)})] \tag{1.4}$$

here $J(\underset{\sim}{U}^{(K)})$ is given by

$$J(\underset{\sim}{U}^{(K)}) = A - \frac{\partial \underset{\sim}{F}}{\partial \underset{\sim}{U}}(\underset{\sim}{U}^{(K)}) \tag{1.5}$$

nder the assumptions given above, $J(\underset{\sim}{U}^{(K)})$ will be symmetric, positive efinite, and sparse (i.e. the number of nonzeroes per row is bounded ndependent of h). As we shall see later, it is these properties which nable our algorithm to be efficient.

We now state the following result concerning the convergence of Newton's Method, cf. [8].

Theorem 1.1: Newton's Method is quadratically convergent:

$$\|\underset{\sim}{U}^{(K+1)} - \underset{\sim}{U}\| \leq c\|\underset{\sim}{U}^{(K)} - \underset{\sim}{U}\|^2$$

Thus convergence to the solution of the discrete problem will be quite rapid, requiring only O(log log N) iterations to reduce the error by a factor of h^2.

As an alternative to this finite difference approximation, we now consider finite element methods for approximating the solution of (1.1). If we let $\frac{\partial\phi}{\partial u} = f(u)$, then (1.1) is equivalent to minimizing the function G[u] given by

$$G[u] = \int_D \{(\frac{\partial u}{\partial x})^2 + (\frac{\partial u}{\partial y})^2 - 2\phi(u)\}\, dx\, dy \tag{1.6}$$

over all functions $u \in W^{1,2}(D)$ satisfying $u = 0$ on ∂D.

Again we must reduce the infinite dimensional continuous problem (1.6) to a finite dimensional discrete one. To do this we introduce S_N, a finite dimensional subspace of $W^{1,2}(D)$ with a linearly independent set of basis functions $\{\psi_i\}_{i=1}^N$. Then we approximate u by

$$\sum_{i=1}^{N} \beta_i \psi_i(x,y)$$

with the coefficients β_i chosen so as to minimize

$$G[\sum_{i=1}^{N} \beta_i\psi_i] = \int_D \{(\frac{\partial}{\partial x} \sum_{i=1}^{N} \beta_i\psi_i)^2 + (\frac{\partial}{\partial y} \sum_{i=1}^{N} \beta_i\psi_i)^2 \tag{1.7}$$

$$-2\phi(\sum_{i=1}^{N} \beta_i\psi_i)\}\, dx\, dy$$

Differentiating, we wish to choose the β_i so that

$$\int_D \psi_i \{-\Delta(\sum_{i=1}^{N} \beta_i \psi_i) - f(\sum_{i=1}^{N} \beta_i \psi_i)\}\, dx\, dy = 0 \tag{1.8}$$

for $1 \leq i \leq N$

Rewriting this is a more convenient form, we wish to solve the system of nonlinear equations given by

$$A\underset{\sim}{U} = \underset{\sim}{F}(\underset{\sim}{U}) \tag{1.9}$$

where $\underset{\sim}{U} = \{\beta_i\}$,

$$A = \{a_{ij}\} = \{\int_D [\frac{\partial \psi_i}{\partial x} \frac{\partial \psi_j}{\partial x} + \frac{\partial \psi_i}{\partial y} \frac{\partial \psi_j}{\partial y}]\, dx\, dy\}$$

$$\text{and } \underset{\sim}{F}(\underset{\sim}{U}) = \{\int_D \psi_i f(\sum_{j=1}^{N} \beta_j \psi_j)\}$$

As before, we use Newton's Method to solve (1.9). In this case, we are given an initial guess $\underset{\sim}{U}^{(0)}$ and obtain a sequence of successive approximations to $\underset{\sim}{U}$

$$\underset{\sim}{U}^{(K+1)} = \underset{\sim}{U}^{(K)} - J(\underset{\sim}{U}^{(K)})^{-1}[A\underset{\sim}{U}^{(K)} - \underset{\sim}{F}(\underset{\sim}{U}^{(K)})] \tag{1.10}$$

where

$$J(\underset{\sim}{U}^{(K)}) = A - \frac{\partial \underset{\sim}{F}}{\partial \underset{\sim}{U}} (\underset{\sim}{U}^{(K)}) = A - \{\int_D \psi_i \psi_j \frac{\partial f}{\partial \underset{\sim}{U}}(\sum_{\ell=1}^{N} \beta_\ell \psi_\ell)\} \tag{1.11}$$

We will assume that $\{\psi_i(x,y)\}_{i=1}^{N}$ is a local basis (i.e. for each i, the support of ψ_i intersects with the support of a bounded number of ψ_j's). Then $J(\underset{\sim}{U})$ will be sparse, symmetric, and positive definite, and the system (1.9) will have the same properties as the system (1.3).

We have seen how we may use Newton's Method with either finite difference or finite element approximations. In practice, we rewrite the iteration equations (1.4) or (1.10) for Newton's Method so as to obtain the sequence of successive approximations to $\underset{\sim}{U}$ by solving

$$J(\underset{\sim}{U}^{(K)})\underset{\sim}{\delta}^{(K)} = A\underset{\sim}{U}^{(K)} - \underset{\sim}{F}(\underset{\sim}{U}^{(K)}) \quad (1.$$

and setting

$$\underset{\sim}{U}^{(K+1)} = \underset{\sim}{U}^{(K)} - \underset{\sim}{\delta}^{(K)} \quad (1.$$

Our problem is then reduced to the solution of a sequence of related, sparse, symmetric, and positive definite systems of linear equations.

2. The Solution of Sparse, Symmetric, Positive Definite Systems of Linear Equations

In this section we consider methods for the solution of an N by N sparse, symmetric, positive definite system of linear equations

$$A\underset{\sim}{z} = \underset{\sim}{b} \tag{2.1}$$

As we saw in Section 1, such systems arise in the use of Newton's Method in the solution of a model nonlinear problem. We may solve (2.1) using either a direct decomposition method or one of several iterative methods, and we will discuss both classes of methods here.

The direct method that we consider is the Choleski decomposition, a symmetric variant of standard Gaussian elimination. To solve (2.1) we first decompose A into the product LL^T, with L a lower triangular matrix. Then we successively solve $L\underset{\sim}{w} = \underset{\sim}{b}$ followed by $L^T\underset{\sim}{z} = \underset{\sim}{w}$ to obtain the desired solution $\underset{\sim}{z}$.

It is well known that this direct method may be equally well applied to PAP^T instead of A for any permutation matrix P [8], and we will use this fact to reorder the variables and equations of the system (2.1) so as to reduce the storage and/or time required for its solution.

In general, the use of the Choleski decomposition to solve an N by N symmetric system of linear equations requires $O(N^2)$ storage and $O(N^3)$ time. However, since our model problem is sparse, we may use any of several techniques to reduce these requirements. All of the direct methods which we will describe are sensitive to the ordering of the variables and equations. It is of great importance to choose a permutation matrix P so that the reordered matrix PAP^T can be decomposed in as little time and with as little storage as possible. It has been shown that our model problem on a square five-point finite difference mesh requires at least $O(N \log N)$ storage and at least $O(N^{3/2})$ time [7]. Moreover, using techniques which we describe later and an ordering due to Birkhoff and George [2], we can actually achieve these lower bounds.

The simplest methods which take advantage of the sparsity in A are the band methods. We define the bandwidth $b(A)$ of A by

$$b(A) = \max \{|i-j| : a_{ij} \neq 0\} \tag{2.2}$$

It is easily verified that if $|i-j| > b(A)$, then $a_{ij} = \ell_{ij} = 0$, so that we need only store and operate on those elements which fall in the band of A (i.e. those elements a_{ij} such that $|i-j| \leq b(A)$). For our model problem the use of a band method with the natural (row by row) variable ordering will reduce the storage and time to $O(N^{3/2})$ and $O(N^2)$, respectively [8].

With more effort we can do even better than this. Band methods assume that all the matrix elements are nonzero within the band. For many variable orderings, though, this is not the case, and we can obtain further reductions in space and time by keeping more careful track of the positions of the nonzeroes in A and L. To do this we store the nonzero matrix elements row by row, keeping track of the number of nonzero elements in each row and the column index of each element. Since column accesses are quite costly, we perform all the eliminations in each row at the same time and insert new nonzeroes where and when they occur during the decomposition process. Using such techniques has the advantage that we can fully utilize our knowledge about the zero structures of A and L in order to reduce the costs of solving (2.1). However, the data structures (linked lists) and programming needed to implement the scheme may be quite complex, the storage requirements are variable so that storage management may be difficult, and the technique may not adapt well to paged, virtual storage systems.

To avoid the difficulties just mentioned, we use an improvement, due to Chang [3], in which we preprocess the system (2.1) in order to determine the exact locations of the nonzeroes in L before we actually do the decomposition. This symbolic factorization, though quite complex in itself, need be done only once for a sequence of problems (2.1) in which A has a fixed zero structure (even if the actual values of the elements of A do change). Thus for an application like Newton's Method, the cost of the symbolic factorization may be spread over the solution to a number of systems (2.1). For each system to be solved, we need perform only a numerical factorization, a process which becomes quite efficient after the symbolic factorization has determined the exact zero structure of L. When this technique is used with the nested dissection variable ordering due to Birkhoff and George [2], the Choleski decomposition requires only $O(N \log N)$ storage and $O(N^{3/2})$ time.

The alternative to using a direct method for the solution of (2.1) would be to use one of the many iterative methods [13] which are available. These have been developed mainly to solve sparse linear systems like our model problem, and they work very well for regular problems on simple areas of the plane. The main advantages of the iterative methods are in the storage requirements (usually O(N)), in the fact that some of them may be faster than direct methods, and in their ability to make use of a good guess at the solution. This last point is especially applicable to our model nonlinear problem, for the solutions to the sequence of linear problems arising from Newton's Method are converging quadratically towards the solution of the nonlinear problem.

However, the good points of the iterative methods may often be outweighed by their negative features. Most important, it is difficult to know when to stop iterating, since a posteriori error bounds are usually not conveniently available. Also, several of the methods have parameters which must be accurately estimated in order for the methods to behave well. Finally, not all of the methods can be extended easily to non-rectangular or irregular domains, so that there are many problems for which the iterative methods may not help at all.

In Table 2.1 (cf. [6]) we summarize the storage and time requirements of several direct and iterative methods. One should be careful about making direct comparisons, however, since the choice of the best method for a problem may depend on the problem size, on the context of the linear system in a larger numerical method, and, especially, on the coefficients of the terms in the expressions for the storage and time requirements.

Table 2.1

Method	Storage	Time
Choleski Decomposition	$O(N^2)$	$O(N^3)$
Band Choleski Decomposition	$O(N^{3/2})$	$O(N^2)$
Sparse Choleski Decomposition	$O(N \log N)$	$O(N^{3/2})$
Gauss-Seidel Iteration	$O(N)$	$O(N^2)$
SOR Iteration (Optimal ω)	$O(N)$	$O(N^{3/2} \log N)$
SSOR Iteration	$O(N)$	$O(N^{5/4})$
ADI Iteration (Optimal α)	$O(N)$	$O(N \log^2 N)$
SIP Iteration	$O(N)$	$O(N \log^2 N)$

3. The Newton-Sparse-Richardson Method

We now return to the model nonlinear problem presented in Section 1 and discuss the numerical solution of the sequence of linear problems generated by Newton's Method. We consider direct decomposition methods an certain variations on them that we will mention later. Although we restrict our analysis to the model problem, it is clear that what we will say is applicable to any numerical method which requires the solution o a sequence of related systems of linear equations.

The most obvious method for solving the sequence of linear syster is to solve each system using the sparse Choleski decomposition presente in Section 2. We shall call this method Newton-Sparse (NS), and it is clear that it has all of the convergence properties of Newton's Method. The advantages of NS are that there are no parameters to estimate and that it is essentially independent of the domain and the source of the linear systems. However, it ignores the relationships which exist betwe the solutions of successive problems, and its requirements for $O(N \log N$ storage and $O(N^{3/2} \log \log N)$ time very likely make it inferior to variations of Newton's Method using SSOR, ADI, or SIP (cf. [8]) to solve the linear systems occurring at each step (although this may depend heavily on implementation efficiency).

In an effort to avoid these difficulties, we will present a modification to the straightforward NS method based on the ideas embodied in the Strongly Implicit Procedures (SIP) of Stone [11], Diamond [4], and others. The basic strategy for the solution of each linear system in the sequence is to use the solution to a different, but related, system in conjunction with an iteration procedure to obtain the solution to the system of interest. In particular, suppose that we wish to solve (2.1) and that for some matrix B (depending on A) we can quickly solve (say in O(N) time) the system

$$(A+B)\underset{\sim}{w} = \underset{\sim}{d} \tag{3.1}$$

We may then obtain the solution of (2.1) by using a Richardson-D'Jakonov iteration (cf. [5, 9, 13]) to generate a sequence of vectors $\{\underset{\sim}{z}^{(K)}\}$ which converge linearly to the desired $\underset{\sim}{z}$. For a given starting guess $\underset{\sim}{z}^{(0)}$, the successive iterates are obtained by solving the system

$$(A+B)\underset{\sim}{z}^{(K+1)} = (A+B)\underset{\sim}{z}^{(K)} + \gamma(A\underset{\sim}{z}^{(K)} - \underset{\sim}{b}) \tag{3.2}$$

or, equivalently, by first solving the system

$$(A+B)\underset{\sim}{\delta}^{(K)} = A\underset{\sim}{z}^{(K)} - \underset{\sim}{b} \tag{3.3}$$

and then setting $\underset{\sim}{z}^{(K+1)} = \underset{\sim}{z}^{(K)} + \gamma\underset{\sim}{\delta}^{(K)}$. The cost of each step of the iteration will be O(N), and the iteration will converge to the solution of (2.1) if $\rho \equiv \|I + \gamma(A+B)^{-1} A\| < 1$. To maximize the rate of convergence, we must choose γ to minimize ρ. If all the eigenvalues of $(A+B)^{-1} A$ lie in the interval $[\alpha,\beta]$, then we will choose $\gamma = 2/(\alpha+\beta)$ and obtain $\rho = (\beta-\alpha)/(\alpha+\beta)$. If A and B are such that α and β are independent of N, then the iteration will converge at a rate independent of the mesh size. Hence, the solution of (2.1) with the above version of SIP would require O(log N) Richardson iterations or O(N log N) total time in order to reduce the initial error by a factor of 1/N. Further, it has been shown by several authors that these requirements may be reduced by using Tchebychev acceleration to choose a sequence of parameters $\{\gamma_K\}$ instead of the single parameter γ [4].

The main problem with SIP is that it has so far defied analysis even for our model problem, and it is not known whether there exists a matrix B for which α and β are independent of N. We now examine a meth similar to SIP which does lead to a rate of convergence independent of N SIP chooses a matrix B so that A+B has a sparse, symmetric decomposition of an especially nice form. What we propose is to choose B so that A+B is a matrix for which we already have a decomposition, even though the decomposition will have O(N log N), rather than O(N) nonzero elements. That is, in order to guarantee a rapid rate of convergence independent o the mesh size, we will sacrifice a factor of log N in the work per iteration. More explicitly, we use a direct decomposition method to decompo $J(\underset{\sim}{U}^{(0)})$ into a product LL^T in the first step of Newton's Method. Then a successive steps of Newton's Method, we define $B = J(\underset{\sim}{U}^{(0)}) - J(\underset{\sim}{U}^{(K)})$ and use the Richardson-D'Jakonov iteration given above to solve (1.12). We shall call this method Newton-Sparse-Richardson (NSR).

There are two main theoretical questions which arise concerning the use of NSR. First, is it possible to maintain the quadratic converg ence of the Newton iteration by using an inner Richardson-D'Jakonov iteration scheme at each step? And if so, how many Richardson-D'Jakonov iterations are required to do it? In answer to these questions, we give the following two theorems, which will be proved elsewhere [10].

Theorem 3.1: If Newton's Method is quadratically convergent for a given semilinear problem (1.1), and if we define $\underset{\sim}{U}^{(K+1)}$ to be the 2^K-th Richardson-D'Jakonov iterate in the K-th step of NSR, then NSR is quadratically convergent independent of N:

$$\|\underset{\sim}{U}^{(K+1)} - \underset{\sim}{U}\| \leq c\|\underset{\sim}{U}^{(K)} - \underset{\sim}{U}\|^2, \; c < 1.$$

Theorem 3.2: If Newton's Method is quadratically convergent for a given semilinear problem (1.1), and if we define $\underset{\sim}{U}^{(K+1)}$ to be the <u>first</u> Richardson-D'Jakonov iterate in the K-th step of NSR, then NSR is linearly convergent independent of N:

$$\|\underset{\sim}{U}^{(K+1)} - \underset{\sim}{U}\| \leq c\|\underset{\sim}{U}^{(K)} - \underset{\sim}{U}\|, \; c < 1.$$

Having determined that NSR will actually converge whenever Newton's method converges, we now wish to know how much effort is required to solve a semilinear problem using NSR. Rather surprisingly, we have the following theorem which shows that if we neglect preprocessing and the cost of the sparse LL^T factorization at the first step, then the NSR methods described in Theorems 3.1 and 3.2 asymptotically require the same amount of time!

Theorem 3.3: Assume that Newton's Method converges quadratically for a given semilinear problem (1.1). Then the NSR methods described in Theorems 3.1 and 3.2 will reduce the initial error $\|\underset{\sim}{U}^{(0)} - \underset{\sim}{U}\|$ by a factor of $1/N$ in $O(N \log^2 N)$ iteration time, if we ignore preprocessing and the cost of the LL^T factorization at the first step.

Proof: Each Richardson-D'Jakonov (RD) iteration requires $O(N \log N)$ time. If we take 2^K RD iterations at the K-th step of NSR, Theorem 3.1 tells us that we can achieve quadratic convergence, so that we need only log log N NSR steps to reduce the initial error by 1/N. The total iteration time required for NSR is then

$$O\left(\sum_{K=1}^{\log \log N} 2^K N \log N\right) = O(N \log^2 N)$$

On the other hand, if we use only one RD iteration at each step of NSR, we will need log N steps to reduce the initial error by 1/N. Since each step of NSR will require $O(N \log N)$ time, we see again that the total time for NSR is $O(N \log^2 N)$.

If we consider the sparse factorization at the first NSR step to be a form of preprocessing, then NSR requires only $O(N \log^2 N)$ time as against the $O(N^{3/2} \log \log N)$ time required for the NS method that we discussed above. Other advantages of the NSR method are that it is independent of both the problem domain and the method used to generate the nonlinear system of equations and that it takes advantage of good guesses to the solution at each iteration step. The only serious disadvantages of NSR are its requirements for $O(N \log N)$ storage locations and $O(N^{3/2})$ preprocessing time.

4. Numerical Experiments

In this section, we shall illustrate our general method by solving the equation

$$(su_x)_x + (tu_y)_y = f(x,y,u)$$

over the domain

$$D = (0,1) \times (0,1) - [\tfrac{3}{8},\tfrac{5}{8}] \times [\tfrac{3}{8},\tfrac{5}{8}]$$

the unit square with a hole cut out. In particular, we consider two problems given by Bartels and Daniel [1]:

$$\{(1 + x^2 + y^2)u_x\}_x + \{(1 + e^x + e^y)u_y\}_y = g(x,y)e^u \qquad (4.1a)$$

in D

$$u(x,y) = x^2 + y^2 \qquad (4.1b)$$

on ∂D

and

$$\{(1 + (x-y)^2)u_x\}_x + \{(10 + e^{xy})u_y\}_y = g(x,y)(1 + u)^3 \qquad (4.2a)$$

in D

$$u(x,y) = x^2 + y^2 \qquad (4.2b)$$

on ∂D

where in each case $g(x,y)$ is chosen to make the unique exact solution $u(x,y) = x^2 + y^2$. For all tests, the initial approximation was chosen to be zero inside the region and to satisfy the boundary conditions. The results are summarized in Table 4.1. By way of comparison, we also give in Table 4.2 the corresponding times for the Nonlinear Conjugate Gradient method (NCG) proposed by Bartels and Daniel [1]. All times listed are in seconds and reflect execution times on a CDC-6600. We note that asymptotically NCG requires $O(N)$ storage, $O(N^{3/2} \log N)$ preprocessing time, and $O(N \log^2 N)$ iteration time, whereas NSR requires $O(N \log N)$ storage, $O(N^{3/2})$ preprocessing time, and $O(N \log^2 N)$ iteration time.

The results are clear: even including domain-dependent preprocess ing time (ordering and symbolic factorization), which accounts for rough half the total execution time, NSR reduces the nonlinear residual by a factor of 10^{-11} in significantly less time than NCG requires to reduce the error by a factor of 10^{-3}. There is little to choose between the tw variants of NSR. Other experiments indicate that letting the number of Richardson iterations vary between the two extremes gives better results

Table 4.1

Newton-Sparse-Richardson Method

	Problem (4.1)				Problem (4.2)			
	h = 1/16		h = 1/32		h = 1/16		h = 1/32	
Domain-dependent preprocessing time	.496	.474	3.901	3.786	.495	.511	3.902	3.908
Problem-dependent preprocessing time	.170	.167	1.211	1.173	.160	.165	1.164	1.162
Number of Richardson iterations per K-th Newton iteration	2^K	1	2^K	1	2^K	1	2^K	1
Richardson iteration time	.032	.032	.169	.165	.032	.034	.170	.174
Number of Newton iterations to reduce nonlinear residual by a factor of 10^{-11}	4	8	4	8	4	12	4	11
Nonlinear residual time	.055	.053	.227	.219	.044	.045	.178	.175
Total iteration time	.669	.650	3.274	2.918	.629	.918	3.086	3.604
Total time	1.335	1.291	8.386	7.877	1.284	1.594	8.152	8.674

Table 4.2

	Nonlinear Conjugate Gradient Method			
	Problem (4.1)		Problem (4.2)	
	h = 1/16	h = 1/32	h = 1/16	h = 1/3
Preprocessing time	.394	2.90	.398	2.82
Number of conjugate gradient iterations to reduce error by factor h^2	6	8	9	12
Time per iteration	.347	1.20	.287	.943
Total iteration time	2.08	9.63	2.58	11.3
Total time	2.57	12.9	3.05	14.4

5. Conclusion

In this paper we have presented a means for improving the efficiency of Newton's Method as used in the numerical solution of semilinear self-adjoint, elliptic partial differential equations. However, our technique of using an SIP-like iteration to solve the linear systems that arise in each step of Newton's Method is not limited to just this one application. Under suitable conditions, the method may be equally useful in the numerical solution of more general nonlinear and time dependent problems. In fact, the technique may prove fruitful whenever the numerical solution of any problem requires the solution of a sequence of related systems of linear equations; the only restriction is that the operators in successive systems be close enough so that the Richardson-D'Jakonov iteration converges quickly.

In theory we have seen that our method is quite efficient. The numerical examples show the practical value of the method, both as a tool for enhancing the efficiency of Newton's Method and as a very efficient means of solving certain semilinear problems for which other methods have been successfully used in the past. We are currently examining the applicability of the method to the numerical solution of more general nonlinear problems, and we fully expect that it will be able to compete quite successfully with methods which are currently in use.

References

[1] R. Bartels and J. W. Daniel. A conjugate gradient approach to nonlinear elliptic boundary value problems in irregular regions. Report #CNA 63, Center for Numerical Analysis, The University of Texas at Austin, 1973.

[2] G. Birkhoff and D. J. Rose. Elimination by nested dissection. Complexity of Sequential and Parallel Numerical Algorithms, edited by J. F. Traub, Academic Press, New York, 1973.

[3] A. Chang. Application of sparse matrix methods in electric power system analysis. Sparse Matrix Proceedings, edited by R. A. Willoughby, IBM Research Report #RA1, Yorktown Heights, New York, 1968.

[4] M. A. Diamond. An Economical Algorithm for the Solution of Finite Difference Equations, PhD dissertation, Department of Computer Science, University of Illinois, 1971.

[5] E. G. D'Jakonov. On certain iterative methods for solving non-linear difference equations. Proceedings of the Conference on the Numerical Solution of Differential Equations, (Scotland, June 1969 Springer-Verlag, Heidelberg, 1969.

[6] F. W. Dorr. The direct solution of the discrete Poisson equation on a rectangle. SIAM Review 12:248-263, 1970.

[7] A. J. Hoffman, M. S. Martin, and D. J. Rose. Complexity bounds fo regular finite difference and finite element grids. SIAM Journal on Numerical Analysis 10:364-369, 1973.

[8] J. M. Ortega and W. C. Rheinboldt. Iterative Solution of Nonlinea Equations in Several Variables, Academic Press, New York, 1970.

[9] L. F. Richardson. The approximate arithmetical solution by finite differences of physical problems involving differential equations with an application to the stresses in a masonry dam. Philosophical Transactions of the Royal Society, London, Series A(210):307-357, 1910.

[10] A. H. Sherman. PhD dissertation, Department of Computer Science, Yale University. To appear.

[11] H. L. Stone. Iterative solution of implicit approximations of

multidimensional partial differential equations. SIAM Journal on Numerical Analysis 10:530-558, 1968.

[12] J. H. Wilkinson. The Algebraic Eigenvalue Problem, Clarendon Press, London, 1965.

[13] D. M. Young. Iterative Solution of Large Linear Systems, Academic Press, New York, 1971.

Collocation solutions of integro-differential equations

Ruben J. Espinosa
Department of Mathematics
University of Pittsburgh
Pittsburgh, Pa. 15260

Abstract. Collocation solutions of mth order nonlinear integro-differential equations are discussed. The solutions are piecewise polynomials and are determined by the requirement that they satisfy the given equation at a finite number of points. Approximations to the collocation equations are also discussed.

1. Introduction. Collocation using piecewise polynomials has been used extensively in the solution of several kinds of linear and nonlinear operator equations. See [4], [6], [7], [9], [11], [12], [13]. We discuss here the application of collocation based on piecewise polynomials to the solution of integro-differential equations on the finite interval [a,b] of the type

$$D^m x(s) - \int_a^b \mathcal{K}(s,t,x(t),\ldots, D^m x(t))\, dt = F(s,x(s),\ldots, D^{m-1}x(s))$$

subject to m boundary conditions. When m = 0 our discussion applies and gives collocation solutions of Urysohn integral equations.

2. Notation and preliminaries. Let $\Delta \equiv \{s_i\}_0^N$ be a strict partition of the finite interval [a,b],

$$a = s_0 < s_1 < \cdots < s_N = b.$$

For each such partition Δ we define $\pi \equiv \max_{0\leq i\leq N-1} (s_{i+1} - s_i)$.

We denote by C [a,b] the space of continuous functions on [a,b] with the supremum-norm, and by $\mathcal{R}$[a,b] the Banach space of bounded Riemann integrable functions on [a,b] with the supremum-norm. Recall that $f \in \mathcal{R}$[a,b] if and only if f is continuous almost everywhere on [a,b] and is bounded.

Let k be a positive integer. We denote by P_k the space of polynomials of degree less than k. For each partition Δ we represent by $P_{k,\Delta}$ the space of functions defined on [a,b] which are polynomials of degree less than k on each subinterval (s_i, s_{i+1}), $i = 0,\ldots, N-1$.

Throughout we shall represent any generic constant by the general symbol K, provided that no ambiguity arises.

We will refer in general to any real valued nondecreasing and subadditive function $\omega(\cdot)$ defined on some interval $(0,\varepsilon)$ and with $\lim_{h\to 0} \omega(h) = 0$, as a modulus of continuity.

3. Collocation. We consider the problem of finding $x \in C^m [a,b]$, $m \geq 0$, so that

$$(3.1) \qquad D^m x = E\,x, \quad \beta_i x = c_i, \quad i = 1,\ldots, m,$$

where

$$E\,x(s) = \int_a^b \mathcal{K}(s,t,\underset{\sim}{x}(t))\,dt + F(s,\underset{\sim}{x}(s)).$$

Here $\mathcal{K}$ and F are maps from some suitable domains in R^{m+3} and R^{m+1} into R, $\{\beta_i\}_1^m$ is a set of continuous linear functionals on $C^{m-1}[a,b]$, $\{c_i\}_1^m$ is a fixed set of constants and $\underset{\sim}{x}(t)$ equals $(x(t),\ldots, D^m x(t))^T$ or $(x(t),\ldots, D^{m-1}x(t))^T$ depending on whether it appears as an argument of $\mathcal{K}$ or F respectively. We assume throughout that (3.1) has a solution that will be denoted by x.

Assume that $\{\beta_i\}_1^m$ is linearly independent over P_m and let $H(s,t)$ denote the Green's function corresponding to the problem

$$D^m\, u = 0, \quad \beta_i\, u = 0, \quad i = 1,\ldots, m.$$

Then x satisfies $D^m\, x = y$, $\beta_i\, x = c_i$, $i = 1,\ldots, m$ if and only if

$$x = p + \int_a^b H(\cdot,t)\, y(t)\, dt,$$

where p is the unique polynomial of degree less than m satisfying $\beta_i\, p = c_i$, $i = 1,\ldots, m$. Define

$$y^{-m+i} \equiv D^i p + \int_a^b H_i(\cdot,t)\, y(t)\, dt, \quad i = 0,\ldots, m-1,$$

where

$$H_i(s,t) = \partial^i/\partial s^i\, H(s,t),$$

then (3.1) is equivalent to finding $y = D^m x$ such that

$$y = Ty \tag{3.2}$$

where

$$Ty \equiv \int_a^b \mathcal{K}(\cdot,t,\underset{\sim}{y}^{-m}(t))\, dt + F(\cdot,\underset{\sim}{y}^{-m}(\cdot)). \tag{3.3}$$

Here $\underset{\sim}{y}^{-m}(t)$ equals $(y^{-m}(t),\ldots, y(t))^T$ or $(y^{-m}(t),\ldots, y^{-1}(t))^T$ depending on whether it appears as an argument of $\mathcal{K}$ or F.

In order to solve (3.1) by collocation, let $\{\xi_i\}_1^k$ be a fixed set of points $-1 \leq \xi_1 < \cdots < \xi_k \leq 1$. For each partition Δ of [a,b] we select the set of collocation points $M_\Delta \equiv \{\tau_i\}_1^{Nk}$ by

$$\tau_{ik+j} = (s_i + s_{i+1} + \xi_j(s_{i+1} - s_i))/2, \quad j = 1,\ldots, k,$$

$i = 0,\ldots, N-1$. Note that if $\xi_1 = -1$ and $\xi_k = 1$, then $\tau_{ik} = \tau_{ik+1}$. In this case we will take as approximating spaces

$$S_\Delta \equiv P_{m+k,\Delta} \cap C^m[a,b]$$

as in Russell and Shampine [13]. If $-1 < \xi_1$ and $\xi_k < 1$ then we will take

$$S_\Delta \equiv P_{m+k,\Delta} \cap C^{m-1}[a,b]$$

as in De Boor and Swartz [6]. Let $S_\Delta^{(m)}$ denote the space of m derivatives of S_Δ. Clearly $S_\Delta^{(m)}$ consists of functions which are not necessarily defined at the interior knots of Δ. For convenience we will assume that their value at these points has been redefined and set equal to the average of the two one sided limits.

We want to determine $x_\Delta \in S_\Delta$ such that

$$(3.4) \quad D^m x_\Delta(\tau_i) - \int_a^b \mathcal{K}(\tau_i,t,\underset{\sim}{x}_\Delta(t))\, dt = F(\tau_i,\underset{\sim}{x}_\Delta(\tau_i)),$$

$$\tau_i \in M_\Delta, \quad \beta_i\, x_\Delta = c_i, \quad i = 1,\ldots, m.$$

Although this is the set of equations that is usually solved, for the analysis of convergence we consider the equivalent problem of determining $y_\Delta = D^m x_\Delta \in S_\Delta^{(m)}$ such that

$$(3.5) \qquad y_\Delta(\tau_i) = T\, y_\Delta(\tau_i),\ \tau_i \in M_\Delta,$$

where T is defined in (3.3). Let Q_Δ denote the projection from C[a,b] into $S_\Delta^{(m)}$ interpolating at the points $\tau_i \in M_\Delta$. Then it can be observed that y_Δ satisfies (3.5) if and only if y_Δ satisfies

(3.6) $$y_\Delta = Q_\Delta \, T \, y_\Delta$$

With respect to properties of Q_Δ we remark that each of them is a continuous projection and its norm is independent of Δ.

A rather general result on collocation solutions of operator equations in one variable is proven in [6, Lemma 3.1]. The following lemma is a slightly modified version of this result, is in a form more amenable to our problem, and has $\mathcal{R}[a,b]$ as the underlying space. The changes are straightforward and the proof will not be reproduced here.

Lemma 3.1. Consider the equations

(3.2') $$y = Ty$$

and

(3.6') $$y_\Delta = Q_\Delta \, T \, y_\Delta$$

where T is an operator from $\mathcal{R}[a,b]$ into C[a,b] and Q_Δ is a projection as described before. Let $y \in C[a,b]$ be a solution of (3.2') and assume:

(i) T is Frechet differentiable at y and for all f and g in an open ball $B(y,\delta) \subset \mathcal{R}[a,b]$ of radius δ and center at y,

$$||E(f,y) - E(g,y)|| \leq K \; ||f-g|| \; \max(||f-y||, \; ||g-y||)$$

and

$$||E(f,y)|| \leq K \; ||f-y||^2$$

where

$$E(f,y) \equiv Tf - Ty - T'(y)\,(f-y).$$

Here $T'(y)$ is the derivative of T at y and K is a constant depending only on T and y.

(ii) For every $f \in \mathcal{R}[a,b]$, $T'(y)\,f \in C[a,b]$ and

$$|T'(y)\,f(s_1) - T'(y)\,f(s_2)| \leq \omega(|s_1 - s_2|)\;||f||$$

where $\omega(\cdot)$ is a modulus of continuity independent of f.

(iii) $I - T'(y)$ as a mapping from $\mathcal{R}[a,b]$ into $\mathcal{R}[a,b]$ has a continuous inverse.

Then, there exists $\varepsilon > 0$ and $d > 0$ such that for all partitions Δ with $\pi \leq d$,

(a) (3.6') has exactly one solution $y_\Delta \in S_\Delta^{(m)} \cap B(y,\varepsilon)$ and this solution satisfies

$$(3.7) \qquad ||y-y_\Delta|| \leq K \; \mathrm{dist}_\infty\,(y,S_\Delta^{(m)}).$$

(b) y is the unique solution of (3.2') in $B(y,\varepsilon)$.

If $y \in C^n[a,b]$, then the use of Jackson's theorem [5], gives that (3.7) can be written in the form

$$||y-y_\Delta|| \leq K \ \pi^{\min(n,k)}$$

where K is a constant independent of Δ.

The lemma that follows shows that Newton's method can be used to determine y_Δ. The details are again a direct modification of a similar lemma proven in [6].

Lemma 3.2. In addition to the assumptions of Lemma 3.1 assume that $T'(g)$ exists and is uniformly bounded for all $g \in B(y,\delta) \subset \mathcal{R}[a,b]$. Also assume that $T'(g)$ is continuous in g at y, that for some constant K independent of f and g

$$||E(f,g)|| \leq K \ ||f-g||^2$$

is valid uniformly for any f, $g \in B(y,\delta)$, and that the assumption (ii) of Lemma 3.1 with $T'(y)$ replaced by $T'(g)$ holds uniformly for $g \in B(y,\delta)$. Then there exists $\hat{\varepsilon} > 0$ and $d > 0$ such that for all partitions Δ with $\pi \leq d$ and $g_0 \in B(y_\Delta,\hat{\varepsilon})$, the Newton iteration

$$g_{i+1} = \hat{N}_\Delta \ g_i, \quad i = 0,1,\ldots, \tag{3.8}$$

with

$$\hat{N}_\Delta(g) \equiv g - (I - Q_\Delta \ T'(g))^{-1} \ (I - Q_\Delta T)g$$

is well defined and converges quadratically to y_Δ. Furthermore $\hat{N}_\Delta$ maps $B(y_\Delta,\hat{\varepsilon}) \cap S_\Delta^{(m)}$ into itself.

We will give conditions that will guarantee the applicability of Lemmas 3.1 and 3.2 to our problem with T defined in (3.3). Let $C_1 \subset R^{m+2}$ and $C_2 \subset R^{m+1}$ denote the curves $C_1 \equiv \{(t,x(t),\ldots, D^m x(t))^T, t \in [a,b]\}$ and $C_2 \equiv \{(t,x(t),\ldots, D^{m-1}x(t))^T, t \in [a,b]\}$. Let M_1 and M_2 denote the closure of ε neighborhoods of $[a,b] \times C_1$ and of C_2 respectively. If $\mathcal{K}$ and F are continuously differentiable at every point of $[a,b] \times C_1$ and of C_2 respectively, then it can be shown that

$$T'(y)\ f(s) = \int_a^b \sum_{i=0}^{m} \mathcal{K}_i(s,t,\underset{\sim}{y}^{-m}(t))\ \hat{f}^{-m+i}(t)\ dt$$

$$+ \sum_{i=0}^{m-1} F_i(s,\underset{\sim}{y}^{-m}(s))\ \hat{f}^{-m+i}(s)$$

where

$$\hat{f}^{-m+i}(s) = \int_a^b H_i(s,t)\ f(t)\ dt$$

$$\mathcal{K}_i(s,t,\underset{\sim}{x}) = \partial/\partial x_i\ \mathcal{K}\ (s,t,\underset{\sim}{x})$$

$$F_i(s,t,\underset{\sim}{x}) = \partial/\partial x_i\ F(s,\underset{\sim}{x}).$$

<u>Theorem 3.1</u>. Let $y = D^m x \in C^n[a,b]$ be a solution of (3.2) and assume that

(i) $\mathcal{K}(s,t,x) \in C^2(M_1)$ and $F(s,x) \in C^2(M_2)$.

(ii) $(I - T'(y))^{-1}$ exists from $\mathcal{R}[a,b]$ onto $\mathcal{R}[a,b]$.

Then there exists $\varepsilon > 0$ and $d > 0$ such that the conclusions of Lemmas 3.1 and 3.2 are valid.

Proof. For the proof of the validity of hypotheses (i) and (ii) of Lemma 3.1 and of the corresponding statements of Lemma 3.2 we refer the reader to the proof of analogous results in [6]. In fact the analysis in the present case follows basically the same lines. The principal tools in this derivation are the triangular inequality, Taylor's formula with remainder and the Lipschitz continuity of $\mathcal{K}$, F, $\mathcal{K}_i$ and F_i. With respect to (iii) of Lemma 3.1, it can be proven that $T'(y)$ is a compact operator from $\mathcal{R}[a,b]$ into $\mathcal{R}[a,b]$ and hence the existence of $(I - T'(y))^{-1}$ implies its continuity. Finally it follows easily that $T'(g)$ is uniformly bounded in a sufficiently small neighborhood of y and is continuous in g at y. Then the results of the theorem follow. Q.E.D.

From the results of this theorem and the fact that

$$D^i(x-x_\Delta) = \int_a^b H_i(\cdot,t)\,(y-y_\Delta)(t)\,dt,\ i = 0,\ldots, m-1,$$

it follows that

(3.9) $\quad ||D^i(x-x_\Delta)|| \leq K \; \pi^{\min(n,k)}, \; i = 0,\ldots, m,$

where K is a constant independent of Δ. With respect to Newton's equations defined in (3.8), we observe that for the present problem they are equivalent to the collocation equatio

(3.10) $$(D^m - U(x_{\Delta,r}))\,(x_{\Delta,r+1} - x_{\Delta,r})\,(\tau_i)$$

$$= -\,(D^m - E)\; x_{\Delta,r}\,(\tau_i),\; \tau_i \in M_\Delta,$$

$$\beta_i\; x_{\Delta,r+1} = c_i,\; i = 1,\ldots, m.$$

where

$$U(x)\; f = \sum_{i=0}^{m} \int_a^b \mathcal{K}_i(\cdot,t,\underset{\sim}{x}(t))\; D^i f(t)\; dt$$

$$+ \sum_{i=0}^{m-1} F_i(\cdot,\underset{\sim}{x}(\cdot))\; D^i f(\cdot),$$

and E is defined in (3.1). The convergence is locally quadratic to x_Δ. We summarize the main results in the following theorem.

<u>Theorem 3.2</u>. If $x \in C^{m+n}$ [a,b] is a solution of (3.1) and if the hypotheses (i) and (ii) of Theorem 3.1 are valid, then there exists $\varepsilon > 0$ and $d > 0$ such that

(a) There is no other solution $\hat{x}$ of (3.1) such that $||D^m(x-\hat{x})|| < \varepsilon$.

(b) For each partition Δ with $\pi \leq d$ (3.4) has a unique solution in this same neighborhood of x.

(c) The error estimates (3.9) are valid.

(d) Newton's method (3.10) for approximately solving (3.4) converges quadratically to x_Δ in $B(x_\Delta, \varepsilon_\Delta) \cap S_\Delta$ for some $\varepsilon_\Delta > 0$.

Remarks. It is shown in [6] that for collocation solutions of two point boundary value problems the error estimates (3.9) are best in general. However it is also shown there that if the collocation points are determined by selecting $\{\xi_i\}_1^k$ coincident with the zeros of the Legendre polynomials of degree k on [-1,1], then those asymptotic error estimates can be notably improved. Although we have not proven this result for the general problem (3.1) several numerical examples tested suggest that selection of the collocation points in the manner described above gives the same improved rates of convergence as [6]. However, for integral equations the following example shows that the error estimates (3.9) are best and are independent of the relative position of the collocation points.

Consider the integral equation

$$x(s) - \int_0^1 x(t)\, dt = s^k - 1/(k+1),$$

where k is a positive integer and whose exact solution is

$x(s) = s^k$. After selecting $\{\xi_i\}_1^k \subset [-1,1]$, arbitrary but fixed, we consider approximating spaces S_Δ consisting of piecewise polynomials of degree less than k. For Δ an arbitrary partition of [0,1], let $x_\Delta \in S_\Delta$ denote the collocation solution of the above equation. Let $p_{(i)}$ denote the monic Chebyshev polynomial of degree k relative to (s_i, s_{i+1}) for $i = 0,\ldots, N-1$ Since $x - x_\Delta$ is a piecewise monic polynomial of degree k on (s_i, s_{i+1}), $i = 0,\ldots, N-1$, it follows that

$$\sup_{s\varepsilon(s_i,s_{i+1})} |x(s) - x_\Delta(s)| \geq \sup_{s\varepsilon(s_i,s_{i+1})} |p_{(i)}(s)|$$

$$= 2^{-k} (s_{i+1} - s_i)^k .$$

Hence $||x - x_\Delta|| \geq 2^{-k} \pi^k$. This proves our claim.

4. <u>Approximations to the collocation equations</u>. In practice it is usually the case that (3.4) or (3.10) cannot be computed explicitely since they could involve integrals which are either to costly to evaluate or which can not be evaluated in closed form. Hence in general one must resort to some approximation. Since this approximation needs to be done only on the integral part of the equation, we will assume for simplicity that $F(s,\underset{\sim}{x}) \equiv z(s)$. In this paper we will discuss some approximatio to (3.4) and (3.10) when $\mathcal{K}$ depends only on $x_0,\ldots, x_{m-1}$ but no on x_m. We will remark briefly at the end on the more general case when $\mathcal{K}$ depends also on x_m.

5. Global approximations to the kernels. Consider first the linear equation

$$D^m x(s) - \int_a^b \sum_{i=0}^{m-1} \mathcal{K}_i(s,t)\, D^i x(t)\, dt = z(s)$$

(5.1)

$$\beta_i x = c_i, \; i = 1,\ldots, m.$$

We assume that each $\mathcal{K}_i$ is continuous on $[a,b] \times [a,b]$ and that this equation has a unique solution $x \in C^m[a,b]$. We associate with each partition Δ of $[a,b]$ a bivariate approximant $\mathcal{K}_{i,\Delta}(s,t)$ of $\mathcal{K}_i(s,t)$ for $i = 0,\ldots, m-1$, and consider the problem of obtaining approximations to x by finding $u_\Delta \in S_\Delta$ satisfying

$$D^m u_\Delta(\tau_i) - \int_a^b \sum_{j=0}^{m-1} \mathcal{K}_{i,\Delta}(\tau_i,t)\, D^j u_\Delta(t)\, dt = z(\tau_i),$$

(5.2)

$$\tau_i \in M_\Delta, \quad \beta_i u_\Delta = c_i, \quad i = 1,\ldots, m.$$

We will assume that these bivariate approximants satisfy the condition

(5.3) $$||\mathcal{K}_{i,\Delta} - \mathcal{K}_i||_S \le \omega(\pi), \; i = 0,\ldots, m-1,$$

where $\omega(\cdot)$ is a modulus of continuity and $||f||_S \equiv \sup\{|f(s,t)|, (s,t) \in [a,b] \times [a,b]\}$. The most common approximants based on bivariate piecewise polynomials and on blending functions [2], [3], [8] satisfy (5.3).

We prove first,

Theorem 5.1. For each partition Δ with π sufficiently small, (5.2) has a unique solution.

Proof. (5.1) and (5.2) are equivalent to solving $y = D^m x$ and $v_\Delta = D^m u_\Delta$ satisfying

$$(I - T)y = z$$

and

$$(I - Q_\Delta T_\Delta)v_\Delta = Q_\Delta z$$

where

$$Ty = \int_a^b \sum_{i=0}^{m-1} \mathcal{K}_i(\cdot,t)\, y^{-m+i}(t)\, dt \tag{5.4}$$

and

$$T_\Delta g = \int_a^b \sum_{i=0}^{m-1} \mathcal{K}_{i,\Delta}(\cdot,t)\, g^{-m+i}(t)\, dt$$

We have that

$$||(I - T) - (I - Q_\Delta T_\Delta)|| \leq ||(I - Q_\Delta) T|| + ||Q_\Delta||\ ||T - T_\Delta|| . \tag{5.5}$$

The facts that Q_Δ converges pointwise to I on $C[a,b]$ and uniformly on the totally bounded set $T\mathcal{B}$ where $\mathcal{B}$ is the unit ball in $\mathscr{R}[a,b]$, imply that $||(I - Q_\Delta) T|| \to 0$ as $\pi \to 0$ and that this convergence is uniform with respect to Δ. From (5.3) and the definition of Q_Δ, it follows that $||Q_\Delta|| \; ||T - T_\Delta|| \to 0$ as $\pi \to 0$ uniformly with respect to Δ. Hence (5.5) together with these results and the fact that $(I - T)$ is boundedly invertible prove the result. Q.E.D.

Now we establish convergence of $D^i u_\Delta$ to $D^i x$ as $\pi \to 0$ for $i = 0,\ldots, m$, and analyze the rate of convergence. First, we have that

$$(5.6) \qquad ||y - v_\Delta|| \leq ||y - Q_\Delta y|| + ||Q_\Delta y - Q_\Delta v_\Delta||.$$

<u>Lemma 5.1</u>. $||Q_\Delta y - Q_\Delta v_\Delta|| \leq K (||(T - T_\Delta)y|| + ||y - Q_\Delta y||)$, where K is a constant independent of Δ.

<u>Proof</u>. Subtracting

$$Q_\Delta v_\Delta - Q_\Delta T_\Delta Q_\Delta v_\Delta = Q_\Delta z$$

from

$$Q_\Delta y - Q_\Delta T_\Delta Q_\Delta y = Q_\Delta z + Q_\Delta Ty - Q_\Delta T_\Delta Q_\Delta y$$

and solving for $Q_\Delta y - Q_\Delta v_\Delta$, we obtain

$$||Q_\Delta y - Q_\Delta v_\Delta|| \leq ||(I - Q_\Delta T_\Delta)^{-1}|| \; ||Q_\Delta|| \; ||Ty - T_\Delta Q_\Delta y||$$

The uniform boundedness of $||(I - Q_\Delta T_\Delta)^{-1}||$ which follows as a consequence of the discussion in Theorem 5.1, the definition of Q_Δ and the use of the triangular inequality give the desired result. Q.E.D.

Now we can write

Theorem 5.2. $D^i u_\Delta$ converges uniformly to $D^i x$, $i = 0,\ldots, m$, as $\pi \to 0$. If in addition $x \in C^{m+n}[a,b]$ and $||\mathcal{K}_i - \mathcal{K}_{i,\Delta}||_S \leq K \pi^p$, $i = 0,\ldots, m-1$, then

$$||D^i(x - x_\Delta)|| \leq K \pi^{\min(n,k,p)}, \quad i = 0,\ldots, m.$$

Proof. The first part follows using Lemma 5.1 in (5.6) and the fact that

$$D^i(x - u_\Delta) = \int_a^b H_i(\cdot,t)\, (y - v_\Delta)\, (t)\, dt, \quad i = 0,\ldots, m-1.$$

The second one follows since $||T - T_\Delta|| \leq K \pi^p$ and $||y - Q_\Delta y|| \leq K \pi^{\min(n,k)}$. Q.E.

Consider now the nonlinear equation

$$(5.7) \qquad D^m x(s) - \int_a^b \mathcal{K}(s,t,x(t),\ldots, D^{m-1}x(t))\, dt = z(s),$$

$$\beta_i x = c_i, \quad i = 1,\ldots, m,$$

satisfying the hypotheses of Theorem 3.2. Here we will discuss approximations to the collocation equations (3.10).

each function g in a neighborhood of y and for each ition Δ, let $\mathcal{K}_{i,\Delta}(s,t,g^{-m}(t))$ denote a bivariate oximation to $\mathcal{K}_i(s,t,g^{-m}(t))$. We will consider approximations that

$$\text{(5.8)}\qquad \max_{(s,t)\in[a,b]\times[a,b]} |\mathcal{K}_i(s,t,\underset{\sim}{g}^{-m}(t)) - \mathcal{K}_{i,\Delta}(s,t,\underset{\sim}{g}^{-m}(t)| \leq \omega(\pi),$$

formly for all $g \in \mathcal{R}[a,b] \cap B(y,\varepsilon)$ for some $\varepsilon > 0$. Here) is a modulus of continuity. Since $\mathcal{K} \in C^2(M_1)$, (5.8) can satisfied using the most common approximants based on ariate piecewise polynomials and on blending functions.

Based on equation (3.10), for each fixed partition Δ we t to determine a sequence $\{u_{\Delta,r}\}_{r=0}^{\infty}$ satisfying the collocation ations

(5.9)

$$(u_{\Delta,r+1} - u_{\Delta,r})(\tau_i) - \int_a^b \sum_{j=0}^{m-1} \mathcal{K}_{j,\Delta}(\tau_i,t,\underset{\sim}{u}_{\Delta,r}(t))D^j(u_{\Delta,r+1} - u_{\Delta,r})(t)dt$$

$$= - D^m u_{\Delta,r}(\tau_i) + \int_a^b \mathcal{K}(\tau_i,t,\underset{\sim}{u}_{\Delta,r}(t))dt + z(\tau_i),$$

$$\tau_i \in M_\Delta, \quad \beta_i u_{\Delta,r+1} = c_i, \quad i = 1,\dots, m, \; r = 0,1,\dots .$$

Remark. It is clear that in practice the integrals on the right nd side of (5.9) will usually have to be approximated too.

However we will not discuss here the effect of such approximations

In order to show that the above sequence of iterates is well defined and convergent to x_Δ, we look at the equivalent problem of finding $v_{\Delta,r+1} = D^m u_{\Delta,r+1}$, satisfying

$$(5.10) \qquad (I - Q_\Delta V_\Delta(v_{\Delta,r}))(v_{\Delta,r+1} - v_{\Delta,r}) = -(I - Q_\Delta T)\, v_{\Delta,r},\ r = 0,1,\ldots,$$

where

$$V_\Delta(g)h = \int_a^b \sum_{i=0}^{m-1} \mathcal{K}_{i,\Delta}(\cdot,t,\underset{\sim}{g}^{-m}(t))\, \hat{h}^{-m+i}(t)dt$$

and

$$(5.11) \qquad Th = \int_a^b \mathcal{K}(\cdot,t,\underset{\sim}{h}^{-m}(t))dt + z.$$

The following lemma follows immediately using the triangular inequality and (5.8).

<u>Lemma 5.2</u>. For all $g \in \mathcal{R}[a,b] \cap B(y,\varepsilon)$ for some $\varepsilon > 0$ and for each partition Δ, $||V_\Delta(g) - T'(g)|| \leq \omega(\pi)$, where $\omega(\cdot)$ is a modulus of continuity.

With the aid of this result we prove now the following lemma.

<u>Lemma 5.3</u>. $(I - Q_\Delta V_\Delta(g))^{-1}$ exists and is uniformly bounded for all $g \in \mathcal{R}[a,b]$ in a sufficiently small neighborhood of y

and all partitions Δ with π sufficiently small.

Proof. We have that

$$||T'(y) - Q_\Delta V_\Delta(g)|| \leq ||(I - Q_\Delta) T'(y)||$$

$$+ ||Q_\Delta|| (||T'(y) - T'(g)|| + ||T'(g) - V_\Delta(g)||)$$

$$\leq ||(I - Q_\Delta) T'(y)|| + K (||y-g|| + ||T'(g) - V_\Delta(g)||),$$

where K is independent of Δ. Since Q_Δ converges uniformly to I on the totally bounded set $T'(y)\mathcal{B}$, then $||(I - Q_\Delta) T'(y)|| \to 0$ as $\pi \to 0$. This result, Lemma 5.2 and the requirement that g be sufficiently near y, yield that for all partitions Δ with π sufficiently small $I - Q_\Delta V_\Delta(g)$ is close to the boundedly invertible map $I - T'(y)$. Hence the result follows. Q.E.D.

Now rewrite (5.10) in the form

$$v_{\Delta,r+1} = G_\Delta v_{\Delta,r} , \quad r = 0,1,\ldots, \tag{5.12}$$

where

$$G_\Delta v = v - (I - Q_\Delta V_\Delta(v))^{-1} (I - Q_\Delta T)v$$

The next two lemmas follow as a consequence of results proven in [10, Prop. 10.1.3 and 10.2.1].

<u>Lemma 5.4</u>. If $||G'_\Delta(y_\Delta)|| < 1$ where y_Δ satisfies $y_\Delta = Q_\Delta\ T\ y_\Delta$, then y_Δ is a point of attraction of (5.12), i.e. for any $v_{\Delta,0} \in S_\Delta^{(m)}$ in a sufficiently small neighborhood of y_Δ the sequence $\{v_{\Delta,r}\}_{r=0}^{\infty}$ is well defined and converges to y_Δ.

<u>Lemma 5.5</u>. If y_Δ is such that $y_\Delta = Q_\Delta\ T\ y_\Delta$ then

$$G'_\Delta(y_\Delta) = I - (I - Q_\Delta\ V_\Delta(y_\Delta))^{-1}\ (I - Q_\Delta\ T'(y_\Delta)).$$

Using these results we prove now the following theorem.

<u>Theorem 5.3</u>. For each Δ with π sufficiently small x_Δ is a point of attraction of (5.9). Furthermore, for all $u_{\Delta,0} \in S_\Delta$ in a sufficiently small neighborhood of x_Δ, $||D^i(u_{\Delta,r} - x_\Delta)|| \to 0$ as $r \to \infty$, for $i = 0,\dots, m$.

<u>Proof</u>. From Lemmas 5.5 and 5.3, from the definition of Q_Δ and since $y_\Delta \to y$ as $\pi \to 0$ it follows that

$$||G'_\Delta(y_\Delta)|| = ||(I - Q_\Delta\ V_\Delta(y_\Delta))^{-1}(I - Q_\Delta\ V_\Delta(y_\Delta) - (I - Q_\Delta\ T'(y_\Delta)))||$$

$$\leq\ K\ ||T'(y_\Delta) - V_\Delta(y_\Delta)||,$$

where K is independent of Δ. Then Lemmas 5.2 and 5.4 show that for each Δ with π sufficiently small y_Δ is a point of attraction of (5.12). Since

$$D^i(x_\Delta - u_{\Delta,r}) = \int_a^b H_i(\cdot,t)\,(y_\Delta - v_{\Delta,r})\,(t)\,dt, \quad i = 0,\ldots, m-1,$$

then $D^i\, u_{\Delta,r}$ converges to $D^i\, x_\Delta$, $i = 0,\ldots, m$. The continuity of D^m from S_Δ to $S_\Delta^{(m)}$ shows that x_Δ is a point of attraction of (5.9).

Q.E.D.

Now we analyze the rate of convergence of the sequence generated by (5.9).

<u>Theorem 5.4</u>. Let $e_{\Delta,r} = u_{\Delta,r} - x_\Delta$. Then

$$(5.13) \quad ||D^i e_{\Delta,r+1}|| \leq K(||D^i e_{\Delta,r}||^2 + \omega(\pi)||D^i e_{\Delta,r}||), \quad i = 0,\ldots, m,$$

where K is a constant in general depending on Δ and $\omega(\cdot)$ is a modulus of continuity.

<u>Proof</u>. By the triangular inequality and Lemma 5.3

$$\begin{aligned} ||v_{\Delta,r+1} - y_\Delta|| \leq\ & K_1\{||(I - Q_\Delta T)y_\Delta - (I - Q_\Delta T)v_{\Delta,r} \\ & - (I - Q_\Delta T'(v_{\Delta,r}))\,(y_\Delta - v_{\Delta,r})|| \\ & + ||Q_\Delta(T'(y_\Delta) - T'(v_{\Delta,r}))\,(y_\Delta - v_{\Delta,r})|| \\ & + ||Q_\Delta(V_\Delta(y_\Delta) - T'(y_\Delta))\,(y_\Delta - v_{\Delta,r})|| \\ & + ||Q_\Delta(V_\Delta(v_{\Delta,r}) - V_\Delta(y_\Delta))\,(y_\Delta - v_{\Delta,r})||\} \end{aligned}$$

It can be easily shown that the first three terms on the right hand side are bounded by $K_2(||v_{\Delta,r} - y_\Delta||^2 + \omega(\pi)\ ||v_{\Delta,r} - y_\Delta||)$ where K_2 is a constant. We can show that this expression is also a bound for the forth term on the right hand side using the triangular inequality and Lemma 5.3. These results prove (5.13) for $i = m$. From the existence of constants $K_3 > 0$, $K_4 > 0$ such that $K_3\ ||D^i(u_{\Delta,r} - x_\Delta)|| \leq ||v_{\Delta,r} - y_\Delta||$ $\leq K_4\ ||D^i(u_{\Delta,r} - x_\Delta)||$, the result of the theorem follows.

Q.E.D.

We observe that for each fixed Δ with π sufficiently small the convergence is, in general, essentially quadratic at the beginning of the iterative process. In fact for the first few iterates we will usually have $||u_{\Delta,r} - x_\Delta|| >> \omega(\pi)$ and hence the first term on the right hand side of (5.13) dominates. However as $u_{\Delta,r}$ approaches x_Δ, then the second term eventually dominates and the convergence becomes basically linear. A similar reduction of the error which in addition is computationally tractable occurs between successive iterates.

<u>Theorem 5.4</u>. Let $\bar{e}_{\Delta,r} = u_{\Delta,r} - u_{\Delta,r-1}$. Then

$$(5.14)\quad ||D^i\bar{e}_{\Delta,r+1}|| \leq K\ (||D^i\bar{e}_{\Delta,r}||^2 + \omega(\pi)\ ||D^i\bar{e}_{\Delta,r}||),$$

$$i = 0,\ldots, m,$$

where K is a constant in general depending on Δ and $\omega(\cdot)$ is a modulus of continuity.

Proof. In the expression for $||v_{\Delta,r+1} - v_{\Delta,r}||$ use the triangular inequality, Taylor's formula with remainder and the results of Lemmas 5.2 and 5.3. Q.E.D.

We remark that some numerical examples tested have exhibited a pattern in the reduction of the error like the one described in this theorem. We also indicate that the form of the expressions (5.13) and (5.14) suggest that if we use approximations $\mathcal{K}_{i,\Delta,r}$ depending on r and satisfying a condition of the kind

$$\max_{(s,t)\varepsilon[a,b]\times[a,b]} |\mathcal{K}_{i,\Delta,r}(s,t,\underset{\sim}{g}^{-m}(t)) - \mathcal{K}_i(s,t,\underset{\sim}{g}^{-m}(t))|$$

$$= 0||u_{\Delta,r} - x_\Delta||,$$

$i = 0,\ldots, m-1$, for all $g \;\varepsilon\; \mathcal{R}[a,b] \cap B(y,\varepsilon)$, then we can at least in theory retain the quadratic convergence of Newton's method.

6. Use of quadrature formulas. Let $\{\Delta(i)\}_{i=1}^{\infty}$ be a sequence of partitions of [a,b] such that $\pi_i \to 0$ as $i \to \infty$ where π_i is the mesh gauge of $\Delta(i)$. With each $\Delta(i)$ associate a quadrature formula B_i of the type

$$(6.1) \qquad B_i f = \sum_{j=1}^{\mu_i} W_{ij}\ f(\eta_{ij}).$$

where $a \leq \eta_{ij} \leq b$. We assume that $B_i f \to \int_a^b f(t)dt$ as $i \to \infty$ for any $f \in C[a,b]$.

Consider first the linear equation (5.1). We want to determine approximations to x by finding $u_j \in S_{\Delta(j)}$ satisfying

$$(6.2) \qquad D^m u_j(\tau_i) - B_j\Big(\sum_{h=0}^{m-1} \mathcal{K}_h(\tau_i,\cdot)\ D^h u_j(\cdot)\Big) = z(\tau_i),$$

$$\tau_i \in M_{\Delta(j)}, \quad \beta_i\ u_j = c_i, \quad i = 1,\ldots, m.$$

This problem is equivalent to finding $v_j = D^m\ u_j \in S_{\Delta(j)}^{(m)}$ satisfying

$$(I - Q_{\Delta(j)}\ T_j)\ v_j = Q_{\Delta(j)}\ z,$$

where

$$T_j\ f(s) = B_j\Big(\sum_{h=0}^{m-1} \mathcal{K}_h(s,\cdot)\ f^{-m+h}(\cdot)\Big).$$

We prove first the following lemma.

Lemma 6.1. $||T_j - T|| \to 0$ as $j \to \infty$, where T is defined in (5.4).

Proof. We have that $T_j f \to Tf$ as $j \to \infty$ for any $f \in \mathcal{R}[a,b]$. Since the collection of functions of $t \in [a,b]$
$\{\mathcal{K}_h(s,\cdot)\ f^{-m+h}(\cdot) \mid h = 0,\ldots, m-1,\ s \in [a,b],\ f \in \mathcal{R}[a,b]$ with $||f|| = 1\}$ is totally bounded, the result follows using the

uniform convergence of T_j to T on totally bounded sets. Q.E.D.

Using Lemma 6.1 and proceeding as in section 5 we obtain the following result.

Theorem 6.1. (6.2) has a unique solution for all j sufficiently large and $||D^i(u_j - x)|| \to 0$ as $j \to \infty$ for $i = 0,\ldots, m$. If in addition

$$(6.3) \qquad \max_{s\varepsilon[a,b]} |B_j(\mathcal{K}_i(s,\cdot)\, y^{-m+i}(\cdot)) - \int_a^b \mathcal{K}_i(s,t)\, y^{-m+i}(t)dt| \leq K\, \pi_j^p ,$$

$i = 0,\ldots, m-1$ and $x \in C^{m+n}[a,b]$, then

$$||D^i(u_j - x)|| \leq K\, \pi_j^{\min(n,k,p)},\ i = 0,\ldots, m.$$

Remark. (6.3) can be satisfied by requiring for example that $|B_j f - \int_a^b f(t)dt| \leq K\, \pi_j^p$ for all $f \in C^p[a,b]$, where K is a constant usually depending on f, and by assuming that $\mathcal{K}_i$ $i = 0,\ldots, m-1$, and y are sufficiently smooth.

Consider now the nonlinear equation (5.7). Based on Newton's method, for each partition $\Delta(j)$ we want to determine a sequence $\{u_{j,r}\}_{r=0}^{\infty}$ satisfying the collocation equations

(6.4)

$$D^m(u_{j,r+1} - u_{j,r})(\tau_i) - B_j(\sum_{h=0}^{m-1} \mathcal{K}_h(\tau_i,\cdot,\underset{\sim}{u}_{j,r}(\cdot))\, D^h(u_{j,r+1} - u_{j,r})$$

$$= - D^m u_{j,r}(\tau_i) + \int_a^b \mathcal{K}(\tau_i,t,\underset{\sim}{u}_{j,r}(t))\, dt + z(\tau_i),$$

$$\tau_i \in M_{\Delta(j)}, \quad \beta_i u_{j,r+1} = c_i, \quad i = 1,\ldots, m, \quad r = 0,1,\ldots$$

where B_j is defined in (6.1). This problem is equivalent to finding $v_{j,r+1} = D^m u_{j,r+1}$ satisfying

$$(I - Q_{\Delta(j)} V_j(v_{j,r}))\,(v_{j,r+1} - v_{j,r}) = -(I - Q_{\Delta(j)}T)\, v_{j,r}$$

$r = 0,1,\ldots$, where

$$V_j(g)\, h(s) = B_j(\sum_{i=0}^{m-1} \mathcal{K}_i(s,\cdot,\underset{\sim}{g}^{-m}(\cdot))\, \hat{h}^{-m+i}(\cdot))$$

and T is defined in (5.11). An analogous proof to the one of Lemma 6.1 gives,

<u>Lemma 6.2</u>. For all $g \in \mathcal{R}[a,b] \cap B(y,\varepsilon)$ for some $\varepsilon > 0$ and for all j's, $||V_j(g) - T'(g)|| \leq \omega(\pi_j)$ where $\omega(\cdot)$ is a modulus of continuity.

This result permits us to proceed as in section 5 and to prove the following theorem.

heorem 6.1. For all j sufficiently large, $x_{\Delta(j)}$ is a point f attraction of (6.4). Furthermore, for all $u_{j,0} \in S_{\Delta(j)}$ in sufficiently small neighborhood of $x_{\Delta(j)}$, $||D^i(u_{j,r} - x_{\Delta(j)})|| \to 0$ s $r \to \infty$, for $i = 0,\ldots, m$.

Also analogous results to Theorems 5.4 and 5.5 can be roven in the present case.

inal remarks. It can be observed that for the more general ase in which $\mathcal{K}$ depends also on x_m, the discussion of section 5 ased on global approximants to the kernels can be extended with o change for the linear equation (5.1). However, for the emaining cases it can be seen that we lose the convergence in orm of the operators discussed in Lemmas 5.2, 6.1 and 6.2. his suggest the use of the theory of collectively compact perators of Anselone [1] to compensate for this disadvantage. n fact, its application permits us to obtain, in this more eneral case, similar results to those we have presented here. In addition, the approximation to Newton's equation as in (6.4) leads to the notion of radially collectively compact operators. This concept will be discussed elsewhere.

Acknowledgments. This paper is part of a Ph.D. dissertation (in progress) of the author at the University of Pittsburgh. The author would like to thank his thesis advisor Dr. George D. Byrne for all the help and encouragement that he provided.

References

[1] P. M. Anselone, Collectively Compact Operator Approximation Theory, Prentice-Hall, Englewood Cliffs N.J., 1971.

[2] G. Birkhoff, M. H. Schultz and R. S. Varga, Piecewise Hermite interpolation in one and two variables with applications to partial differential equations, Numer. Math., 11 (1968), 232-256.

[3] R. Carlson and C. Hall, On piecewise polynomial interpolatio in rectangular polygons, J. Approximation Theory, 4 (1971), 37-53.

[4] J. C. Cavendish and C. A..Hall, L_∞-convergence of collocatio and Galerkin approximations to linear two point parabolic problems, Aequationes Math, to appear.

[5] E. W. Cheney, Introduction to Approximation Theory, Mc Graw-Hill, New York, 1966.

[6] C. DeBoor and B. Swartz, Collocation at Gaussian points, SIAM J. Numer. Anal. 10 (1973), 582-606.

[7] J. Douglas, Jr. and T. Dupont, A finite element collocation method for quasilinear parabolic equations, Math. of Comp. 27 (1973), 17-28.

[8] W. J. Gordon, Blending-function methods for bivariate and multivariate interpolation and approximation, SIAM J. Numer. Anal. 8 (1971), 158-177.

[9] T. R. Lucas and G. W. Reddien, Jr., Some collocation methods for nonlinear boundary value problems, SIAM J. Numer. Anal 9 (1972), 341-356.

[10] J. M. Ortega and W. C. Rheinboldt, Iterative Solution of Nonlinear Equations in Several Variables, Academic Press, New York, 1970.

[11] J. L. Phillips, The use of collocation as a projection method for solving linear operator equations, SIAM J. Numer. Anal. 9 (1972), 14-28.

2] P. M. Prenter, A collocation method for the numerical solution of integral equations, SIAM J. Numer. Anal. 10 (1973), 570-581.

3] R. D. Russell and L. F. Shampine, A collocation method for boundary value problems, Numer. Math. 19 (1972), 1-28.

ON DIRICHLET'S PROBLEM FOR QUASI-LINEAR ELLIPTIC EQUATIONS[‡]

by

Robert P. Gilbert* and George C. Hsiao**

ABSTRACT

This paper concerns the existence and uniqueness as well as the approximations of the solutions to the Dirichlet problem for the second-order quasi-linear elliptic equation of n variables in a simply connected domain with Lyapunov boundary. Sufficient conditions on the coefficients of the equation are established for the existence proofs of the solution so that Vekua's function theoretic method for treating Dirichlet's problems for equations in two variables can be employed for the case $n \geq 3$ variables. Based on Warschawski's work, an iterative scheme is also presented for constructing an approximating solution in the three-dimensional case. It is shown that the approximants converge to the actual solution geometrically.

[‡]An invited address at Constructive and Computational Methods for Differential and Integral Equations, Systems Analysis Institute, Research Center for Applied Science, Indiana University, February 17-20, 1974. This research was supported in part by the Air Force Office of Scientific Research through AF-AFOSR Grant No. 74-2592.

*Department of Mathematics, Indiana University, Bloomington, Indiana, 47401.

**Department of Mathematics, University of Delaware, Newark, Delaware, 19711.

1. Introduction

In [9], Vekua showed by using Green's function of the Laplace equation for a unit circle, one can reduce the interior Dirichlet problem for quasi-linear elliptic equations in two independent variables to a non-linear functional equation from which existence and uniqueness of the solution to the problem can be established by the generalized principle of contraction mappings. In this paper, we shall discuss the feasibility for adapting Vekua's approach to equations in higher dimensions. More precisely, we consider the Dirichlet problem consisting of quasi-linear equations of the form

$$(E) \qquad L[u] \equiv \sum_{i,j=1}^{n} a_{ij}\left(x_1,\ldots,x_n,u,\frac{\partial u}{\partial x_1},\ldots,\frac{\partial u}{\partial x_n}\right)\frac{\partial^2 u}{\partial x_i \partial x_j} = f\left(x_1,\ldots,x_n,u,\frac{\partial u}{\partial x_1},\ldots,\frac{\partial u}{\partial x_n}\right)$$

in a simply-connected domain Ω with Lyapunov boundary $\partial\Omega$ together with the homogeneous boundary condition

$$(B) \qquad u = 0 \quad \text{on} \quad \partial\Omega .$$

Here the functions a_{ij} and f are assumed to satisfy certain properties (to be specified later), and under these assumptions, we are able to extend Vekua's results at least to the case for $n = 3$ and 4 .

We shall first show that by introducing several complex variables, Eq. (E) can be rewritten in a form similar to the one used in [9]. We refer to this form as a normal form of (E) (see Section 2). Then, we shall confine ourselves to the case where the domain Ω is a unit sphere; in this case, Green's function of the Laplace equation is known explicity, and Vekua's existence theorem can be generalized. We present these results for $n = 3$ and 4 in Section 3; it is readily seen that in the same manner these results can be established in general. Based on these results for the unit sphere, it is not difficult to see, in general, how to carry out the analysis for an arbitrary domain Ω, if one knows Green's function in Ω for the Laplace equation. We shall give a brief indication how to treat this situation in Section 4. Finally, we consider the case when we do not know Green's function. We present a new iterative scheme for solving the problem (E) and (B). Our procedure is a variation of a technique due to Warschawski[10], which differs from the one used in [1] where Green's function is expressed in terms of the approximated kernel function. Again, we confine ourselves for the case of $n = 3$, since the approach is no different in the higher dimensional case.

In what follows, we shall assume that the following assumptions are satisfied:

(1) a_{ij} is bounded and measurable for

$$(x_1,\ldots,x_n) \in \Omega+\partial\Omega \ , \quad u^2+\sum_{i=1}^{n}\left(\frac{\partial u}{\partial x_i}\right)^2 \leq M \ ,$$

where M is an arbitrary fixed positive number;

(2) for the same values of the arguments

$$\sum_{i,j=1}^{n} a_{ij}\xi_i\xi_j \geq \Delta_o(M) \sum_{i=1}^{n}\xi_i^2 \quad \text{for all real}$$

$\xi_1,\ldots,\xi_n$, where $\Delta_o(M) > 0$ is a constant;

(3) $f\left(x_1,\ldots,x_n,u,\frac{\partial u}{\partial x_1},\ldots,\frac{\partial u}{\partial x_n}\right)$ is a measurable function and $f(x_1,\ldots,x_n,0,0,\ldots,0) \in L_p(\Omega+\partial\Omega)$, $p > n$;

(4) a_{ij} satisfies the Lipschitz condition with respect to the arguments $u,\left(\frac{\partial u}{\partial x_1},\ldots,\frac{\partial u}{\partial x_n}\right)$,

$$\left| a_{ij}\left(x_1,\ldots,x_n,\hat{u},\frac{\partial \hat{u}}{\partial x_1},\ldots,\frac{\partial \hat{u}}{\partial x_n}\right) - a_{ij}\left(x_1,\ldots,x_n,u,\frac{\partial u}{\partial x_1},\ldots,\frac{\partial u}{\partial x_n}\right)\right|$$

$$\leq M_1\left(|\hat{u}-u| + \sum_{i=1}^{n}\left|\frac{\partial \hat{u}}{\partial x_i} - \frac{\partial u}{\partial x_i}\right|\right) \ ,$$ where M_1 is

a constant independent of x_i ;

(5) $f\left(x_1,\ldots,x_n,u,\frac{\partial u}{\partial x_1},\ldots,\frac{\partial u}{\partial x_n}\right)$ satisfies a condition of the form

$$\left|f\left(x_1,\ldots,x_n,\hat{u},\frac{\partial \hat{u}}{\partial x_1},\ldots,\frac{\partial \hat{u}}{\partial x_n}\right) - f\left(x_1,\ldots,x_n,u,\frac{\partial u}{\partial x_1},\ldots,\frac{\partial u}{\partial x_n}\right)\right|$$

$$\leq f_o(x_1,\ldots,x_n)\left(|\hat{u}-u| + \sum_{i=1}^{n}\left|\frac{\partial \hat{u}}{\partial x_i} - \frac{\partial u}{\partial x_i}\right|\right),$$

where $f_o(x_1,\ldots,x_n) \in L_p(\Omega+\partial\Omega)$, $p > n$.

2. Reduction to the Normal Form

In terms of complex variables $z = x_1+ix_2$ and $\bar{z} = x_1-ix_2$, for $n = 2$, Eq. (E) can be rewritten in the complex form [9]

$$L_2[u] \equiv \frac{\partial^2 u}{\partial z\partial\bar{z}} + \mathrm{Re}\left[A(z,u,u_z)\frac{\partial^2 u}{\partial z^2}\right] + B(z,u,u_z) = 0 \tag{2.1}$$

where

$$A = \frac{a_{11}-a_{22}+(a_{12}+a_{21})i}{2(a_{11}+a_{22})}, \quad B = \frac{-f}{2(a_{11}+a_{22})}.$$

(In [9], the number 2 is missing in the denominator of A and B.) We refer to (2.1) as the normal form of (E) for $n = 2$. In the following we shall see how (2.1) can be generalized in higher dimensions.

To illustrate the idea, we consider separately the even and odd dimensions and write $n = 2m$ or $n = 2m+1$ accordingly. We set, in the even case, $z_k = x_{2k-1}+ix_{2k}$, $\zeta_k = x_{2k-1}-ix_{2k}$ for $k = 1,\dots,m$, and in addition, $X = x_{2m+1}$ in the odd case. The formal derivatives with respect to z_k and ζ_k are defined by the relations

$$\frac{\partial}{\partial z_k} = \frac{1}{2}\left(\frac{\partial}{\partial x_{2k-1}} - i\frac{\partial}{\partial x_{2k}}\right) , \quad \frac{\partial}{\partial \zeta_k} = \frac{1}{2}\left(\frac{\partial}{\partial x_{2k-1}} + i\frac{\partial}{\partial x_{2k}}\right) .$$

A simple computation then shows that (E) can be written in the form:

$$\begin{aligned}(2.2)\qquad & \sum_{j=1}^{m}\left\{B_j \frac{\partial^2 u}{\partial z_j \partial \zeta_j} + A_j \frac{\partial^2 u}{\partial z_j^2} + \bar{A}_j \frac{\partial^2 u}{\partial \zeta_j^2}\right\} \\ & + \sum_{\substack{k,j=1\\ k\neq j}}^{m}\left\{A_{kj}\frac{\partial^2 u}{\partial z_k \partial z_j} + B_{kj}\frac{\partial^2 u}{\partial z_k \partial \zeta_j} + \bar{B}_{kj}\frac{\partial^2 u}{\partial z_j \partial \zeta_k}\right. \\ & \left. + \bar{A}_{kj}\frac{\partial^2 u}{\partial \zeta_k \partial \zeta_j}\right\} + \mathcal{O}[u] - f = 0 ,\end{aligned}$$

where

$$A_j \equiv -a_{2j,2j}+2ia_{2j,2j-1}+a_{2j-1,2j-1} ,$$

$$B_j \equiv 2a_{2j,2j}+2a_{2j-1,2j-1} ,$$

(2.3)

$$A_{kj} \equiv -a_{2k,2j}+2ia_{2k,2j-1}+a_{2k-1,2j-1} ,$$

$$B_{kj} \equiv a_{2k,2j}+2ia_{2k,2j}+a_{2k-1,2j-1}$$

The operator O is defined so that $O[u] = 0$ for $n = 2m$, and for $n = 2m+1$,

$$O[u] \equiv a_{2m+1,2m+1} \frac{\partial^2 u}{\partial x^2} + \sum_{j=1}^{m} \left\{ C_j \frac{\partial^2 u}{\partial x \partial z_j} + \overline{C}_j \frac{\partial^2 u}{\partial x \partial \zeta_j} \right\} ,$$

where

(2.4) $$C_j \equiv a_{2j-1,2m+1}+ia_{2j,2m+1} .$$

In (2.2), a bar above the letter denotes the complex conjugate, and the symmetrical condition on the coefficients,

(2.5) $$a_{2k,2j-1} = a_{2j-1,2k} , \quad k,j = 1,2,\ldots,m ,$$

has been assumed. Comparing (2.2) and (2.1), we now assume that

$$B_j = 4a_{2m+1,2m+1} = \delta \tag{2.6}$$

$$(\text{or } a_{2j,2j}+a_{2j-1,2j-1} = 2a_{2m+1,2m+1} = \delta) \quad ,$$

$$j = 1,2,\ldots,m \quad .$$

Note that this corresponds to m conditions which can be satisfied in the <u>linear</u> case, by a coordinate transformation. If x_{2k-1}, x_{2k} are real, i.e. $\zeta_k = \bar{z}_k$ then under the assumptions (2.5) and (2.6) we have the normal form:

$$L_{2m}[u] \equiv \Delta_{2m}u+\mathrm{Re}\left\{\sum_{j=1}^{m} A_j \frac{\partial^2 u}{\partial z_j^2} + \sum_{\substack{k,j=1\\k\neq j}}^{m} \left[A_{kj} \frac{\partial^2 u}{\partial z_k \partial z_j} + B_{kj} \frac{\partial^2 u}{\partial z_k \partial \zeta_j}\right]\right\}+F = 0 \quad ,$$

(2.7)

$$L_{2m+1}[u] \equiv \Delta_{2m+1}u+\mathrm{Re}\left\{\sum_{j=1}^{m} A_j \frac{\partial^2 u}{\partial z_j^2} + \sum_{\substack{k,j=1\\k\neq j}}^{m} \left[A_{kj} \frac{\partial^2 u}{\partial z_k \partial z_j} + B_{kj} \frac{\partial^2 u}{\partial z_k \partial \zeta_j}\right] + \sum_{j=1}^{m} C_j \frac{\partial^2 u}{\partial x \partial z_j}\right\} +F = 0$$

for n even and odd respectively. Here Δ_n denotes the

Laplacian operator in the n-dimensional space; $F \equiv -f$ in terms of variables z_k and ζ_k. It is understood that in (2.7), δ has been normalized equal to 4 (otherwise the coefficients in (2.7) should be multiplied by a factor of $\frac{4}{\delta}$).

3. Existence Theorem

In this section, we consider the Dirichlet problem, (E) and (B) when Ω is the interior of a unit sphere. Here, and henceforth, we denote Ω by S_n. As is well known, the Green's function $G(\underset{\sim}{x},\underset{\sim}{\xi})$ for $\Delta_n u = 0$ in S_n has the form

$$(3.1) \qquad G(\underset{\sim}{x},\underset{\sim}{\xi}) = \frac{1}{(n-2)\omega_n}\left\{\frac{1}{|\underset{\sim}{x}-\underset{\sim}{\xi}|^{n-2}} - \left(\frac{1}{|\underset{\sim}{\xi}|}\right)^{n-2}\frac{1}{|\underset{\sim}{x}-\underset{\sim}{\xi}^*|^{n-2}}\right\} .$$

Here $\xi^* = \dfrac{\underset{\sim}{\xi}}{|\underset{\sim}{\xi}|^2}$ denotes the symmetric point for $\underset{\sim}{\xi} = (\xi_1,\ldots,\xi_n)$, $|\underset{\sim}{\xi}|^2 = \sum_{i=1}^{n} \xi_i^2$, and ω_n is the surface area of S_n. We shall look for solutions u of (E) in the Sobolev space $W_p^2(S_n+\partial S_n)$, $p > n$, and hence, u belongs to class $C_\alpha'(S_n+\partial S_n)$ with exponent $\alpha = \frac{p-n}{p}$. (Recall that $u \in C_\alpha^m(D)$ if it has derivatives up to the order m which satisfy Holder conditions with exponent $\alpha < 1$.)

Following Vekua, we seek a solution of (E) (or Eq. (2.7)) in the form

$$(3.2) \qquad u(\underset{\sim}{x}) = \int_{S_n} G(\underset{\sim}{x},\underset{\sim}{\xi})\rho(\underset{\sim}{\xi})d\underset{\sim}{\xi} \equiv (\underset{\sim}{\pi}_o\rho)(\underset{\sim}{x}) \quad ,$$

where $\rho \in L_p(S_n+\partial S_n)$, $p > n$.

Remark. We note that every function v in the class of $W_p^2(S_n+\partial S_n)$ satisfying the boundary condition (B) can be represented in the form (3.2). This can be seen as follows. If we set $\rho = \Delta_n v$, clearly $v-\underset{\sim}{\pi}_o\rho$ is harmonic in S_n and vanishes on the boundary ∂S_n . Hence, by the uniqueness theorem of Dirichlet's problem for the Laplace equation, we conclude that $v = \underset{\sim}{\pi}_o\rho$ in $S_n+\partial S_n$.

From the theory of integral operators whose kernel has a weak singularity (see Mikhlin [6], p. 136), it is seen that u defined in (3.2) is in the Sobolev space $W_p^2(S_n)$. Hence, if $p > n$ (∂S_n is a Lyapunov boundary), we can conclude that $u \in C^m(S_n+\partial S_n)$ for some non negative integer $m < \left[2-\frac{n}{p}\right]$ (see Friedman [2], p. 30). Vekua's approach for the existence proof was first to reduce the differential equation (2.1) to a functional equation for ρ and then to establish the existence of its solution by the usual contraction mapping principle. In our case, we reduce (2.7) to a functional equation for ρ. The functional equation for ρ, in general, takes the form $\rho - \underset{\sim}{P}\rho = 0$ for some nonlinear operator $\underset{\sim}{P}$ which will be defined later. Here, of course, the first term in the equation is deduced from the Laplacian term in (2.7). In order to insure the existence of a solution to this nonlinear functional equation, the fact that $\underset{\sim}{P}$ maps $L_p(S_n+\partial S_n)$ into $L_p(S_n+\partial S_n)$ for $p > n$ should be established. To do so, it is necessary to investigate the properties $(\underset{\sim}{\pi}_0\rho)(\underset{\sim}{x})$ in (3.2)

It is a standard result that if $\rho \in L_p(S_n+\partial S_n)$ then the first derivatives of $(\underset{\sim}{\pi}_0\rho)(\underset{\sim}{x})$ can be obtained by the direct differentiation of $G(\underset{\sim}{x},\underset{\sim}{\xi})$ under the integral sign. We define the operators $\underset{\sim}{\pi}_i$, $i = 1,\ldots,n$ so that

$$\text{(3.2)} \qquad \frac{\partial u(\underset{\sim}{x})}{\partial x_i} = \int_{S_n} \frac{\partial}{\partial x_i} G(\underset{\sim}{x},\underset{\sim}{\xi})\rho(\underset{\sim}{\xi})d\xi \equiv (\underset{\sim}{\pi}_i\rho)(\underset{\sim}{x}) \quad .$$

We remark that the derivatives in (3.2) are to be interpreted in a generalized or Sobolev sense. From the definition of $G(\underset{\sim}{x},\underset{\sim}{\xi})$ in (3.1), a simple computation then yields the formula

$$(\pi_i \rho)(\underset{\sim}{x}) = -\frac{1}{\omega_n}\int_{S_n}\left\{\frac{x_i-\xi_i}{|\underset{\sim}{x}-\underset{\sim}{\xi}|^n} - \frac{1}{|\underset{\sim}{\xi}|^{n-2}}\frac{x_i-\xi_i^*}{|x-\underset{\sim}{\xi}^*|^n}\right\}\rho(\underset{\sim}{\xi})d\underset{\sim}{\xi}$$

$$= -\frac{1}{\omega_n}\left\{\int_{|\underset{\sim}{\xi}|\le 1}\frac{x_i-\xi_i}{|\underset{\sim}{x}-\underset{\sim}{\xi}|^n}\rho(\underset{\sim}{\xi})d\underset{\sim}{\xi}\right.$$

$$\left. - \int_{|\xi|\ge 1}\frac{x_i-\xi_i}{|\underset{\sim}{x}-\underset{\sim}{\xi}|^n}|\underset{\sim}{\xi}|^{-n-2}\rho\left(\frac{\underset{\sim}{\xi}}{|\underset{\sim}{\xi}|^2}\right)d\xi\right\}$$

To show that $\underset{\sim}{\pi}_i$ maps $L_p(S_n+\partial S_n)$ into $L_p(S_n+\partial S_n)$ for $p > n$, we extend ρ to ρ_* defined in the whole space E_n such that

$$(3.3)^* \qquad \rho_*(\underset{\sim}{\xi}) \equiv \begin{cases} \rho(\underset{\sim}{\xi}) & , \ |\underset{\sim}{\xi}| \le 1 \\ -|\underset{\sim}{\xi}|^{-n-2}\rho\left(\frac{\underset{\sim}{\xi}}{|\underset{\sim}{\xi}|^2}\right) & , \ |\underset{\sim}{\xi}| \ge 1 \end{cases}$$

To this end, let $L_{p,\upsilon}(E_n)$ be the Banach space of functions such that if $\rho \in L_p(S_n+\partial S_n)$ then $\rho^{(\upsilon)} \in L_p(S_n+\partial S_n)$ where $\rho^{(\upsilon)}(\underset{\sim}{\xi}) = |\underset{\sim}{\xi}|^{-\upsilon}\rho\left(\frac{\underset{\sim}{\xi}}{|\underset{\sim}{\xi}|^2}\right)$. Then we desire that $\rho_* \in L_{p,n+2}(E_n)$ for our case. We can then write

$$(3.4) \qquad (\underset{\sim}{\pi}_i\rho)(\underset{\sim}{x}) = -\frac{1}{\omega_n}\int_{E_n}\frac{x_i-\xi_i}{|\underset{\sim}{x}-\underset{\sim}{\xi}|^n}\rho_*(\underset{\sim}{\xi})d\underset{\sim}{\xi}$$

and consider the derivatives of $(\underset{\sim}{\pi}_i\rho)(\underset{\sim}{x})$. We introduce,

---

*For $n = 2$, the definition of ρ_* coincides with the one in [9].

when they exist, the formal operators

$$(3.5) \qquad (\pi_{ij}\rho)(x) \equiv \frac{n}{\omega_n}\int_{E_n} \frac{(x_i-\xi_i)(x_j-\xi_j)}{|x-\xi|^{n+2}}\rho_*(\xi)d\xi \quad .$$

We shall show that some special combinations of the operators defined by (3.5) are meaningful. They satisfy the so-called cancellation conditions [8, p. 35] and thus provide the desirable properties for our existence proofs. To simplify computations, we consider here only the cases when $n = 3$ and 4, since the generalization to higher dimensions is straight forward.

We return for a moment to the differential equation (2.7) for $n = 3$. For convenience, we rewrite the equation in the form

$$(3.6) \qquad \Delta_3 u + \mathrm{Re}\{Au_{zz} + Bu_{xz}\} + F = 0 \quad ,$$

with the substitutions of $z_1 = z$, $A_1 = A$, and $C_1 = B$. In this case, the coefficients satisfy the conditions:

H_1: for any arbitrary fixed $M > 0$ there exists constants $q_1 = q_1(M) < 1$ and $q_2 = q_2(M) < 1$

such that

$$|A(\underset{\sim}{x},u,u_x,u_z)| \leq q_1(M) \ , \quad |B(\underset{\sim}{x},u,u_x,u_z)| \leq q_2(M)$$

for $\underset{\sim}{x} \in S_3+\partial S_3$, $|u|+|u_x|+|u_z| \leq M$;

H_2: A and B satisfy Lipschitz condition with coefficients M_1 , M_2 such that

$$|A(\underset{\sim}{x},\hat{u},\hat{u}_x,\hat{u}_z)-A(\underset{\sim}{x},u,u_x,u_z)|$$

$$\leq M_1(|\hat{u}-u|+|\hat{u}_x-u_x|+|\hat{u}_z-u_z|) \quad \text{and}$$

$$|B(\underset{\sim}{x},\hat{u},\hat{u}_x,\hat{u}_z)-B(\underset{\sim}{x},u,u_x,u_z)|$$

$$\leq M_2(|\hat{u}-u|+|\hat{u}_x-u_x|+|\hat{u}_z-u_z|) \quad ;$$

H_3: $F(\underset{\sim}{x},u,u_x,u_z)$ is measurable, and

$F(\underset{\sim}{x},0,0,0) \in L_p(S_3+\partial S_3)$, $p > 3$;

H_4: $|F(\underset{\sim}{x},\hat{u},\hat{u}_x,\hat{u}_z)-F(\underset{\sim}{x},u,u_x,u_z)|$

$\leq F_o(\underset{\sim}{x})[|\hat{u}-u|+|\hat{u}_x-u_x|+|\hat{u}_z-u_z|]$ where

$F_o(\underset{\sim}{x}) \in L_p(S_3+\partial S_3)$, $p > 3$.

From (3.6), we are led to consider terms such as

$$\frac{\partial u}{\partial x} = \underset{\sim}{\pi}_3\rho \equiv \underset{\sim}{\pi}_x\rho \quad , \quad \frac{\partial u}{\partial z} = \frac{1}{2}\,[\underset{\sim}{\pi}_1\rho - i\underset{\sim}{\pi}_2\rho] \equiv \underset{\sim}{\pi}_z\rho$$

$$(3.7)\qquad \frac{\partial^2 u}{\partial z^2} = \frac{1}{4}\,[(\underset{\sim}{\pi}_{11}\rho - \underset{\sim}{\pi}_{22}\rho) - 2i\underset{\sim}{\pi}_{12}\rho] \equiv \underset{\sim}{T}_1\rho$$

$$\frac{\partial^2 u}{\partial x \partial z} = \frac{1}{2}\,[\underset{\sim}{\pi}_{13}\rho - i\underset{\sim}{\pi}_{23}\rho] \equiv \underset{\sim}{T}_2\rho \ ,$$

in terms of which equation (3.6) may be replaced by the operator equation:

$$(3.8)\qquad \begin{aligned} &\rho(\underset{\sim}{x}) + \mathrm{Re}\{A(\underset{\sim}{x},\underset{\sim}{\pi}_o\rho,\underset{\sim}{\pi}_x\rho,\underset{\sim}{\pi}_z\rho)\underset{\sim}{T}_1\rho + B(\underset{\sim}{x},\underset{\sim}{\pi}_o\rho,\underset{\sim}{\pi}_x\rho,\underset{\sim}{\pi}_z\rho)\underset{\sim}{T}_2\rho\} \\ &\quad + F(\underset{\sim}{x},\underset{\sim}{\pi}_o\rho,\underset{\sim}{\pi}_x\rho,\underset{\sim}{\pi}_z\rho) = 0 \ . \end{aligned}$$

We define the nonlinear operator $\underset{\sim}{P}$ such that

$$(3.9)\qquad \underset{\sim}{P}\rho \equiv -\mathrm{Re}\{\ldots\} - F(\underset{\sim}{x},\underset{\sim}{\pi}_o\rho,\underset{\sim}{\pi}_x\rho,\underset{\sim}{\pi}_z\rho) \ ,$$

and we shall show indeed $\underset{\sim}{P}$ is a nonlinear operator from a closed convex subset of $L_p(S_3+\partial S_3)$, $p > 3$ into itself.

Let us introduce the notation

$$(3.10)\qquad \Omega_{ij} \equiv \int_{\partial S_n} \frac{x_i x_j}{|\underset{\sim}{x}|^2}\, d\omega \ ,$$

and we say that $\underset{\sim}{\pi}_{ij}$ (or a linear combination of π_{ij}'s) satisfies the cancellation condition if Ω_{ij} (or the corresponding linear combination of π_{ij}'s) is equal to zero (see Stein [8], p. 39). For $n = 3$, it is easy to verify that $\Omega_{12} = \Omega_{13} = \Omega_{23} = 0$ and $\Omega_{11} = \Omega_{22} = \frac{4\pi}{3}$. Thus, the operators $\underset{\sim}{T}_k$, $k = 1,2$, defined in (3.7) satisfy the cancellation condition. Furthermore, one can easily show, in general, when $p \geq 2\left(\frac{n-1}{n}\right)$, $L_{p,n+2}(E_n) \subset L_p(E_n)$. Therefore, using a theorem in Stein [8, p. 39], we realize that for $p \geq \frac{4}{3}$ there exists constants $\hat{A}_k(p)$, $(k = 1,2)$, independent of ρ_* such that

$$||(\underset{\sim}{T}_k\rho_*)(\underset{\sim}{x})||_{L_p,E_3} \leq \hat{A}_k||\rho_*||_{L_p,E_3} .$$

Consequently, there exist constant $A_k(p)$, $(k = 1,2)$,

$$(3.11)\quad \begin{aligned} ||(\underset{\sim}{T}_k\rho)(\underset{\sim}{x})||_{L_p,S_3} &\leq ||(\underset{\sim}{T}_k\rho_*)(\underset{\sim}{x})||_{L_p,E_3} \\ &\leq \hat{A}_k||\rho_*||_{L_p,E_3} \leq A_k||\rho||_{L_p,S_3} . \end{aligned}$$

The last inequality can be shown as follows:

$$||\rho_*||_{L_p,E_n} = \left\{\int_{S_n} |\rho|^p d\underset{\sim}{x} + \int_{S_n^c} |\rho_*|^p d\underset{\sim}{x}\right\}^{1/p}$$

$$= \left\{\int_{|\underset{\sim}{\xi}|\leq 1} |\rho|^p d\underset{\sim}{\xi} + \int_{|\underset{\sim}{\xi}|\leq 1} |\underset{\sim}{\xi}|^{-2n}(|\underset{\sim}{\xi}|^{(n+2)}|\rho|)^p d\underset{\sim}{\xi}\right\}^{1}$$

$$= \left\{\int_{|\underset{\sim}{\xi}|\leq 1} |\rho|^p\left(1+|\underset{\sim}{\xi}|^{p(n+2)-2n}\right)d\underset{\sim}{\xi}\right\}^{1/p}$$

$$\leq 2\left\{\int_{|\underset{\sim}{\xi}|\leq 1} |\rho|^p d\underset{\sim}{\xi}\right\}^{1/p}$$

$$= 2||\rho||_{L_p,S_n} \quad , \quad \text{for} \quad p \geq n \quad .$$

Because of the weak singularity of the Green's function defined in (3.1), it is easy to see that the following estimates hold for $\rho \in L_p(S_3+\partial S_3)$:

$$(3.12) \qquad \begin{aligned} &|\underset{\sim}{\pi}_o\rho| \leq K||\rho||_{L_p,S_3} \quad , \quad |\underset{\sim}{\pi}_x\rho| \leq K||\rho||_{L_p,S_3} \quad , \\ &|\underset{\sim}{\pi}_z\rho| \leq K||\rho||_{L_p,S_3} \quad , \end{aligned}$$

where K is some constant. We note also that from a

lemma due to Hile [4, p. 98], the functions $\underset{\sim}{\pi}_x\rho$ and $\underset{\sim}{\pi}_z\rho$ are of class $C_\alpha(S_3+\partial S_3)$, $\alpha = \frac{p-3}{p}$; that is Hölder continuous in $S_3+\partial S_3$ with exponent α .

Now, let us consider the operator $\underset{\sim}{P}$ defined by (3.9). From the definition of the operator $\underset{\sim}{P}$, and properties H_1-H_4 , we have, for $\rho_1,\rho_2 \in L_p(S_3+\partial S_3)$, $p > 3$,

$$|\underset{\sim}{P}\rho_1-\underset{\sim}{P}\rho_2| \leq q_1(M)|\underset{\sim}{T}_1(\rho_1-\rho_2)|+q_2(M)|\underset{\sim}{T}_2(\rho_1-\rho_2)|$$

$$+3K[M_1|\underset{\sim}{T}_1\rho_1|+M_2|\underset{\sim}{T}_2\rho_1|+F_o(\underset{\sim}{x})]\,||\rho_1-\rho_2||_{L_p,S_3} \;.$$

We choose M such that

$$|\underset{\sim}{\pi}_o\rho_2|+|\underset{\sim}{\pi}_x\rho_2|+|\underset{\sim}{\pi}_z\rho_2| \leq 3K||\rho_2||_{L_p,S_3} \leq M \;.$$

Then from the Minkowski inequality and estimates of (3.11), we have

$$||\underset{\sim}{P}\rho_1-\underset{\sim}{P}\rho_2||_{L_p,S_3} \leq K_p||\rho_1-\rho_2||_{L_p,S_3} \;,$$

where

$$K_p = \left\{ q(M)\Lambda_p + 3K(M_1+M_2)\Lambda_p ||\rho_1||_{L_p,S_3} + 3K||\underset{\sim}{F}_o||_{L_p,S_3} \right\}$$

Here $q(M) \equiv q_1(M)+q_2(M)$ and $\Lambda_p \equiv \max_{k=1,2} \{A_k\}$ which is a fixed quantity. Now let $q(M)$ be such that $q(M)\Lambda_p < 1$. (This can always be done provided $q_i(M)$, $i = 1,2$, are sufficiently small.) If positive constants ε and r are chosen so that

$$3Kr < M$$

(3.14)

$$\alpha \equiv q(M)\Lambda_p + 3K(M_1+M_2)\Lambda_p r + 3K\varepsilon < 1$$

then for $||F_o||_{L_p,S_3} < \varepsilon$ and for ρ_1, ρ_2 satisfying $||\rho_k||_{L_p,S_3} < r$, $(k = 1,2)$, we have

$$K_p \leq \alpha < 1 \quad .$$

Thus, taking into account the fact that the operator $\underset{\sim}{P}$ acting on the zero element $\theta = 0$ yields $F(\underset{\sim}{x},0,0,0)$ let us assume that

$$(3.15) \qquad \|\underset{\sim}{P}\theta\|_{L_p,S_3} \equiv \|\underset{\sim}{F}(x,0,0,0)\|_{L_p,S_3} < (1-\alpha)r \quad .$$

Under these conditions, clearly $\underset{\sim}{P}$ maps the sphere $\|\rho\|_{L_p,S_3} < r$ into itself, and by the Banach contraction mapping principle, $\underset{\sim}{P}$ has a unique fixed point ρ in $L_p(S_3,\partial S_3)$ for $\|\rho\|_{L_p,S_3} < r$.

We treat next the four variable case, which we may put (2.7) in the form

$$(3.16) \qquad \Delta_4 u+\mathrm{Re}\{Au_{zz}+Bu_{yy}+Cu_{zy}+Du_{z\bar{y}}\}+F = 0$$

by setting $z_1 = z$, $z_2 = y$, $\zeta_1 = \bar{z}$, $\zeta_2 = \bar{y}$; $A_1 = A$, $A_2 = B$, $A_{12}+A_{21} = C$ and $B_{12}+\bar{B}_{21} = D$. All the coefficients are functions of $(\underset{\sim}{x},u,u_z,u_y)$. We assume as before

H_1: $|A| \leq q_1(M) < 1$, $|B| \leq q_2(M) < 1$,

$|C| \leq q_3(M) < 1$, $|D| \leq q_4(M) < 1$ for

$\underset{\sim}{x} \in S_4+\partial S_4$, $|u|+|u_z|+|u_y| \leq M$;

H_2: A , B , C , D satisfy a Lipschitz condition,

with coefficients M_1, M_2, M_3, M_4 for $|u|+|u_z|+|u_y| \le M$;

H_3: $F(\underset{\sim}{x},u,u_z,u_y)$ is measurable and $F(\underset{\sim}{x},0,0,0) \in L_p(S_4+\partial S_4)$, $p > 4$; and

H_4: $|F(\underset{\sim}{x},\hat{u},\hat{u}_z,\hat{u}_y)-F(\underset{\sim}{x},u,u_z,u_y)|$

$\le F_o(\underset{\sim}{x})[|\hat{u}-u|+|\hat{u}_z-u_z|+|\hat{u}_y-u_y|]$ with

$F_o(\underset{\sim}{x}) \in L_p(S_4+\partial S_4)$, $p > 4$.

The operators $\underset{\sim}{\pi}_i$ and $\underset{\sim}{\pi}_{ij}$, $(i,j = 1,2,3,4)$, are defined by (3.4) and (3.5) with $n = 4$, while $\underset{\sim}{\pi}_z\rho$ and $\underset{\sim}{\pi}_y\rho$ take the form:

$$\frac{\partial u}{\partial z} = \frac{1}{2}[\underset{\sim}{\pi}_1\rho - i\underset{\sim}{\pi}_2\rho] \equiv \underset{\sim}{\pi}_z\rho \quad \text{and}$$
(3.17)
$$\frac{\partial u}{\partial y} = \frac{1}{2}[\underset{\sim}{\pi}_3\rho - i\underset{\sim}{\pi}_4\rho] \equiv \underset{\sim}{\pi}_y\rho \; .$$

As before, we define formally the operators $\underset{\sim}{T}_k$, $k = 1,2,3,4$,

$$\underset{\sim}{T}_1 = \underset{\sim}{\pi}_{zz} = \frac{1}{4}[(\underset{\sim}{\pi}_{11}-\underset{\sim}{\pi}_{22})-2i\underset{\sim}{\pi}_{12}]$$

$$\underset{\sim}{T}_2 = \underset{\sim}{\pi}_{yy} = \frac{1}{4}[(\underset{\sim}{\pi}_{33}-\underset{\sim}{\pi}_{44})-2i\underset{\sim}{\pi}_{34}]$$

(3.18)

$$\underset{\sim}{T}_3 = \underset{\sim}{\Pi}_{zy} = \frac{1}{4} [(\underset{\sim}{\Pi}_{13}-\underset{\sim}{\Pi}_{24})-i(\underset{\sim}{\Pi}_{14}+\underset{\sim}{\Pi}_{23})]$$

$$\underset{\sim}{T}_4 = \underset{\sim}{\Pi}_{z\bar{y}} = \frac{1}{4} [(\underset{\sim}{\Pi}_{13}+\underset{\sim}{\Pi}_{24})+i(\underset{\sim}{\Pi}_{14}-\underset{\sim}{\Pi}_{23})] \quad .$$

In order to see whether these operators are well defined, we check to see whether they satisfy the cancellation conditions. To this end, we compute $\Omega_{ij} = \int_{\partial S_4} \frac{x_i - x_j}{r^2} d\omega$. One easily finds that $\Omega_{11} = \Omega_{22} = \frac{\pi^2}{4}$, $\Omega_{33} = \Omega_{44} = \pi$, and $\Omega_{ij} = 0$, for $i \neq j$. Using these in the above formal definitions of $\underset{\sim}{T}_k$, we see that they have meaning in the principal value sense and indeed we have, as in the case $n = 3$,

$$(3.19) \qquad ||\underset{\sim}{T}_k \rho||_{L_p,S_4} \leq A_k(p)||\rho||_{L_p,S_4} \quad , \quad p \geq \frac{3}{2} \quad .$$

As in the case $n = 3$, the differential equation (3.16) may then be replaced by the functional equation $\rho - \underset{\sim}{P}\rho = 0$, where the operator $\underset{\sim}{P}$ is defined by

$$(3.20) \qquad \underset{\sim}{P}\rho \equiv -\mathrm{Re}\{A(\underset{\sim}{x},\underset{\sim}{\Pi}_o\rho,\underset{\sim}{\Pi}_z\rho,\underset{\sim}{\Pi}_y\rho)\underset{\sim}{T}_1\rho+B(\ldots)T_2\rho+C(\ldots)\underset{\sim}{T}_3\rho$$

$$+D(\ldots)\underset{\sim}{T}_4\rho\}-F(\ldots) \quad .$$

In the same manner as in the case $n = 3$, one can show $\underset{\sim}{P}$ is a nonlinear mapping from a closed subset of $L_p(S_4+\partial S_4)$, $p > 4$ into itself. Under the conditions H_1-H_4, if M is chosen so that for $\rho_1, \rho_2 \in L_p(S_4+\partial S_4)$, $p > 4$,

$$|\underset{\sim}{\pi}_o\rho_2|+|\underset{\sim}{\pi}_z\rho_2|+|\underset{\sim}{\pi}_y\rho_2| \leq 3K||\rho_2||_{L_p,S_4} \leq M$$

then we have,

$$||\underset{\sim}{P}\rho_1-\underset{\sim}{P}\rho_2||_{L_p,S_4} \leq K_p||\rho_1-\rho_2||_{L_p,S_4} ,$$

with

$$K_p = \Big\{q(M)\Lambda_p+3K(M_1+M_2+M_3+M_4)\Lambda_p||\rho_1||_{L_p,S_4}$$

$$+3K||F_o||_{L_p,S_4}\Big\} .$$

Here $q(M) = \sum_{k=1}^{4} q_k(M)$ and $\Lambda_p = \max_{k=1,2,3,4} \{A_k\}$.

The remainder of the proof is identical to the three variable cases. The details are omitted. It is clear that one can follow the same procedure used here to study the case for higher dimensions. We now summarize our results for the case $n = 3$ and 4 in the following theorem.

Theorem 1. *For* $n = 3$ *and* 4, *under the conditions* H_1-H_4, *if* $||F_o(\underset{\sim}{x})||_{L_p,S_n}$ *and* $||\underset{\sim}{F}(\underset{\sim}{x},0,0,0)||_{L_p,S_n}$ *are sufficiently small and* $q(M)\Lambda_p < 1$, *then the Dirichlet problem for the quasi-linear equation* (2.7) *with homogeneous boundary conditions has a unique solution in the class* $W_p^2(S_n+\partial S_n)$, *which is consequently in the class* $C_\alpha(S_n+\partial S_n)$, $\alpha = \frac{p-n}{p}$. *Moreover, the solution* $u(\underset{\sim}{x})$ *can be represented in the form*

$$u(\underset{\sim}{x}) = \int_{S_n} G(\underset{\sim}{x},\underset{\sim}{\xi})\rho(\underset{\sim}{\xi})d\underset{\sim}{\xi} \quad ,$$

where $G(\underset{\sim}{x},\xi)$ *is the Green's function for* $\Delta_n u = 0$, *and* $\rho(\xi)$ *is a uniquely determined function in* $L_p(S_n+\partial S_n)$, $p > n$.

4. Remarks on Existence Proofs

We now turn our attention to the Dirichlet problem (2.7) and (B) for a regular domain Ω with a Lyapunov boundary $\partial\Omega$. We shall indicate how one can carry out the existence proof as in the case of a sphere.

The operators $\underset{\sim}{\pi}_o$, $\underset{\sim}{\pi}_i$, $i = 1,2,\ldots n$ are defined as before

$$u(\underset{\sim}{x}) = (\underset{\sim}{\pi}_o \rho)(\underset{\sim}{x}) \equiv \int_\Omega G(\underset{\sim}{x},\xi)\rho(\xi)d\underset{\sim}{\xi}$$

(4.1)

$$\frac{\partial u}{\partial x_i} = (\underset{\sim}{\pi}_i \rho)(\underset{\sim}{x}) \equiv \int_\Omega \frac{\partial}{\partial x_i} G(\underset{\sim}{x},\underset{\sim}{\xi})\rho(\underset{\sim}{\xi})d\underset{\sim}{\xi} \quad ,$$

where $\rho \in L_p(\Omega+\partial\Omega)$, $p > n$ and $G(\underset{\sim}{x},\underset{\sim}{\xi})$ is the Green's function of $\Delta_n u = 0$ in Ω . The derivatives in (4.1) are to be interpreted in a generalized or Sobolev sense. Since the Green's function $G(\underset{\sim}{x},\xi)$ is dominated by its singular part, $\frac{1}{\omega_n(n-2)} |\underset{\sim}{x}-\underset{\sim}{\xi}|^{-n+2}$, for convenience, we shall henceforth denote $\hat{\underset{\sim}{\pi}}$ the corresponding singular part of the operator $\underset{\sim}{\pi}$ defined by G . Notice that the inequalities (for $p > n$),

$$|(\hat{\underset{\sim}{\pi}}_o \rho)(\underset{\sim}{x})| \equiv \frac{1}{\omega_n(n-2)} \left| \int_\Omega \frac{1}{|\underset{\sim}{x}-\underset{\sim}{\xi}|^{n-2}} \rho(\xi)d\underset{\sim}{\xi} \right|$$

$$\leq \frac{1}{\omega_n(n-2)} \left[\int_\Omega |\rho(\underset{\sim}{\xi})|^p d\underset{\sim}{\xi} \right]^{1/p}$$

$$\cdot \left[\int_\Omega |\underset{\sim}{x}-\underset{\sim}{\xi}|^{-(n-2)q} d\underset{\sim}{\xi} \right]^{1/q} \quad ,$$

(4.2)

$$|(\hat{\pi}_i \rho)(\underset{\sim}{x})| \equiv \frac{1}{\omega_n} \left| \int_\Omega \frac{x_i - \xi_i}{|\underset{\sim}{x} - \underset{\sim}{\xi}|^n} \rho(\underset{\sim}{\xi}) d\underset{\sim}{\xi} \right|$$

$$\leq \frac{1}{\omega_n} \left[\int_\Omega |\rho(\underset{\sim}{\xi})|^p d\underset{\sim}{\xi} \right]^{1/p}$$

$$\cdot \left[\int_\Omega \left(\frac{|x_i - \xi_i|}{|\underset{\sim}{x} - \underset{\sim}{\xi}|^n} \right)^q d\underset{\sim}{\xi} \right]^{1/q} ,$$

where $\frac{1}{p} + \frac{1}{q} = 1$. It is clear that Eqs. (4.2) imply the existence of bounded positive constants K_i $(i = 0,1,\ldots,n)$ such that

$$\sup_{\underset{\sim}{x} \in \Omega+\partial\Omega} |(\underset{\sim}{\pi}_o \rho)(\underset{\sim}{x})| \leq \sup_{\underset{\sim}{x} \in \Omega+\partial\Omega} ||G||_{L_q} ||\rho||_{L_p}$$

$$= K_o ||\rho||_{L_p}$$

$$\sup_{\underset{\sim}{x} \in \Omega+\partial\Omega} |(\pi_i \rho)(x)| \leq \sup_{\underset{\sim}{x} \in \Omega+\partial\Omega} \left|\left| \frac{\partial G}{\partial x_i} \right|\right|_{L_q} ||\rho||_{L_p}$$

$$= K_i ||\rho||_{L_p} .$$

Let $K = \max_{i=0,1,\ldots,n} \{K_i\}$. Then we have

(4.3) $$|(\underset{\sim}{\pi}_i \rho)| \leq K ||\rho||_{L_p} \quad (i = 0,1,2,\ldots,n) , \quad p > n .$$

We extend $\rho(\underset{\sim}{\xi})$ to $\rho_*(\underset{\sim}{\xi})$ defined in the whole space such that

$$\rho_*(\underset{\sim}{x}) = \begin{cases} \rho \,, & \underset{\sim}{x} \in \Omega \\ 0 \,, & \underset{\sim}{x} \notin \Omega \end{cases} .$$

As in the case of the sphere, for $n = 3$, we introduce the formal operators

$$(4.4) \qquad \hat{\underset{\sim}{T}}_1\rho \equiv \hat{\underset{\sim}{\pi}}_{zz}\rho \quad \text{and} \quad \hat{\underset{\sim}{T}}_2\rho \equiv \hat{\underset{\sim}{\pi}}_{xz}\rho$$

and we can show that $\hat{\underset{\sim}{T}}_i$, $i = 1,2$ satisfy the cancellation condition (see (3.10)) which implies $\rho_* \in L_p(E_3)$, $\hat{\underset{\sim}{\pi}}_{zz}\rho_*$ and $\hat{\underset{\sim}{\pi}}_{xz}\rho_*$ are of class $L_p(E_3)$. Consequently, there are constant $\hat{\underset{\sim}{A}}_k$ $(k = 1,2)$ such that

$$(4.5) \qquad ||(\hat{T}_k\rho)(\underset{\sim}{x})||_{L_p,\Omega} = ||(\hat{T}_k\rho_*)(\underset{\sim}{x})||_{L_p,E_3}$$
$$\leq \hat{A}_k||\rho_*||_{L_p,E_3} = \hat{A}_k||\rho||_{L_p,\Omega}$$

The operators $\underset{\sim}{T}_k$, $(k = 1,2)$ are then defined by

$$\underset{\sim}{T}_1\rho \equiv \hat{\underset{\sim}{T}}_1\rho + \int_\Omega \frac{\partial^2}{\partial z^2}\, g(\underset{\sim}{x},\underset{\sim}{\xi})\,\rho(\underset{\sim}{\xi})\,d\underset{\sim}{\xi}$$

(4.6)

$$\underset{\sim}{T}_2\rho \equiv \hat{\underset{\sim}{T}}_2\rho + \int_\Omega \frac{\partial^2}{\partial x\partial z}\, g(\underset{\sim}{x},\underset{\sim}{\xi})\,\rho(\underset{\sim}{\xi})\,d\underset{\sim}{\xi} \quad ,$$

where $g(\underset{\sim}{x},\underset{\sim}{\xi})$ is the regular part of the Green's function $G(\underset{\sim}{x},\underset{\sim}{\xi})$ in Ω. One sees that $(\underset{\sim}{T}_k - \hat{\underset{\sim}{T}}_k)$ is a compact operator on $L_p(\Omega+\partial\Omega)$ and we must have

$$\left\{\int_\Omega |\underset{\sim}{T}_k\rho|^p d\underset{\sim}{x}\right\}^{1/p} = \left\{\int_{E_3} |\underset{\sim}{T}_k\rho_*|^p d\underset{\sim}{x}\right\}^{1/p}$$

$$\leq A_k ||\rho||_{L_p,\Omega} \quad .$$

Let $\Lambda_p \;\max_{k=1,2} \{A_k\}$. The proof proceeds then as in the case of the sphere <u>if</u> we know the Green's function for Laplace's equation.

5. <u>An Iterative Scheme</u>

It is seen that applying Vekua's procedure to construct a solution of the Dirichlet problem, one needs an explicit form of the Green's function for the Laplace equation in the domain under consideration. From the viewpoint of numerical computation, it is rather impractical.

As is well known, to find the Green's function, in general, is a difficult task, although one can show theoretically the existence of the Green's functions for regular domains. To circumvent this difficulty, we present here a modified procedure which depends only upon the fundamental solution,

$$(5.1) \qquad E_n(\underset{\sim}{x},\underset{\sim}{\xi}) = \frac{1}{\omega_n(n-2)} \frac{1}{|\underset{\sim}{x}-\underset{\sim}{\xi}|^{n-2}} ,$$

to the Laplace equation. We give an iterative procedure for solving the Dirichlet problem for Eq. (2.7). Our approach follows Warschawski's idea for treating the Lichtenstein-Gershgorin integral [10]. For simplicity, we again confine ourselves to the three-dimensional case. It is not difficult to see that our results should also extend to higher dimensions.

For convenience, we relabel Eq. (3.6) as (5.2). We seek a solution of the equation

$$(5.2) \qquad \Delta_3 u+\mathrm{Re}\{Au_{zz}+Bu_{xz}\}+F = 0$$

in a Lyapunov domain Ω, satisfying the homogeneous data $u = 0$ on the boundary $\partial\Omega$. To this end we establish the following lemma.

<u>Lemma</u> 5.1. <u>Each solution of</u> (5.2) <u>in</u> $W_p^2(\Omega+\partial\Omega)$,

$p > 3$, *such that* $u = 0$ *on* $\partial\Omega$ *can be represented in the form*

$$(5.3) \qquad u(\underset{\sim}{x}) = -\int_{\Omega} E_3(\underset{\sim}{x},\underset{\sim}{\xi})\rho(\underset{\sim}{\xi})d\underset{\sim}{\xi} + \int_{\partial\Omega} \frac{\partial}{\partial\nu_{\underset{\sim}{\xi}}} E_3(\underset{\sim}{x},\underset{\sim}{\xi})\mu(\underset{\sim}{\xi})d\sigma_{\underset{\sim}{\xi}}$$

for $\rho \in L_p(\Omega+\partial\Omega)$ *and* $\mu \in L_p(\partial\Omega)$, $p > 3$, *where* $\frac{\partial}{\partial\nu_{\underset{\sim}{\xi}}}$ *denotes differentiation with respect to the inward normal at* $\underset{\sim}{\xi} \in \partial\Omega$.

Proof. If $\rho \in L_p(\Omega+\partial\Omega)$, $p > 3$, the first integral is in $W_p^2(\Omega+\partial\Omega)$ (see Mikhlin [6], p. 135). The second integral is analytic in Ω . Consequently,

$$\Delta_3 u = \rho(\underset{\sim}{x}) \in L_p(\Omega+\partial\Omega) \quad , \quad p > 3$$

and

$$u(\underset{\sim}{x}) + \int_{\Omega} \rho(\underset{\sim}{\xi})E_3(\underset{\sim}{x},\underset{\sim}{\xi})d\underset{\sim}{\xi} \equiv h(\underset{\sim}{x})$$

is harmonic. Since $p > 3$, $h(\underset{\sim}{x}) \in C^1(\Omega+\partial\Omega)$, and hence may be represented as a double layer potential, which proves the results (5.3).

We now indicate the idea how one can modify Vekua's

approach by using the representation of $u(\underset{\sim}{x})$ in (5.3) to reduce Eq. (5.2) to an operator equation such as Eq. (3.8). Observe first that from the homogeneous boundary condition $u = 0$ on $\partial\Omega$ and the double layer potential as $\underset{\sim}{x}$ approaches $\partial\Omega$, Eq. (5.3) reads

$$(5.4) \qquad 0 = -\int_{\Omega} E_3(\underset{\sim}{x},\underset{\sim}{\xi})\rho(\underset{\sim}{\xi})d\underset{\sim}{\xi}+\mu(\underset{\sim}{x})+\int_{\partial\Omega} \frac{\partial}{\partial\upsilon_{\underset{\sim}{\xi}}} E_3(\underset{\sim}{x},\underset{\sim}{\xi})\mu(\underset{\sim}{\xi})d\sigma_{\underset{\sim}{\xi}}$$

which we rewrite in operator notation as

$$(5.5) \qquad -(\underset{\sim}{\pi}_E\rho)(\underset{\sim}{x})+(\underset{\sim}{I}+\underset{\sim}{K})\mu(\underset{\sim}{x}) = 0 \ , \quad \underset{\sim}{x} \in \partial\Omega$$

with

$$(\underset{\sim}{\pi}_E\rho)(\underset{\sim}{x}) \equiv \int_{\Omega} E_3(\underset{\sim}{x},\underset{\sim}{\xi})\rho(\underset{\sim}{\xi})d\underset{\sim}{\xi} \quad ;$$

$$(\underset{\sim}{K}\mu)(\underset{\sim}{x}) \equiv \int_{\partial\Omega} K(\underset{\sim}{x},\underset{\sim}{\xi})\mu(\underset{\sim}{\xi})d\sigma_{\underset{\sim}{\xi}} \quad ,$$

where $K(\underset{\sim}{x},\underset{\sim}{\xi}) \equiv \frac{\partial}{\partial\upsilon_{\underset{\sim}{\xi}}} E_3(\underset{\sim}{x},\underset{\sim}{\xi})$.

We note that $W_p^2(\Omega+\partial\Omega) \subset C^1(\Omega+\partial\Omega)$; hence, if $(\underset{\sim}{I}+\underset{\sim}{K})^{-1}$ exists on $C^1(\Omega+\partial\Omega)$ we may formally solve Eq. (5.5) for the density $\mu(\underset{\sim}{x})$, i.e.,

$$(5.6) \qquad \mu(\underset{\sim}{x}) = (\underset{\sim}{I}+\underset{\sim}{K})^{-1}(\underset{\sim}{\pi}_E\rho)(\underset{\sim}{x}) \quad , \quad \underset{\sim}{x} \in \partial\Omega \ .$$

Then Eq. (5.3) may be written as

$$(5.7) \qquad u(\underset{\sim}{x}) = -(\underset{\sim}{\pi}_E\rho)(\underset{\sim}{x}) + \int_{\partial\Omega} K(\underset{\sim}{x},\xi)[(\underset{\sim}{I}+\underset{\sim}{K})^{-1}(\underset{\sim}{\pi}_E\rho)(\underset{\sim}{\xi})]ds_\xi$$

$$\equiv (\hat{\underset{\sim}{\pi}}_0\rho)(\underset{\sim}{x}) \quad .$$

It is clear then by introducing the operators $\hat{\underset{\sim}{\pi}}_0\rho$, $\hat{\underset{\sim}{\pi}}_x\rho$, $\hat{\underset{\sim}{\pi}}_z\rho$, $\hat{\underset{\sim}{T}}_1\rho$ and $\hat{\underset{\sim}{T}}_2\rho$ as we did in Section 3 (see (3.7)), we reduce Eq. (5.3) to the operator equation

$$(5.8) \qquad \rho - \hat{\underset{\sim}{P}}\rho = 0 \quad ,$$

where

$$\hat{\underset{\sim}{P}}\rho \equiv -\mathrm{Re}\{A(\underset{\sim}{x},\hat{\underset{\sim}{\pi}}_0\rho,\hat{\underset{\sim}{\pi}}_z\rho,\hat{\underset{\sim}{\pi}}_x\rho)\hat{\underset{\sim}{T}}_1\rho + B(\ldots)\hat{\underset{\sim}{T}}_2\rho\} - F(\ldots) \quad .$$

It should be noted that so far what we have done is purely formal, but it is not difficult to see that by a similar analysis as in Section 3, one can indeed show that Eq. (5.8) is the corresponding operator equation for (5.2) and the nonlinear operator $\hat{\underset{\sim}{P}}$ possesses the desired properties so that one can apply the contraction mapping principle to establish the existence results.

Let us return to Eq. (5.6), and consider a more general operator equation

$$(5.9) \qquad \mu(\underset{\sim}{x}) - \lambda(\underset{\sim}{K}\mu)(\underset{\sim}{x}) = f(\underset{\sim}{x}) \quad , \quad \underset{\sim}{x} \in \partial\Omega$$

with the kernel $K(\underset{\sim}{x},\underset{\sim}{\xi}) = \frac{\partial}{\partial \upsilon_{\underset{\sim}{\xi}}} E_3(\underset{\sim}{x},\underset{\sim}{\xi})$ as before. The cases $\lambda = 1$ and $\lambda = -1$ of (5.9) correspond to the exterior and interior Dirichlet problem for Laplace's equation. It is well known that for continuous $f(\underset{\sim}{x})$ and Lyapunov Ω these problems are well posed; hence the representation (5.6) may be shown to be valid. To see this, one requires some information about the resolvent kernel $R(\underset{\sim}{x},\underset{\sim}{\xi},\lambda)$ of Eq. (5.9). The resolvent kernel takes the form:

$$(5.10) \qquad R(\underset{\sim}{x},\underset{\sim}{\xi},\lambda) \equiv \sum_{n=0}^{\infty} \lambda^n K_{n+1}(\underset{\sim}{x},\underset{\sim}{\xi}) \quad ,$$

where $K_{n+1}(\underset{\sim}{x},\underset{\sim}{\xi})$ are the iterated kernels. We note that the only pole of the resolvent occurs for $\lambda = 1$; furthermore, this is a simple pole, (see Goursat [3], p. 192-3). Since this is a simple pole, it is also well known that the principal part of the resolvent at $\lambda = 1$ is $\sum_{j=1}^{q} \psi_j(\underset{\sim}{x}) \overline{\phi_j(\underset{\sim}{\xi})}$ where ψ_j and ϕ_j are the eigenfunctions of $\underset{\sim}{K}$ and $\underset{\sim}{K}^*$ respectively. For the potential equation, it is also known that $q = 1$, $\psi_1(\underset{\sim}{x}) = 1$, and $\phi_1(x)$ is the so-called "equilibrium distribution" (see Hoheisel [5], p. 54). Therefore, the function $\Gamma(\underset{\sim}{x},\underset{\sim}{\xi},\lambda)$ defined by

$$(5.11) \qquad \Gamma(\underset{\sim}{x},\underset{\sim}{\xi},\lambda) \equiv R(\underset{\sim}{x},\underset{\sim}{\xi},\lambda) - \frac{\phi_1(\underset{\sim}{\xi})}{1-\lambda} = \sum_{\upsilon=0}^{\infty} \lambda^{\upsilon}[K_{\upsilon+1}(\underset{\sim}{x},\underset{\sim}{\xi})-\phi_1(\underset{\sim}{\xi})]$$

converges for all λ such that $|\lambda| < |\lambda_2|$, where λ_2 is the eigenvalue of $K(\underset{\sim}{x},\underset{\sim}{\xi})$, the absolute value of which is nearest to 1, and, of course, $|\lambda_2| > 1$. It is easily seen that the series

$$(5.12) \qquad f(\underset{\sim}{x}) + \sum_{\upsilon=0}^{\infty} \lambda^{\upsilon+1} \int_{\partial\Omega} [K_{\upsilon+1}(\underset{\sim}{x},\underset{\sim}{\xi})-\phi_1(\underset{\sim}{\xi})]f(\underset{\sim}{\xi})\,d\sigma_{\underset{\sim}{\xi}}$$

converges for $|\lambda| < |\lambda_2|$. For the case $\lambda = -1$ which is of particular interest to us, (5.12) represents precisely a solution of (5.9) which we obtain by the iteration process.

We next turn to the problem of obtaining estimates on the error for the approximate solution of (5.9). To this end, we introduce some function space. Our approach is an n-dimensional analogue of the ideas given by Warschawski [10].

Consider $C(\partial\Omega)$, the set of all continuous functions on $\partial\Omega$. For $f,g \in C(\partial\Omega)$, we introduce the inner product

$(\, , \,)$ such that

$$(5.13) \qquad (f,g) \equiv \int_{\partial\Omega} \int_{\partial\Omega} f(\underset{\sim}{x}) g(\underset{\sim}{y}) \frac{D}{|\underset{\sim}{x}-\underset{\sim}{y}|^{n-2}} d\sigma_{\underset{\sim}{x}} d\sigma_{\underset{\sim}{y}} \qquad (n \geq 3)$$

where D is a fixed number such that the diameter of $\partial\Omega$ does not exceed D. Clearly, $||f||^2 = (f,f)$ is positive definite, and the completion of $C(\partial\Omega)$ under $(\, , \,)$ is a Hilbert space. We denote it by $\mathcal{H}$. The purpose of introducing such a particular form (5.13) for the inner product is that our kernel $K(\underset{\sim}{x},\underset{\sim}{\xi}) = \frac{\partial}{\partial \upsilon_{\underset{\sim}{\xi}}} E_3(\underset{\sim}{x},\underset{\sim}{\xi})$ in Eq. (5.3) is symmetrizable by means of $\frac{D}{|\underset{\sim}{x}-\underset{\sim}{y}|^{n-2}}$, i.e., the function

$$(5.14) \qquad H(\underset{\sim}{x},\underset{\sim}{y}) \equiv \int_{\partial\Omega} K(\underset{\sim}{x},\underset{\sim}{\xi}) \frac{D}{|\underset{\sim}{\xi}-\underset{\sim}{y}|^{n-2}} d\sigma_{\underset{\sim}{\xi}}$$

is symmetrical in $(\underset{\sim}{x},\underset{\sim}{y})$: $H(\underset{\sim}{x},\underset{\sim}{y}) = H(\underset{\sim}{y},\underset{\sim}{x})$. From the theory of symmetrizable kernels (see Riesz-Nagy [7], p. 241) it is known that if the symmetrizable kernel is compact, there exists at least one eigenvalue, and at most a denumerable number of eigenvalues; moreover, the eigenvalues are real. The eigenvalues of $K(\underset{\sim}{x},\underset{\sim}{y})$ are

$$\lambda_1 = 1, \lambda_2, \lambda_3, \ldots,$$

where

$$|\lambda_1| < |\lambda_2| \leq |\lambda_3| < \dots \quad .$$

(Also $\underset{\sim}{K}$ may be extended to the Hilbert space $\mathcal{H}$.)
To each λ_i we associate the characteristic function $\phi_i(\underset{\sim}{x}) \not\equiv 0$ such that

$$\text{(5.15)} \qquad \phi_i(\underset{\sim}{y}) = \lambda_i \int_{\partial\Omega} K(\underset{\sim}{x},\underset{\sim}{y})\phi_i(\underset{\sim}{x})\,d\sigma_{\underset{\sim}{x}} \equiv \lambda_i \underset{\sim}{K}^*\phi_i \quad ,$$

and to these functions we associate the function

$$\text{(5.16)} \qquad \psi_i(\underset{\sim}{y}) \equiv \int_{\partial\Omega} \phi_i(\underset{\sim}{x}) \frac{D}{|\underset{\sim}{x}-\underset{\sim}{y}|^{n-2}}\,d\sigma_{\underset{\sim}{x}} \quad .$$

By making use of the symmetrical property of $H(\underset{\sim}{x},\underset{\sim}{y})$ in (5.14), it is easy to see that ψ_i's are the eigenfunctions of $\underset{\sim}{K}$,

$$\psi_i(\underset{\sim}{y}) = \lambda_i(\underset{\sim}{K}\psi_i)(\underset{\sim}{y}) \quad ,$$

and

$$[\phi_j,\psi_i] = \delta_{ij} \quad ,$$

where $[f,g] \equiv \int_{\partial\Omega} f(\underset{\sim}{x})g(\underset{\sim}{x})ds_{\underset{\sim}{x}}$ and δ_{ij} is the Kronecker symbol. Consequently, if $\psi_1(\underset{\sim}{x}) \equiv 1$, then

(5.17) $$\int_{\partial\Omega} \phi_i(\underset{\sim}{x})ds_{\underset{\sim}{x}} = \delta_{i1} \quad , \quad i = 1,2,\ldots \quad .$$

We now state two lemmas which will be needed later. For easy reading, we present the proofs in the Appendix.

<u>Lemma</u> 5.2. <u>Suppose</u> $g \in C(\partial\Omega)$, <u>and</u>

$g_{m+1}(\underset{\sim}{x}) \equiv \int_{\partial\Omega} K_{m+1}(\underset{\sim}{x},\underset{\sim}{y})g(\underset{\sim}{y})d\sigma_{\underset{\sim}{y}}$. <u>Then</u>,

(5.18) $$\lim_{m \to \infty} g_{m+1}(\underset{\sim}{x}) = \gamma = \int_{\partial\Omega} g(\underset{\sim}{y})\phi_1(\underset{\sim}{y})d\phi_{\underset{\sim}{y}} \quad ,$$

<u>and</u>,

(5.19) $$|g_{m+1}(\underset{\sim}{x})-\gamma| \leq \frac{M}{|\lambda_2|^m} \quad , \quad |\lambda_2| > 1 \quad ,$$

<u>where</u> M <u>is a constant depending on</u> g .

<u>Lemma</u> 5.3. <u>Let</u> $f(\underset{\sim}{x})$ <u>be any function of class</u> $C(\partial\Omega)$. <u>Then for</u> $|\lambda| < |\lambda_2|$, <u>the sequence</u> $\mu_n(\underset{\sim}{x})$, <u>defined by</u>

$$\mu_0(\underset{\sim}{x}) = f(\underset{\sim}{x})$$

(5.20)

$$\mu_{n+1}(\underset{\sim}{x}) = \lambda \int_{\partial\Omega} K(\underset{\sim}{x},\underset{\sim}{y})\mu_n(\underset{\sim}{y})\,d\sigma_{\underset{\sim}{y}} + f(\underset{\sim}{x}) \ , \quad n = 0,1,2,\ldots \ ,$$

satisfies the estimates

$$(5.21) \qquad |\mu_{n+1}(\underset{\sim}{x}) - \mu_n(\underset{\sim}{x}) - \gamma\lambda^{n+1}| \leq M\left|\frac{\lambda}{\lambda_2}\right|^n$$

where $\gamma = \int_{\partial\Omega} f(\underset{\sim}{x})\phi_1(\underset{\sim}{x})\,d\sigma_{\underset{\sim}{x}}$ *and* M *is a constant depending on* f . *Furthermore, for* $\lambda \neq 1$ *the sequence* $\mu_{n+1}(\underset{\sim}{x}) - \frac{\gamma}{\lambda-1}\lambda^{n+2}$ *converges uniformly to the solution* $\mu(\underset{\sim}{x})$ *of* (5.9) *such that*

$$(5.22) \qquad |\mu_{n+1}(\underset{\sim}{x}) - \frac{\gamma}{\lambda-1}\lambda^{n+2} - \mu(\underset{\sim}{x})| \leq \left|\frac{\lambda}{\lambda_2}\right|^{n+1} \frac{M|\lambda_2|}{|\lambda_2|-|\lambda|} \ .$$

For the case $\lambda = -1$ which is of interest to us, the above results can be simplified by introducing the sequence

$$(5.23) \qquad \tau_n = \frac{\mu_{n+1}+\mu_n}{2} \ .$$

Observe that from (5.22), one can obtain the estimates

$$(5.24)\qquad \left|\tau_n(\underset{\sim}{x}) - \frac{\gamma\lambda^{n+1}(\lambda+1)}{2(\lambda-1)} - \mu(x)\right| \le \frac{1}{2} M \left|\frac{\lambda}{\lambda_2}\right|^n \left\{\frac{|\lambda_2|+|\lambda|}{|\lambda_2|-|\lambda|}\right\}$$

so that the term involving γ is seen to be eliminated for $\lambda = -1$. Also, from the definition of τ_n , one has from (5.21) for $\lambda = -1$,

$$(5.25)\qquad |\tau_n - \tau_{n-1}| = \frac{1}{2} \left| (\mu_{n+1}+\mu_n) - (\mu_n+\mu_{n-1}) \right|$$

$$= \frac{1}{2} \left| [\mu_{n+1}-\mu_n-\gamma(-1)^{n+1}] + [\mu_n-\mu_{n-1}-\gamma(-1)^n] \right|$$

$$\le \frac{1}{2} M \left|\frac{1}{\lambda_2}\right|^{n-1} \left\{\frac{1+|\lambda_2|}{|\lambda_2|}\right\} .$$

This gives an estimate for τ_n , since one can write τ_n as

$$\tau_n = \tau_o + \sum_{k=1}^{n} (\tau_k - \tau_{k-1})$$

$$= \tau_o + \frac{1}{2} \sum_{k=1}^{n} (\mu_{k+1}-\mu_k) + \frac{1}{2} \sum_{k=1}^{n} (\mu_k-\mu_{k-1}) ,$$

which implies that

$$(5.26)\qquad \tau_n = \frac{1}{2}[2f-\underset{\sim}{K}f] + \frac{1}{2} \sum_{\ell=1}^{n} (-1)^{\ell+1} \underset{\sim}{K}_{\ell+1} f + \frac{1}{2} \sum_{\ell=1}^{n} (-1)^{\ell} \underset{\sim}{K}_{\ell} f$$

$$\equiv \underset{\sim}{T}_n f .$$

From the above discussions, it is clear that we have the following estimates:

$$|||\underset{\sim}{T}_n||| \leq 1+ \sum_{\ell=1}^{n} \frac{M}{|\lambda_2|^\ell} \tag{5.27}$$

$$|||\underset{\sim}{T}_n-(\underset{\sim}{I}+\underset{\sim}{K})^{-1}||| \leq \sum_{\ell=n}^{\infty} \frac{M}{|\lambda_2|^\ell} \tag{5.28}$$

Here $|||\cdot|||$ denotes the norm defined by

$$|||\underset{\sim}{T}||| = \sup_{f \,\varepsilon\, C(\partial\Omega),\,||f||_o = 1} ||\underset{\sim}{T}f||_o \quad ,$$

$$||f||_o = \sup_{x \,\varepsilon\, \partial\Omega} |f(\underset{\sim}{x})| \quad .$$

The sequence τ_n converges to $\mu(\underset{\sim}{x})$ in the sup norm. We shall approximate the solution of (5.6) by $\underset{\sim}{T}_n(\underset{\sim}{\pi}_E\rho)$.

Now returning to Eq. (5.7), we define the operator

$$\hat{\underset{\sim}{\pi}}_o^{(n)}\rho \equiv -(\underset{\sim}{\pi}_E\rho)(\underset{\sim}{x})+\int_{\partial\Omega} K(\underset{\sim}{x},\underset{\sim}{\xi})(\underset{\sim}{T}_n\underset{\sim}{\pi}_E\rho)(\underset{\sim}{\xi})d\sigma_{\underset{\sim}{\xi}} \quad , \tag{5.29}$$

and the corresponding operators such as $(\hat{\underset{\sim}{\pi}}_x^{(n)}\rho)(\underset{\sim}{x})$, $(\hat{\underset{\sim}{\pi}}_z^{(n)}\rho)(\underset{\sim}{x})$, $(\hat{\underset{\sim}{T}}_1^{(n)}\rho)(\underset{\sim}{x})$ and $(\hat{\underset{\sim}{T}}_2^{(n)}\rho)(\underset{\sim}{x})$. We also define the operator $\hat{\underset{\sim}{p}}^{(n)}$ from the differential equation such that

$$\rho = \hat{\underset{\sim}{p}}^{(n)}\rho \tag{5.30}$$

where

$$(\hat{\underset{\sim}{P}}^{(n)}\rho) \equiv -\mathrm{Re}\{A(\underset{\sim}{x},\hat{\underset{\sim}{\pi}}_o^{(n)}\rho,\hat{\underset{\sim}{\pi}}_z^{(n)}\rho,\hat{\underset{\sim}{\pi}}_x^{(n)}\rho)\hat{\underset{\sim}{T}}_1^{(n)}\rho$$

$$+B(\ldots)\hat{\underset{\sim}{T}}_2^{(n)}\rho\}-F(\ldots) \quad .$$

Here $\rho \in L_p(\Omega+\partial\Omega)$, $p > 3$ and all the derivatives are to be interpreted in a generalized or Sobolev sense. We note that in (5.29), $\underset{\sim}{\pi}_E\rho \in C\,(\Omega+\partial\Omega)$, $\alpha = \frac{p-3}{p}$, and

$$\sup_{\underset{\sim}{x}\,\in\,\Omega+\partial\Omega}\left|\int_{\partial\Omega} K(\underset{\sim}{x},\underset{\sim}{\xi})(\underset{\sim}{T}_n\underset{\sim}{\pi}_E\rho)(\underset{\sim}{\xi})\,d\sigma_{\underset{\sim}{\xi}}\right|$$

$$\le \text{const.} \sup_{\underset{\sim}{\xi}\,\in\,\partial\Omega} |(\underset{\sim}{\pi}_E\rho)(\underset{\sim}{\xi})| \quad . \tag{5.31}$$

Then as before, it is easy to show that the operators $\hat{\underset{\sim}{T}}_k^{(n)}$, $k = 1,2$ have meaning, and that if Ω is sufficiently small, $\hat{\underset{\sim}{P}}^{(n)}$ is a contraction mapping from a closed ball in $L_p(\Omega+\partial\Omega)$ into itself, and hence has a unique fixed point $\rho^{(n)}(\underset{\sim}{x})$ in $L_p(\Omega+\partial\Omega)$, $p > 3$. Our candidate for an approximation to our original Dirichlet problem is now given by

$$u^{(n)}(\underset{\sim}{x}) = (\hat{\underset{\sim}{\pi}}_o^{(n)} \rho^{(n)})(\underset{\sim}{x}) \quad . \tag{5.32}$$

In the following we wish to obtain an error estimate for the difference $u - u^{(n)}$.

We begin with the definition of u and $u^{(n)}$. From Eqs. (5.7) and (5.32), we have

$$\begin{aligned} |u - u^{(n)}| &= |\hat{\underset{\sim}{\pi}}_o \rho - \hat{\underset{\sim}{\pi}}_o^{(n)} \rho^{(n)}| \\ &\leq |\hat{\underset{\sim}{\pi}}_o^{(n)}(\rho^{(n)} - \rho)| + |(\hat{\underset{\sim}{\pi}}_o^{(n)} - \hat{\underset{\sim}{\pi}}_o)\rho| \quad . \end{aligned}$$

The first term on the right takes the form:

$$\begin{aligned} |\hat{\underset{\sim}{\pi}}_o^{(n)}(\rho^{(n)} - \rho) &= |-\underset{\sim}{\pi}_E(\rho^{(n)} - \rho) + \underset{\sim}{K}\,\underset{\sim}{T}_n \underset{\sim}{\pi}_E(\rho^{(n)} - \rho)| \\ &\leq \sup_{\underset{\sim}{x}\in\Omega+\partial\Omega} ||E_3(\underset{\sim}{x},\cdot)||_{L_q} ||\rho^{(n)} - \rho||_{L_p} \\ &\quad + \left(\sup_{\underset{\sim}{x}\in\partial\Omega} \int_{\partial\Omega} |K(\underset{\sim}{x},\underset{\sim}{\xi})|\, d\sigma_{\underset{\sim}{\xi}}\right)\left(|||\underset{\sim}{T}_n|||\right) \\ &\quad \cdot \left(\sup_{\underset{\sim}{x}\in\partial\Omega} ||K(\underset{\sim}{x},\cdot)||_{L_q}\right) ||\rho^{(n)} - \rho||_{L_p} \end{aligned} \tag{5.33}$$

The second term satisfies the estimate

$$
(5.34)\quad \begin{aligned}
|(\hat{\pi}_o^{(n)} - \hat{\pi}_o)\rho| &\leq |\underset{\sim}{K}[\underset{\sim}{T}_n - (\underset{\sim}{I}+\underset{\sim}{K})^{-1}]\underset{\sim}{\pi}_E\rho| \\
&\leq \left(\sup_{\underset{\sim}{x}\in\partial\Omega} \int_{\partial\Omega} |K(\underset{\sim}{x},\underset{\sim}{\xi})|d\sigma_{\underset{\sim}{\xi}}\right)\left(|||\underset{\sim}{T}_n-(\underset{\sim}{I}+\underset{\sim}{K})^{-1}|\right. \\
&\quad \cdot \left(\sup_{\underset{\sim}{x}\in\partial\Omega} ||K(\underset{\sim}{x},\cdot)||_{L_q}\right)||\rho||_{L_p} .
\end{aligned}
$$

Setting

$$
(5.35)\quad \begin{aligned}
E_{x_i}(\overline{\Omega}) &= \sup_{\underset{\sim}{x}\in\Omega+\partial\Omega} || \frac{\partial}{\partial x_i} E_3(\underset{\sim}{x},\cdot)||_{L_q} , \\
&\left(E_o(\overline{\Omega}) = \sup_{\underset{\sim}{x}\in\Omega+\partial\Omega} ||E_3(\underset{\sim}{x},\cdot)||_{L_q}\right) \\
\Gamma_{x_i}(\partial\Omega) &= \sup_{\underset{\sim}{x}\in\partial\Omega} \int_{\partial\Omega} |\frac{\partial}{\partial x_i} K(\underset{\sim}{x},\xi)|d\sigma_{\underset{\sim}{\xi}} , \\
&\left(\Gamma_o(\partial\Omega) = \sup_{\underset{\sim}{x}\in\partial\Omega} \int_{\partial\Omega} |K(\underset{\sim}{x},\underset{\sim}{\xi})|d\sigma_{\underset{\sim}{\xi}}\right) ,
\end{aligned}
$$

we have from (5.33) and (5.34),

$$
(5.36)\quad \begin{aligned}
|u-u^{(n)}| &\leq A_o^{(n)} ||\rho^{(n)} - \rho||_{L_p} \\
&\quad + M_o^{(n)} \frac{1}{|\lambda_2|^n} ||\rho||_{L_p} .
\end{aligned}
$$

Here,

$$A_o^{(n)} = E_o(\overline{\Omega}) + |||\underset{\sim}{T}_n||| \Gamma_o(\partial\Omega) \left(\sup_{\underset{\sim}{x}\varepsilon\partial\Omega} ||K(\underset{\sim}{x},\cdot)||_{L_q} \right)$$

(5.37)

$$\leq E_o(\overline{\Omega}) + \left[1 + \frac{M(|\lambda_2|^n-1)}{|\lambda_2|^n(|\lambda_2|-1)} \right]$$

$$\cdot\ \Gamma_o(\partial\Omega) \left(\sup_{\underset{\sim}{x}\varepsilon\partial\Omega} ||K(\underset{\sim}{x},\cdot)||_{L_q} \right)$$

from (5.27) and

$$(5.38) \qquad M_o^{(n)} = \frac{M|\lambda_2|}{|\lambda_2|-1} \Gamma_o(\partial\Omega) \left(\sup_{\underset{\sim}{x}\varepsilon\partial\Omega} ||K(\underset{\sim}{x},\cdot)||_{L_q} \right)$$

following from the estimate (5.28).

Next we wish to establish an estimate for $||\rho^{(n)} - \rho||_{L_p}$. From the definition of ρ and $\rho^{(n)}$ and the conditions $H_1 - H_4$ (for Ω) as in Section 3, it follows that

$$|\rho-\rho^{(n)}| \leq q_1(M)|\underset{\sim}{T}_1\rho - \underset{\sim}{T}_1^{(n)}\rho^{(n)}| + q_2(M)|\underset{\sim}{T}_2\rho-\underset{\sim}{T}_2^{(n)}\rho^{(n)}|$$

$$+ \left[M_1|\underset{\sim}{T}_1\rho| + M_2|\underset{\sim}{T}_2\rho| + F_o(\underset{\sim}{x}) \right] \left\{ |\hat{\underset{\sim}{\pi}}_o\rho-\hat{\underset{\sim}{\pi}}_o^{(n)}\rho^{(n)}| \right.$$

$$\left. + |\hat{\underset{\sim}{\pi}}_x\rho - \hat{\underset{\sim}{\pi}}_x^{(n)}\rho^{(n)}| + |\hat{\underset{\sim}{\pi}}_z\rho - \hat{\underset{\sim}{\pi}}_z^{(n)}\rho^{(n)}| \right\}$$

A similar analysis as (5.33)-(5.34) shows that there exist constants $A_x^{(n)}$, $A_z^{(n)}$, $A_{zz}^{(n)}$, $A_{zx}^{(n)}$; $M_x^{(n)}$, $M_z^{(n)}$, $M_{zz}^{(n)}$, and $M_{zx}^{(n)}$ such that

$$
(5.39)\qquad
\begin{aligned}
|\underset{\sim}{T}_1\rho - \underset{\sim}{T}_1^{(n)}\rho^{(n)}| &\le A_{zz}^{(n)}||\rho - \rho^{(n)}||_{L_p} + M_{zz}^{(n)}\frac{1}{|\lambda_2|^n}||\rho||_{L_p} \\
|\underset{\sim}{T}_2\rho - \underset{\sim}{T}_2^{(n)}\rho^{(n)}| &\le A_{zx}^{(n)}||\rho - \rho^{(n)}||_{L_p} + M_{zx}^{(n)}\frac{1}{|\lambda_2|^n}||\rho||_{L_p} \\
|\hat{\underset{\sim}{\pi}}_x\rho - \hat{\underset{\sim}{\pi}}_x^{(n)}\rho^{(n)}| &\le A_x^{(n)}||\rho - \rho^{(n)}||_{L_p} + M_x^{(n)}\frac{1}{|\lambda_2|^n}||\rho||_{L_p} \\
|\hat{\underset{\sim}{\pi}}_z\rho - \hat{\underset{\sim}{\pi}}_z^{(n)}\rho^{(n)}| &\le A_z^{(n)}||\rho - \rho^{(n)}||_{L_p} + M_z^{(n)}\frac{1}{|\lambda_2|^n}||\rho||_{L_p}
\end{aligned}
$$

The constants are defined in the same way as $A_o^{(n)}$ and $M_o^{(n)}$ in (5.37)-(5.38) with $E\ (\overline{\Omega})$ and $\Gamma_o(\partial\Omega)$ replaced by the corresponding $E_{x_i}\ (\overline{\Omega})$ and $\Gamma_{x_i}\ (\partial\Omega)$. From (5.39), it follows that

$$
\begin{aligned}
|\rho - \rho^{(n)}| \le{} & \left(q_1(M)A_{zz}^{(n)} + q_2(M)A_{zx}^{(n)}\right)||\rho - \rho^{(n)}||_{L_p} \\
& + \left(q_1(M)M_{zz}^{(n)} + q_2(M)M_{zx}^{(n)}\right)\frac{1}{|\lambda_2|^n}\ ||\rho||_{L_p} \\
& + \left[M_1|\underset{\sim}{T}_1\rho| + M_2|\underset{\sim}{T}_2\rho| + F_o(x)\right] \\
& \cdot\left\{\left(A_o^{(n)} + A_x^{(n)} + A_z^{(n)}\right)||\rho-\rho^{(n)}||_{L_p}\right. \\
& \left. + \left(M_o + M_x + M_z\right)\ \frac{1}{|\lambda_2|^n}\ ||\rho||_{L_p}\right\}\ .
\end{aligned}
$$

Consequently,

$$||\rho - \rho^{(n)}||_{L_p} \leq \left\{q(M)\Lambda_p^{(n)} + 3K^{(n)}\left[(M_1 + M_2)\Lambda_p||\rho||_{L_p} + ||F_o||_{L_p}\right]\right\}||\rho - \rho^{(n)}||_{L_p} + \left\{q(M)\overline{\Lambda}_p^{(n)} + 3\overline{K}^{(n)}\left[(M_1 + M_2)\Lambda_p||\rho||_{L_p} + ||F_o||_{L_p}\right]\right\} \cdot \frac{1}{|\lambda_2|^n}||\rho||_{L_p} \quad .$$

Here we use the notation:

$$q(M) = q_1(M) + q_2(M) \quad ,$$

$$\Lambda_p^{(n)} = \max\left\{A(\Omega)A_{zz}^{(n)}, A(\Omega)A_{zx}^{(n)}\right\} \ , \ A(\Omega) = (\text{area of } \Omega)^{\frac{1}{p}} \quad ,$$

$$K^{(n)} = \max\left\{A_o^{(n)}, A_x^{(n)}, A_z^{(n)}\right\} \quad ,$$

$$\overline{\Lambda}_p^{(n)} = \max\left\{A(\Omega)M_{zz}^{(n)}, A(\Omega)M_{zx}^{(n)}\right\} \quad ,$$

$$\overline{K}^{(n)} = \max\left\{M_o^{(n)}, M_x^{(n)}, M_z^{(n)}\right\} \quad ,$$

$$\Lambda_p = \max\left\{A_1, A_2\right\} \quad ,$$

and recall that $|\underset{\sim}{T}_i\rho| \leq A_i \ ||\rho||_{L_p}$ $(i = 1,2)$. In consistence with the previous notation, we set

$$K_p^{(n)} = q(M)\Lambda_p^{(n)} + 3K^{(n)}(M_1 + M_2)\Lambda_p||\rho||_{L_p} + 3K^{(n)}||F_o||_{L_p}$$

$$\overline{K}_p^{(n)} = q(M)\overline{\Lambda}_p^{(n)} + 3\overline{K}^{(n)}(M_1 + M_2)\Lambda_p||\rho||_{L_p} + 3\overline{K}^{(n)}||F_o||_{L_p}$$

It is easy to see that for $K_p^{(n)} < 1$,

$$(5.40) \qquad ||\rho - \rho^{(n)}||_{L_p} \leq \left(\frac{\overline{K}_p^{(n)}||\rho||_{L_p}}{1 - K_p^{(n)}}\right) \frac{1}{|\lambda_2|^n} .$$

We remark that it is possible to make $K_p^{(n)} < 1$ for Ω sufficiently small (see (3.14)). Equations (5.36) and (5.40) then yield the following estimate:

$$(5.41) \qquad |u - u^{(n)}| \leq \frac{r}{|\lambda_2|^n}\left\{\frac{K^{(n)}\overline{K}_p^{(n)}}{1 - K_p^{(n)}} + \overline{K}^{(n)}\right\}, \quad ||\rho||_{L_p} < r .$$

In view of the fact that $|\lambda_2| > 1$ we can now conclude that the approximating sequence $\{u^n(\underset{\sim}{x})\}$ tends geometrically to the exact solution of the Dirichlet problem for Eq. (5.2). We have thus established the following result.

<u>Theorem</u> <u>2</u>. <u>Let</u> $u(x)$ <u>be</u> <u>the</u> <u>unique</u> <u>solution</u> <u>to</u> <u>the</u> <u>Dirichlet</u> <u>problem</u> <u>for</u> Eq. (5.2), <u>which</u> <u>has</u> <u>the</u>

representation defined by Eq. (5.3). If the sequence $\{u^{(n)}(\underset{\sim}{x})\}$, $n = 1,2,3,\ldots$, is defined by Eqs. (5.29), (5.30) and (5.32), then for Ω sufficiently small $u^{(n)}$ tends to u such that

$$\max_{\underset{\sim}{x}\in\Omega+\partial\Omega} |u(\underset{\sim}{x}) - u^{(n)}(\underset{\sim}{x})| = O\left(\frac{1}{\lambda_2{}^n}\right)$$

where λ_2 is the second eigenvalue of the Fredholm integral operator in (5.9) (that is, the absolute value of which is nearest to one, of course, $|\lambda_2| > 1$).

APPENDIX

Proof of Lemma 5.2. We first assume the existence of a function on $\partial\Omega$, $h(x)$, with the properties:

$$g(x) = \gamma + \int_{\partial\Omega} h(y) \frac{D}{|x-y|^{n-2}} \, d\sigma_y \; , \text{ where } \int_{\partial\Omega} h(y)\, d\sigma_y =$$

If $K^*h = \int_{\partial\Omega} K(t,x) h(t)\, d\sigma_t$, and $K^*_m h = \int_{\partial\Omega} K_m(t,x) h(t)\, d\sigma_t$,

we have, considering x as a parameter,

$$\left(K^*_m h,\; K(x,u)\right) = \left(h, K_m\, K(x,u)\right) = \left(h, K_{m+1}(x,y)\right)$$

$$= \int_{\partial\Omega} \int_{\partial\Omega} h(u) K_{m+1}(x,y) \frac{D}{|u-y|^{n-2}} \, d\sigma_u d\sigma_y$$

$$= \int_{\partial\Omega} K_{m+1}(x,y) \int_{\partial\Omega} h(u) \frac{D}{|u-y|^{n-2}} \, d\sigma_u d\sigma_y$$

$$= \int_{\partial\Omega} K_{m+1}(x,y) [g(y) - \gamma] \, d\sigma_y \; .$$

However, since $\gamma = \int_{\partial\Omega} g(y) \phi_1(y)\, d\sigma_y$, and $\psi_1 = 1$ has the eigenvalue $\lambda_1 = +1$, we have

$$\int_{\partial\Omega} K_{m+1}(x,y) [g(y)-\gamma]\, d\sigma_y = g_{m+1}(x) - \gamma \; .$$

Hence

$$\left(K^*_m h, K(x,y)\right) = g_{m+1}(x) - \gamma \; .$$

On the other hand

$$\left|\left(\underset{\sim}{K}_m^* h, K(\underset{\sim}{x},\underset{\sim}{u})\right)\right| \leq ||\underset{\sim}{K}_m^* h|| \; ||K(\underset{\sim}{x},\cdot)||$$

and, if we set $h_\upsilon = (h,\phi_\upsilon)$, we then conclude that

$$|g_{m+1}(\underset{\sim}{x}) - \gamma|^2 \leq \sum_{\upsilon=1}^{\infty} \frac{h_\upsilon^{\,2}}{\lambda_\upsilon^{\,2m}} ||K(\underset{\sim}{x},\cdot)||^2$$

$$= \sum_{\upsilon=2}^{\infty} \frac{h_\upsilon^{\,2}}{\lambda_\upsilon^{\,2m}} ||K(\underset{\sim}{x},\cdot)||^2 ,$$

which follows from the fact that $\phi_1(\underset{\sim}{x}) \equiv 1$ and hence $h_1 = 0$. Consequently,

$$|g_{m+1}(\underset{\sim}{x}) - \gamma| \leq \left\{\sum_{\upsilon=2}^{\infty} \frac{h_\upsilon^{\,2}}{\lambda_\upsilon^{\,2m}}\right\}^{\frac{1}{2}} ||K(\underset{\sim}{x},\cdot)||$$

$$\leq \frac{1}{|\lambda_2|^m} \left(\sum_{\upsilon=1}^{\infty} h_\upsilon^{\,2}\right)^{\frac{1}{2}} ||K(\underset{\sim}{x},\cdot)|| .$$

We must next remove the assumption that there exists an $h \in C(\partial\Omega)$, having the previous cited properties. We mollify $g(\underset{\sim}{x})$ on $\partial\Omega$ to obtain a sequence of C^∞ - functions $g_\varepsilon(\underset{\sim}{x})$ on $\partial\Omega$. Let $u_\varepsilon^e(\underset{\sim}{x})$ be the solution of the exterior Dirichlet problem for $\partial\Omega$ with data $g_\varepsilon(\underset{\sim}{x})$ and bounded at infinity. If $\underset{\sim}{\upsilon}$ is the exterior normal, then $\frac{\partial u_\varepsilon^e}{\partial \upsilon}$ exists.

Now suppose $U(x)$ is the solution of the exterior Neumann problem for $\partial\Omega$ with data $\frac{\partial U}{\partial \nu} = \frac{\partial U_\varepsilon^e}{\partial \nu}$ on $\partial\Omega$, and which vanishes at infinity. Then there exists a function $h_\varepsilon(\underset{\sim}{x})$ such that

$$\int_{\partial\Omega} h_\varepsilon(\underset{\sim}{x})\, d\sigma_{\underset{\sim}{x}} = 0 \quad \text{and} \quad U(\underset{\sim}{x}) = \int_{\partial\Omega} \frac{D}{|\underset{\sim}{x}-\underset{\sim}{y}|} h_\varepsilon(\underset{\sim}{y})\, d\sigma_{\underset{\sim}{y}}$$

We conclude $U(\underset{\sim}{x}) = u_\varepsilon^e(\underset{\sim}{x}) + k$. The constant k is determined by noting that

$$0 = \int_{\partial\Omega} \phi_1(\underset{\sim}{x})\; U(\underset{\sim}{x})\, d\sigma_{\underset{\sim}{x}} = \int_{\partial\Omega} \phi_1(\underset{\sim}{x}) u^e(\underset{\sim}{x})\, d\sigma_{\underset{\sim}{x}} + k$$

$$= \int_{\partial\Omega} g_\varepsilon(\underset{\sim}{x}) \phi_1(\underset{\sim}{x})\, d\sigma_{\underset{\sim}{x}} + k ,$$

so

$$k = - \int_{\partial\Omega} g_\varepsilon(\underset{\sim}{x}) \phi_1(\underset{\sim}{x})\, d\sigma_{\underset{\sim}{x}} \equiv - \gamma_\varepsilon$$

From this we have $U(\underset{\sim}{x}) + \gamma_\varepsilon = u_\varepsilon^e(\underset{\sim}{x})$, and hence, for $\underset{\sim}{x} \in \partial\Omega$

$$g_\varepsilon(\underset{\sim}{x}) = \gamma_\varepsilon + \int_{\partial\Omega} h_\varepsilon\; (\underset{\sim}{y}) \frac{D}{|\underset{\sim}{x}-\underset{\sim}{y}|^{n-2}}\, d\sigma_{\underset{\sim}{y}} \quad .$$

Since $U(\underset{\sim}{x}) + \gamma_\varepsilon$ is the solution also to the interior

Dirichlet problem with boundary values $g_\varepsilon(x)$, we have

$$\lim_{\substack{\underset{\sim}{x}\to\partial\Omega \\ \underset{\sim}{x}\in\Omega}} \frac{\partial U}{\partial \upsilon} = \frac{\partial u_\varepsilon^i}{\partial \upsilon} \quad \text{on} \quad \partial\Omega \quad \text{and} \quad h_\varepsilon(\underset{\sim}{x}) = \frac{1}{4\pi}\left(\frac{\partial u_\varepsilon^e}{\partial \upsilon} - \frac{\partial u_\varepsilon^i}{\partial \upsilon}\right) .$$

Letting the mollifer parameter $\varepsilon \to 0$, our results are seen to hold for continuous data. This completes the proof of Lemma (5.2).

<u>Proof of Lemma 5.3</u>. The estimates (5.21) follow immediately from Lemma 5.2 with $g(\underset{\sim}{x})$ replaced by $f(\underset{\sim}{x})$. To see (5.22), following Warschawski's idea, we introduce, for $\lambda \neq 1$, the sequence

$$\mu_n^*(\underset{\sim}{x}) = \mu_n(\underset{\sim}{x}) - \gamma \sum_{\upsilon=1}^{\infty} \lambda^\upsilon \equiv \mu_n(\underset{\sim}{x}) - \gamma\lambda \frac{\lambda^{n-1}}{\lambda - 1} , \; n \geq 1 .$$

This can be seen to converge uniformly to a function $\mu^*(\underset{\sim}{x})$ satisfying the integral equation

$$\mu^*(\underset{\sim}{x}) = \lambda \int_{\partial\Omega} K(\underset{\sim}{x},\underset{\sim}{y})\mu^*(y)\,d\sigma_{\underset{\sim}{y}} + f(\underset{\sim}{x}) - \gamma\lambda .$$

Using the estimates (5.21) it can then be shown that

$$|\mu^*_{n+1}(\underset{\sim}{x}) - \mu^*(\underset{\sim}{x})| = |\mu_{n+1}(x) - \gamma \sum_{\upsilon=1}^{n+1} \lambda^\upsilon - \mu - \frac{\gamma\lambda}{\lambda-1}|$$

$$\leq M \sum_{\upsilon=n+1}^{\infty} \left|\frac{\lambda}{\lambda_2}\right|^\upsilon$$

which proves the results (5.22).

References

[1] D. Colton and R. P. Gilbert, Rapidly convergent approximations to Dirichlet's problem for semi-linear elliptic equations, Applicable Analysis, 1972, Vol. 2, pp. 229-240.

[2] A. Friedman, Partial differential equations, Holt, Rinehart and Winston, New York, 1969.

[3] E. Goursat, A course in mathematical analysis, Vol. III, part two, Dover Pub., New York, 1964.

[4] G. N. Hile, Generalized hyperanalytic function theory, Doctoral dissertation, Indiana University, Bloomington, Indiana 47401.

[5] G. Hoheisel, Integral Equations, Frederick Ungar Pub. Co., New York, 1968.

[6] S. G. Mikhlin, Multidimensional singular integrals and integral equations, Pergamon Press, London, 1965.

[7] F. Riesz and B. Sz-Nagy, Functional Analysis, Frederick Ungar Pub. Co., New York, 1965.

[8] E. M. Stein, Singular integrals and differentiability properties of functions, Princeton Univ. Press, Princeton, New Jersey, 1970.

[9] I. N. Vekua, Generalized analytic functions, Pergamon Press, London, 1962.

[10] S. E. Warschawski, On the solution of the Lichtenstein-Gershgorin integral equation in conformal mapping: I theory, Nat. Bureau of Standards, Applied Math. Series 42, (1955) pp. 7-29.

Department of Mathematics
Indiana University
Bloomington, Indiana 47401

and

Department of Mathematics
University of Delaware
Newark, Delaware 19711

The Numerical Solution of Some Elliptic Boundary Value Problems by Integral Operator Methods*

by

R. P. Gilbert, Department of Mathematics, Indiana University

Peter Linz, Department of Mathematics, University of California at Davis

Abstract: An integral representation for the solution of elliptic equations of the form $\Delta_n u - P(r^2)u = 0$, developed by Gilbert, is used to construct approximate solutions for problems of this type. Properties of the G-function, needed in the integral representation, are discussed and a numerical scheme for its computation is given. The approximate G-function is used to represent the solution and minimization techniques are used to satisfy the boundary conditions. A discussion of the practical usefulness of methods of this type is given.

1. Introduction.

One of the methods for treating certain partial differential equations is the so-called method of particular solutions. Let

$$Lu = 0 \quad \text{on } D \tag{1}$$

$$u = f \quad \text{on } \partial D \tag{2}$$

be an elliptic boundary value problem on a bounded, simply connected domain D, with prescribed boundary conditions on ∂D, and let $\{\varphi_i\}$ be a set of functions satisfying (1) and complete on $D + \partial D$. The method of particular solutions then consists of finding an approximate solution of the form

*This research was supported in part by the Air Force Office of Scientific Research through grant AFOSR-71-2205.

$$u_m = \sum_{i=0}^{m} \alpha_i \varphi_i , \tag{3}$$

where the α_i are selected so as to satisfy the boundary conditions approximately. The success of this approach depends on our ability to find a complete set of particular solutions and, from a practical point of view, on the ease with which the φ_i can be computed. For the important class of equations

$$Lu = \Delta_n u - P(r^2)u = 0, \quad P(r^2) \geq 0, \tag{4}$$

$$\Delta_n u = \frac{\partial^2 u}{\partial x_1^2} + \cdots \frac{\partial^2 u}{\partial x_n^2} ,$$

$$r = ||\underset{\sim}{x}|| = (x_1^2 + x_2^2 + \cdots x_n^2)^{1/2},$$

Gilbert [6] has obtained a simple representation for the particular solutions in terms of the so-called G-function. Given a complete set $\{\psi_i\}$ of solutions of

$$\Delta_n u = 0 \tag{5}$$

a complete set $\{\varphi_i\}$ of solutions of (4) is given by

$$\varphi_i(\underset{\sim}{x}) = \psi_i(\underset{\sim}{x}) + \int_0^1 \sigma^{n-1} G(r, 1-\sigma^2) \psi_i(\sigma^2 \underset{\sim}{x}) d\sigma, \tag{6}$$

where $G(r,t)$ is the solution of

$$2(1-t)G_{rt} - G_r + r(G_{rr}-PG) = 0, \tag{7}$$

$$G(0,t) = 0,$$

$$G(r,0) = \int_0^r rP(r^2)dr.$$

Methods of this type have been studied by a number of authors ([2], [3], [4], [5]). Recently Schryer [11] considered the general case of an elliptic equation in two dimensions with analytic coefficients. There are, however, a number of limitations in Schryer's approach which can be eliminated for the case considered here by using the G-function representation. First of all, the numerical solution of the $\{\varphi_i\}$ has been much simplified. Secondly, the restriction of analyticity of the coefficients can be removed. Finally, and perhaps most importantly, the formulation is essentially independent of the dimension and so is suitable for handling three-dimensional problems.

2. Properties and Numerical Computation of the G-function.

When $P(r^2)$ is entire then $G(r,t)$ has a known series expansion [6]. For the general case we want to consider the computation of $G(r,t)$ in $0 \leq r \leq a$, $0 \leq t \leq 1$ from equation (7). Introduce $W(\rho,t)$ by

$$G(r,t) = W(\rho,t)/(1-t), \tag{8}$$

$$\rho = r\sqrt{1-t} \ .$$

Then $W(\rho,t)$ satisfies the equation

$$W_{\rho t} = \frac{\rho P(\rho^2/1-t)}{2(1-t)^2} W, \tag{9}$$

$$W(0,t) = 0,$$

$$W(\rho,0) = \int_0^\rho rP(r^2)dr = \Phi(\rho), \quad 0 \leq t \leq 1, \quad 0 \leq \rho \leq a\sqrt{1-t} \ .$$

Theorem 1. Let $0 \leq P(r^2) \leq \lambda$. Then,

$$W(\rho,t) \leq \rho\lambda\sqrt{\frac{1-t}{t}} \; I_1\left(\lambda\rho\sqrt{\frac{t}{1-t}}\right), \tag{10}$$

and $$G(r,t) \leq \frac{r\lambda}{\sqrt{t}} I_1(\lambda r\sqrt{t}). \tag{11}$$

Proof: Integrating (9) we get

$$W(\rho,t) = W(\rho,0) + \int_0^\rho \int_0^t \frac{\rho' P(\rho'^2/1-\tau')W(\rho',\tau')}{2(1-\tau')^2} d\tau' d\rho'. \tag{12}$$

Let $W_\lambda(\rho,t)$ denote the solution of this equation with $P(r^2) = \lambda$. Then

$$W_\lambda(\rho,t) = \rho\lambda\sqrt{\frac{1-t}{t}}\; I_1\left(\lambda\rho\sqrt{\frac{t}{1-t}}\right).$$

Choose $\lambda \geq P(r^2)$, and set $\Delta(\rho,\tau) = W_\lambda(\rho,\tau) - W(\rho,\tau)$. Then

$$\Delta(\rho,\tau) = \Delta(\rho,0) + \int_0^\rho\int_0^t \frac{\rho' P(\rho'^2/1-\tau')\Delta(\rho',\tau')}{2(1-\tau')^2}\, d\tau'\, d\rho'$$

$$+ \int_0^\rho\int_0^t \frac{\rho'\{\lambda - P(\rho'^2/1-\tau')\}W_\lambda(\rho',\tau')}{2(1-\tau')^2}\, d\tau'\, d\rho' .$$

For $0 \leq t \leq 1-\epsilon$, this is a Volterra equation with positive kernel and positive free term, so

$$\Delta(\rho,t) \geq 0.$$

Since ϵ can be made as small as desired (10) holds, and (11) follows from the definition of W.

In the region of interest $(0 \leq t \leq 1,\ 0 \leq \rho \leq a\sqrt{1-t}\,)$ $W(\rho,t)$ is bounded (from eqn (10)), and a numerical approximation can be found by integrating (9). General methods for problems of this type have been studied by several authors (e.g. [7], [8]); however it is convenient here to develop a special method for integrating (9). On the one hand, the linearity of (9) simplifies matters somewhat, while on the other hand the presence of the factor $(1-t)^{-2}$ causes difficulties near $t = 1$ and has to be treated specially.

Introduce the mesh-points $\{\rho_i, t_j\}$ with $\rho_o = t_o = 0$ and integrate (9) over one mesh. Then

$$W(\rho_{i+1},t_{j+1}) = W(\rho_{i+1},t_j) + W(\rho_i,t_{j+1}) - W(\rho_i,t_j)$$

$$+ \int_{\rho_i}^{\rho_{i+1}} \int_{t_j}^{t_{j+1}} \frac{\rho' P(\rho'^2/1-t')}{2(1-t')^2} W(\rho',t')dt'd\rho' . \qquad (12)$$

To compute an approximate solution the integral is replaced by an appropriate numerical quadrature. For simplicity we will use quadratures involving only values of W at the mesh corners. The approximate solution $\overline{W}_{ij} \simeq W(\rho_i,t_j)$ is then computed by

$$\overline{W}_{i+1,j+1} = \alpha_{ij}\overline{W}_{i+1,j+1} + (1+\beta_{ij})\overline{W}_{i+1,j} + (1+\gamma_{ij})\overline{W}i,j+1$$

$$- (1-\delta_{ij})\overline{W}_{i,j} , \qquad (13)$$

where the coefficients α_{ij}, β_{ij}, γ_{ij}, δ_{ij} depend on the quadrature chosen. If $\alpha_{ij} \neq 1$, then

$$\overline{W}_{i+1,j+1} = (1-\alpha_{ij})^{-1}\{(1+\beta_{ij})\overline{W}_{i+1,j} + (1+\gamma_{ij})\overline{W}_{i,j+1} - (1-\delta_{ij})\overline{W}_{i,j}\} \qquad (14)$$

can be used to compute the approximate solution at successive mesh-points starting with

$$\overline{W}_{o,j} = 0, \qquad \overline{W}_{i,o} = \Phi(\rho_i). \qquad (15)$$

We now investigate the convergence of the approximate solution to the true solution under appropriate conditions. For this, we will need the following

<u>Lemma</u>: Let ξ_{ij} be the solution of

$$\xi_{i+1,j+1} = \alpha_{ij}\xi_{i+1,j+1} + (1+\beta_{ij})\xi_{i+1,j} + (1+\gamma_{ij})\xi_{i,j+1} - (1-\delta_{ij})\xi_{ij} + Q_{ij} \tag{16}$$

for $i=0,1,\ldots N$; $j=0,1,\ldots M$; $|\alpha_{ij}| < 1$, and $\xi_{i,o}$, $\xi_{o,j}$ given. If

$$B = \max_{i,j} (|\alpha_{ij}| + |\beta_{ij}| + |\gamma_{ij}| + |\delta_{ij}|) < 1,$$

$$C = 3 \max_{i,j} |\xi_{i,o}, \xi_{o,j}| ,$$

$$R = \frac{1}{1-B} \max_{i,j} \left(\sum_{k=0}^{i} \sum_{\ell=0}^{j} |Q_{k\ell}| \right) + C ,$$

$$A = \frac{B}{1-B} ,$$

then

$$|\xi_{ij}| \leq (1+A)^{ij} R . \tag{17}$$

Proof: The proof is straightforward; we will omit some of the details. The solution of (16) also satisfies

$$\xi_{i+1,j+1} = \sum_{k=0}^{i} \sum_{\ell=0}^{j} (\alpha_{k\ell}\xi_{k+1,\ell+1} + \beta_{k\ell}\xi_{k+1,\ell} + \gamma_{k\ell}\xi_{k,\ell+1} + \delta_{k\ell}\xi_{k\ell} + Q_{k\ell})$$

from which it follows that

$$|\xi_{i+1,j+1}| \leq B\{ |\xi_{i+1,j+1}| + |\xi_{i+1,j}| + |\xi_{i,j+1}| + |\xi_{ij}|\} + B \sum_{k=0}^{i} \sum_{\ell=0}^{j} |\xi_{k\ell}| + R(1-B)$$

Let η_{ij} be the solution of

$$\eta_{i+1,j+1} = B\eta_{i+1,j+1} + B\sum_{k=0}^{i}\sum_{\ell=0}^{j}\eta_{k\ell} + R(1-B)$$

with $\eta_{io} = |\xi_{io}|$, $\eta_{oj} = |\xi_{oj}|$. Then, by induction,

$$|\xi_{ij}| < \eta_{ij}$$

Then, $\eta_{i+1,j+1} = A\sum_{k=0}^{i}\sum_{\ell=0}^{j}\eta_{k\ell} + R$

Since $\eta_{11} = A\eta_{01} + A\eta_{10} + A\eta_{00} + R \leq (1+A)R$, we obtain by induction, that

$$\eta_{i,j} \leq (1+A)^{ij}R \quad \text{and (17) is proved.}$$

Corollary: If there exist positive numbers c, h, k, r such that $rhN \geq a$, $rkM \geq 1$, $A \leq chk$, then

$$|\xi_{ij}| \leq e^{cr^2a}R, \quad \text{for } 0 \leq i \leq N, \quad 0 \leq j \leq M.$$

Let $\epsilon_{ij} = W(\rho_i,\tau_j) - \overline{W}_{ij}$. Then (13) from (12) we have

$$\epsilon_{i+1,j+1} = \alpha_{ij}\epsilon_{i+1,j+1} + (1+\beta_{ij})\epsilon_{i+1,j} + (1+\gamma_{ij})\epsilon_{i,j+1}$$
$$- (1-\delta_{ij})\epsilon_{ij} + Q_{ij},$$

where Q_{ij} represents the local quadrature error

$$ij = \int_{\rho_i}^{\rho_{i+1}} \int_{t_j}^{t_{j+1}} \frac{\rho' P(\rho'^2/1-t')}{2(1-t')^2} W(\rho',t')dt'd\rho' - \alpha_{ij}W(\rho_{i+1},t_{j+1})$$

$$- \beta_{ij}W(\rho_{i+1},t_{j+1}) - \gamma_{ij}W(\rho_i,t_{j+1}) - \delta_{ij}W(\rho_i,t_j) . \tag{18}$$

f α_{ij}, β_{ij}, γ_{ij}, δ_{ij} satisfy the conditions of the previous lemma and the orollary, and if the sum of the absolute values of the local quadrature rrors goes to zero, as the grid-size goes to zero, then $\overline{W}_{ij} \to W(\rho_i,t_i)$.

We will require numerical values for W in $0 \leq t \leq 1-\eta$, $0 \leq \rho \leq a$ 1-t η is small, but > 0). Since all quantities in the integrand are bounded, ne could use a simple quadrature rule like the trapezoidal method. However, he presence of the factor $(1-t')^{-2}$ causes difficulties neat $t' = 1$. To alleviate this problem, we use a product-integration technique; in $[\rho_i,\rho_{i+1}] \times [t_j,t_{j+1}]$ we approximate $\rho'P(\rho'^2/1-t')W(\rho',t')$ by a bilinear function and integrate the result. For the resulting quadrature we then get coefficients

$$\alpha_{ij} = \frac{1}{2}\left\{\frac{1}{1-t_{j+1}} + \frac{1}{t_{j+1}-t_j} \ln\left(\frac{1-t_{j+1}}{1-t_j}\right)\right\} \rho_{i+1}(\rho_{i+1}-\rho_i)P\left(\frac{\rho_{i+1}^2}{1-t_{j+1}}\right)$$

$$\beta_{ij} = \frac{1}{2}\left\{-\frac{1}{1-t_j} - \frac{1}{t_{j+1}-t_j} \ln\left(\frac{1-t_{j+1}}{1-t_j}\right)\right\} \rho_{i+1}(\rho_{i+1}-\rho_i)P\left(\frac{\rho_{i+1}^2}{1-t_j}\right)$$

$$\gamma_{ij} = \frac{1}{2}\left\{\frac{1}{1-t_{j+1}} + \frac{1}{t_{j+1}-t_j} \ln\left(\frac{1-t_{j+1}}{1-t_j}\right)\right\} \rho_i(\rho_{i+1}-\rho_i)P\left(\frac{\rho_i^2}{1-t_{j+1}}\right)$$

$$\delta_{ij} = \frac{1}{2}\left\{-\frac{1}{1-t_j} - \frac{1}{t_{j+1}-t_j} \ln\left(\frac{1-t_{j+1}}{1-t_j}\right)\right\} \rho_i(\rho_{i+1}-\rho_i)P\left(\frac{\rho_i^2}{1-t_j}\right)$$

It can be shown without difficulty that if

$$h = \max|\rho_{i+1}-\rho_i|, \qquad h_o = \min|\rho_{i+1}-\rho_i|$$

$$k = \max|t_{i+1}-t_i|, \qquad k_o = \min|t_{i+1}-t_i|$$

$$h/h_o \leq r, \qquad k/k_o \leq r$$

then the conditions of the lemma are satisfied and the approximate solution converges to the true solution. It is also easy to show that the order of convergence is two.

It is worth noticing at this point that the computation of the G-function is independent of the dimension. It is also independent of D and the prescribed boundary values.

3. Computation of the Approximate Solution.

An approximate solution $\hat{u}$ of (4) can be computed by expressing $\hat{u}$ as a linear combination of φ_i

$$\hat{u}(\underset{\sim}{x}) = \sum_{i=1}^{m} \alpha_i \varphi_i(\underset{\sim}{x}). \tag{19}$$

The α_i are then selected so as to approximately satisfy the boundary conditions. Typically, we select points $z_1, z_2, \ldots z_M$ on ∂D and determine the α_i such that

$$\max_{1 \leq i \leq M} |\hat{u}(z_i) - f(z_i)| \tag{20}$$

is minimized. This process can either be formulated as a linear programming problem ([4], [10]), or solved by using an algorithm for finding the minimax solution to an overdetermined system of linear equations.

Approximate values for $\{\varphi_i\}$ can be computed by numerically integrating

) if a complete set $\{\psi_i\}$ is known. We will outline the procedure for n=2.

In two-dimensions the harmonic polynomials are a complete set of

olutions of (4). In polar coordinates

$$\psi_j(r,\theta) = r^k \cos k\theta, \qquad j = \text{odd}$$

$$= r^k \sin k\theta \qquad j = \text{even}, \quad k = \text{integer part of } j/2$$

hen

$$\varphi_j(r,\theta) = r^k \begin{Bmatrix} \cos k\theta \\ \sin k\theta \end{Bmatrix} (1 + m_k(r))$$

here

$$m_k(r) = \int_0^1 \sigma^{2k+1} G(r, 1-\sigma^2) d\sigma$$

$$= \int_0^1 \sigma^{2k-1} W(r\sigma, 1-\sigma^2) d\sigma. \tag{21}$$

rom (10), $|\sigma^{2k-1} W(r\sigma, 1-\sigma^2)| \leq Mr^2\sigma^{2k+1}$, so the moment integrals $m_k(r)$

exist for k=0,1,... and can be approximated numerically, once numerical

values for W are available.

The approximate solution, eqn (19) is then computed by minimizing (20).

Once this is done the error in the approximate solution can be estimated simply.

Let

$$e(\underset{\sim}{x}) = u(x) - \hat{u}(\underset{\sim}{x}),$$

then

$$Le(\underset{\sim}{x}) = 0 \ .$$

If $|e(x)| \leq \epsilon$ for $x \in \partial D$, then from standard maximum principle arguments ([9])

$$|e(\underset{\sim}{x})| \leq \epsilon \qquad \text{for } x \in D$$

To obtain ϵ one can compute $u(x) - \hat{u}(x)$ at discrete points along the boundary and use simple analytical arguments to get a bound everywhere. To get a reasonable estimate one can take

$$\epsilon \simeq \epsilon_B = \max_{1\leq i\leq M} |\hat{u}(z_i)-f(z_i)|$$

if M is not too small.

4. Some Numerical Results.

Numerical results for some typical two-dimensional problems are summarized below with various values of m (eqn. 19) and M (eqn. 20). An algorithm of Bartels and Golub [1] was used to solve the minimization problem.

Example 1.

$$\Delta_2 u - u = 0$$

$$u = e^{\sin \Theta} \qquad \text{for } x^2 + y^2 = 1$$

(m,M) / (r,θ)	(6,31)	(8,31)	(12,31)
(0,0)	.9999	.9999	.9999
(.1,0)	1.0007	1.0007	1.0007
(.1,π/2)	1.1057	1.1058	1.1058
(.5,0)	1.0004	1.0008	1.0007
(.25,π/2)	1.2848	1.2848	1.2848
ϵ_B	5.5×10^{-3}	5.3×10^{-4}	3.0×10^{-6}

Example 2.

$$\Delta_2 u - u = 0$$

$$u = 1 \qquad \text{for } x^2 + 4y^2 = 1$$

(m,M) / (r,θ)	(4,31)	(8,31)	(12,31)
(0,0)	.9062	.9064	.9063
(.1,0)	.9078	.9079	.9079
(.1,π/2)	.9105	.9107	.9107
(.5,0)	.9297	.9295	.9295
(.25,π/2)	.9298	.9301	.9300
ϵ_B	7×10^{-4}	2×10^{-4}	1×10^{-4}

5. Discussion.

The method presented here is a convenient approach to the solution of certain boundary value problems. A meaningful comparison with the more usual methods such as finite difference or finite element methods is difficult to make since no realistic criteria for such a comparison exist. One can, however, single out cases where our method can be considered a promising alternative. Clearly, our method is less general than the standard ones, but when applicable it has some significant advantages

(1) Curved boundaries present no problems. In fact the method works best when the boundary is smooth. Corners slow down the convergence, although this problem can be overcome by using special expansion functions, as pointed out by Schryer.

(2) The computation of G does not depend on D or on the boundary conditions. If the solution is required for changing boundaries or varying boundary conditions, a considerable amount of work can be saved.

(3) The formulation is essentially independent of n. Thus the method is a promising approach to three-dimensional problems.

References

[1] R. H. Bartels and G. H. Golub, Chebyshev solution to an overdetermined system, Comm. ACM 11(1968), pp. 428-430.

[2] S. Bergman and J. G. Herriot, Application of the method of the kernel function for solving boundary value problems, Num. Math. 3(1961), pp. 209-225.

[3] S. Bergman and J. G. Herriot, Numerical solution of boundary value problems by the method of integral operators, Num. Math. 7(1965), pp. 42-65.

[4] J. R. Cannon, The numerical solution of the Dirichlet problem for Laplace's equation by linear programming, SIAM J. Appl. Math. 12(1964), pp. 233-237.

[5] J. R. Cannon and M. M. Cecchi, The numerical solution of some biharmonic problems by mathematical programming techniques, SIAM J. Numer. Anal. 3(1966), pp. 451-466.

[6] R. P. Gilbert, Integral operator methods for approximating solutions of Dirichlet problems, Proceedings of the Conference on "Numerische Methoden der Approximationstheorie, Oberwolfach, 1969.

[7] M. K. Jain and R. D. Sharma, Cubature method for the solution of the characteristic initial value problem $u_{xy} = f(x,y,u, u_x,u_y)$, J. Aust. Math. Soc. 8(1968), pp. 355-368.

[8] R. H. Moore, A Runge-Kutta procedure for the Goursat problem in hyperbolic partial differential equations, Arch. Rat. Mech. Anal. 7(1961), pp. 37-63.

[9] M. Protter and H. Weinberger, Maximum Principles in Differential Equations, Prentice-Hall, Englewood Cliffs, N.J. 1967.

[10] P. Rabinowitz, Applications of linear programming to numerical analysi SIAM Rev. 10(1968), pp. 121-159.

[11] N. L. Schryer, Constructive approximation of solutions to linear elliptic boundary value problems, SIAM J. Numer. Anal. 9(1972), pp. 546-572.

Iterative Schemes for Elliptic Systems

Robert P. Gilbert* — Indiana University

Richard J. Weinacht — University of Delaware

In an earlier work [6] the authors developed a theory of reproducing kernels for elliptic systems

$$\Delta U - CU = 0 , \tag{1}$$

where $C = C^*$ (conjugate transpose) is a positive definite $n \times n$ matrix, and U is an $n \times n$ matrix. The results for the system (1) are analogous to those obtained by Bergman and Schiffer [1] for the scalar case. Particular attention must be paid to the order of the matrix entries in the various representation formulae, however, because of the general noncommutivity of these terms. By introducing an "inner product" structure for $n \times n$ matrices $V, W \in C'(\bar{D})$ by means of

$$E\{V,W\} \equiv \int_D \left[V_x^* W_x + V_y^* W_y + V^* C W\right] dxdy , \tag{2}$$

and using the following norm[≠]

$$\|V\| \equiv \sup_{|\xi|=1} |\xi^* V \xi| \qquad \xi \in \underset{\sim}{C}^n , \tag{3}$$

one is able to obtain estimates for the error in approximating

---

[≠] The structure we have introduced here is a Hilbert module [5,8]. The matrix norm used for Hilbert modules, however, is not as convenient for obtaining our estimates.

*This research was supported in part by the Air Force Office of Scientific Research through AF-AFOSR Grant No. 74-2592.

the kernel function by truncated series representations.

We use these results here to consider the Dirichlet problem for the semilinear **sys**tem

$$(4) \qquad \Delta \underset{\sim}{u} = \underset{\sim}{f}(x,y,\underset{\sim}{u},\underset{\sim}{u}_x,\underset{\sim}{u}_y)$$

in the domain D with vanishing data prescribed on $\dot{D}$. Here $\underset{\sim}{u}$ and $\underset{\sim}{f}$ are n-vectors $\underset{\sim}{u} = \text{col.}(u_1,u_2,\ldots,u_n)$ etc. Our results extend those of Colton and Gilbert [3,4] for the scalar case $n = 1$. The region D is bounded and regular.

It is assumed that

$$H_1 : \quad f_k(x,y,\underset{\sim}{0},\underset{\sim}{0},\underset{\sim}{0}) \in L_p(D) \;, \quad p > 2$$

$$H_2 : \quad |f_k(x,y,\underset{\sim}{\xi}^0,\underset{\sim}{\eta}^0,\underset{\sim}{\zeta}^0) - f_k(x,y,\underset{\sim}{\xi}^1,\underset{\sim}{\eta}^1,\underset{\sim}{\zeta}^1)|$$

$$\leq f_k^{(0)}(x,y)\{\|\underset{\sim}{\xi}^0-\underset{\sim}{\xi}^1\|+\|\underset{\sim}{\eta}^0-\underset{\sim}{\eta}^1\|+\|\underset{\sim}{\zeta}^0-\underset{\sim}{\zeta}^1\|\}$$

for $\|\underset{\sim}{\xi}^i\| + \|\underset{\sim}{\eta}^i\| + \|\underset{\sim}{\zeta}^i\| \leq R, i=0,1,$ where R is a sufficiently large fixed positive constant and where $f_k^{(0)}$ is a non-negative element of $L_p(D)$. We use the vector norm $\|\underset{\sim}{\xi}\| = \max_{\ell} |\xi_\ell|$.

Now let C be any positive-definite Hermitian matrix belonging to $C'(\overline{D})$ and consider in place of (4)

$$(5) \qquad \Delta \underset{\sim}{u} - C(x,y)\underset{\sim}{u} = \underset{\sim}{f} - C(x,y)\,\underset{\sim}{u} \equiv \underset{\sim}{g}\ .$$

The vector function $\underset{\sim}{g}$ will satisfy the hypotheses H_1', H_2' obtained by replacing f by g throughout H_1 and H_2.

To attack this problem we introduce on $L_p(D)$ Green's function G for (1) and the operators

$$(6\text{-a}) \qquad \underset{\sim}{\Pi}_0[\underset{\sim}{\rho}](P) \equiv \int_D G(P,Q)\,\underset{\sim}{\rho}(Q)\,d\tau_Q$$

$$(6\text{-b}) \qquad \underset{\sim}{\Pi}_1[\underset{\sim}{\rho}](P) \equiv \int_D \frac{\partial}{\partial x_p} G(P,Q)\,\underset{\sim}{\rho}(Q)\,d\tau_Q$$

$$(6\text{-c}) \qquad \underset{\sim}{\Pi}_2[\underset{\sim}{\rho}](P) \equiv \int_D \frac{\partial}{\partial y_p} G(P,Q)\,\underset{\sim}{\rho}(Q)\,d\tau_Q\ .$$

By estimating first the singular parts of these operators, obtained by replacing G by $(2\pi)^{-1} I \log (1/r)$ in (6), it is easy to see that there exist constants a_i such that

$$(7) \qquad \max_{P \in D} \| \underset{\sim}{\Pi}_i[\underset{\sim}{\rho}](P) \| \le a_i \| \underset{\sim}{\rho} \|_p\ , \quad (i = 0,1,2)$$

where

$$\| \underset{\sim}{\rho} \|_p \equiv \max_{\ell} \left(\int_D |\rho_\ell|^p d\tau \right)^{1/p}$$

and the a_i are constant multiples of the maximum over P in $\bar{D}$ of the scalar L_q norms (in the variable Q) of the norm of the Green's function and the first derivatives thereof. Therefore, letting

$$(8) \quad \underset{\sim}{\Pi}[\underset{\sim}{\rho}](P) \equiv \underset{\sim}{g}(P, \underset{\sim}{\Pi}_0[\underset{\sim}{\rho}](P), \underset{\sim}{\Pi}_1[\underset{\sim}{\rho}](P), \underset{\sim}{\Pi}_2[\underset{\sim}{\rho}](P))$$

one sees from H_2' and (7) that

$$\|\underset{\sim}{\Pi}[\underset{\sim}{\rho}^{(1)}] - \underset{\sim}{\Pi}[\underset{\sim}{\rho}^{(2)}]\|_p \leq a\|\underset{\sim}{g}^{(0)}\|_p \|\underset{\sim}{\rho}^{(1)} - \underset{\sim}{\rho}^{(2)}\|_p$$

where $a = a_1+a_2+a_3$. Now suppose that $\|\underset{\sim}{g}^{(0)}\|_p < \delta$ where δ is chosen so that $\delta a < 1$ and R is chosen so that

$$\|g(P, \underset{\sim}{0}, \underset{\sim}{0}, \underset{\sim}{0})\|_p \leq (1-\delta a)R .$$

Then $\underset{\sim}{\Pi}$ is a contraction mapping of the closed ball $|\underset{\sim}{\rho}| \leq R$ into itself and hence has a unique fixed point $\underset{\sim}{\rho}^{(0)}$ in $L_p(D)$. Thus from the theory of integral operators with weak singularities [7;p. 136], $\underset{\sim}{u} = \underset{\sim}{\Pi}_0[\underset{\sim}{\rho}^{(0)}]$ is a generalized solution in $W_p^2(D)$ of our Dirichlet problem for (4).

We now obtain an approximation for $\underset{\sim}{u}$ with estimates of the error. Let $\underset{\sim}{\Pi}_i^{(N)}$ (i=0,1,2) denote the operators obtained by replacing G in (6) by G_N , where the approximate Green's matrix

$$(9) \qquad G_N(P,Q) = \frac{1}{2\pi}\left[I \log r_{PQ}^{-1} + \int_{\dot{D}} \log r_{QT}^{-1} \frac{\partial}{\partial \nu_T} K_N(P,T)\,ds_T\right]$$

is defined in terms of the truncated matrix [6]

$$(10) \qquad K_N(P,Q) \equiv \sum_{\mu=0}^{N} \sum_{\nu=0}^{\mu} (-1)^{\nu}\binom{\mu}{\nu} i_{\nu+1}(P,Q) ,$$

where the matrices $i_\nu(P,Q)$ are "geometric entities" defined recursively as follows

$$i_1(P,Q) = 4I(P,Q)$$

$$(11) \qquad i_\nu(P,Q) = E\{i^*_{\nu-1}(P,T), i_1(T,Q)\} , \qquad \nu \geq 0 ,$$

with

$$I(P,Q) = \int_{\dot{D}} \frac{\partial S^*}{\partial \nu_T}(T,P)S(T,Q)\,ds_T .$$

We note that if K_N is replaced by K in (9) the left hand side of (9) becomes $G(P,Q)$. Now the existence of fixed points $\underset{\sim}{\rho}^{(N)}$ of $\underset{\sim}{\Pi}^{(N)}$ (defined by replacing $\underset{\sim}{\Pi}_i$ in (8) by $\underset{\sim}{\Pi}_i^{(N)}$ etc.) follows as for $\underset{\sim}{\rho}^{(0)}$ and yields an approximate solution $\underset{\sim}{u}^{(N)} \equiv \underset{\sim}{\Pi}^{(N)}[\underset{\sim}{\rho}^{(N)}]$.

The estimates for the difference $\underset{\sim}{u}-\underset{\sim}{u}^N$ will now be obtained. From the expression (9) for G_N and the corresponding expression for G follows the estimate

$$(12) \qquad \|(\underset{\sim}{\Pi}_0^{(N)} - \underset{\sim}{\Pi}_0)[\underset{\sim}{\rho}](P)\| \leq$$

$$\leq [\text{CONSTANT}]\|\underset{\sim}{\rho}\|_p \cdot \{\|K-K_n\|_q + \|\int_{\dot{D}} \|K-K_N\| \,|\frac{\partial}{\partial \nu_T} \log r_{QT}\,|ds_T\|_q\}$$

and making use of the estimate [6]

$$(13) \quad \|K(P,Q)-K_N(P,Q)\| \leq \frac{1}{\lambda_1^{2N}}\|M(P,P)\|^{\frac{1}{2}} \cdot \|M(Q,Q)\|^{\frac{1}{2}} ,$$

where the matrix functions $M(P,Q)$ are defined by [6]

$$(14) \qquad M(P,Q) \equiv K(P,Q) - 4I(P,Q) ,$$

we conclude there exists a constant M_0, independent of ρ, such that

$$(15) \qquad \|(\underset{\sim}{\Pi}_0^{(N)} - \underset{\sim}{\Pi}_0)[\underset{\sim}{\rho}](P)\| \leq \frac{M_0}{\lambda_1^{2N}}\|\underset{\sim}{\rho}\|_p .$$

Similarly, as in [3], for $i = 1,2$

$$\max_{P \in \overline{D}} \|(\underset{\sim}{\Pi}_i^{(N)} - \underset{\sim}{\Pi}_i)[\underset{\sim}{\rho}]\| \leq \frac{M_i}{\lambda_1^{2N}}\|\underset{\sim}{\rho}\|_p .$$

Moreover, with $M = M_1 + M_2 + M_3$

$$\|\underset{\sim}{\rho}^0(P) - \underset{\sim}{\rho}^{(N)}(P)\| \leq g^0(P)\left\{\frac{M}{\lambda_i^{2N}}\|\rho^{(0)}\|_p + a^{(N)}\|\rho^0 - \rho^{(N)}\|_p\right\}$$

where $a^{(N)} = a_1^{(N)} + a_2^{(N)} + a_3^{(N)}$ and the $a_i^{(N)}$ are exactly the constants for bounds on $G^{(N)}$ as the a_i for G in (7). Hence, for D small enough such that $a^{(N)} \|\underset{\sim}{g}^0\|_p < 1$

$$(16) \qquad \|\rho^0 - \rho^{(N)}\|_p \le \frac{M\|g^{(0)}\|_p \|\rho\|_p}{\lambda_1^{2N}(1-a^{(N)}\|g^{(0)}\|_p)}$$

and so, collecting these estimates

$$\|\underset{\sim}{u}^{(N)}(P) - \underset{\sim}{u}(P)\| \le \|\underset{\sim}{\Pi}_0^{(N)}[\rho_0^{(N)} - \underset{\sim}{\rho}](P)\| + \|\underset{\sim}{\Pi}_0^{(N)}[\underset{\sim}{\rho}](P) - \underset{\sim}{\Pi}_0[P(0)\|$$

$$\le \frac{R}{\lambda_1^{2N}}\left[\frac{a_0^{(N)} M\|g\|_p}{1-a^{(N)}\|g\|_p} + M_0\right].$$

This yields geometric convergence of our approximation scheme since [6] $\lambda_1 > 1$. We summarize our results as

<u>Theorem</u>: <u>Let the vector function</u> $\underset{\sim}{f}$ <u>of the semilinear system</u> (4) <u>satisfy the hypotheses</u> H_k, $(k = 1,2)$. <u>Then for</u> D <u>sufficiently small there exists a unique solution</u> $\underset{\sim}{u}(P)$ <u>to the Dirichlet problem with vanishing data on</u> $\dot{D}$, <u>and the sequence</u> $\underset{\sim}{u}^{(N)}(P) = (\underset{\sim}{\Pi}^{(N)}\underset{\sim}{\rho})(P)$ <u>of vector functions converges geometrically to this unique solution</u>, i. e.,

$$\|\underset{\sim}{u}^{(N)}(P)-\underset{\sim}{u}(P)\| = 0(1/\lambda_1^{2N}) , \quad \lambda_1 > 1 .$$

References

[1] Bergman, S. and Schiffer, M.: "Kernel Functions and Elliptic Differential Equations in Mathematical Physics", Academic Press, New York, (1953).

[2] _________: "Kernel functions in the theory of partial differential equations of elliptic type", Duke J. Math., 15, 1948, 535-566.

[3] Colton, D. L. and Gilbert, R. P.: "Rapidly convergent approximations to Dirichlet's problem for semilinear elliptic equations", Applicable Analysis, 2, 1972, 229-240.

[4] _________: "New results on the approximation of solutions to partial differential equations" in Analytic Theory of Differential Equations, Lecture Notes in Mathematics vol. 183, Springer-Verlag, Heidelberg, (1971), 213-220.

[5] Goldstine, H. H. and Horwitz, L. P.: "Hilbert space with non-associative scalars II", Math. Annalen, 164, 1966, 291-316.

[6] Gilbert, R. P. and Weinacht, R. J.: Reproducing kernels for elliptic systems (to appear).

[7] Mikhlin, G.: Multidimensional Singular Integrals and Integral Equations, Pergamon Press, London, (1965).

[8] Saworotnow, P. P.: "A generalized Hilbert space", Duke J. Math., 35, 1968, 191-197.

Extrapolation in the Finite Element Method with Penalty

by

J. Thomas King*

Abstract: Consider the model problem $\Delta u = f$ in Ω, $u = 0$ on $\partial\Omega$. Here Ω is a bounded open subset of R^n with smooth boundary, $\partial\Omega$. The penalty method provides a method for obtaining an approximate solution without requiring the approximant to satisfy boundary conditions. Unfortunately, we pay a price for this convenience, namely loss of accuracy. We show that this difficulty may be alleviated by a particular type of extrapolation process. For a particular choice of boundary weight in the penalty method we show that repeated extrapolation always yields "optimal" error estimates in the energy norm.

Introduction

In the past few years many researchers have investigated finite element methods where the approximants are not required to satisfy essential boundary conditions [1], [2], [6], [7], [8], and [11]. Others [4], [15] have considered methods using approximants which "nearly" satisfy boundary conditions.

* Department of Mathematical Sciences, University of Cincinnati, Cincinnati, Ohio 45221

The penalty method of Babuška [2] is essentially a finite element version of a method proposed by Courant [10] in 1941. A fundamental difficulty of this method has been the nonoptimal nature of the error estimates. Hence in order to circumvent the problem of satisfying essential boundary conditions we are penalized by loss of accuracy.

The purpose of this paper is to survey some recent results of the author [13] where the penalty method is combined with a certain extrapolation process. It is shown that, for a particular choice of boundary weight, repeated extrapolation yields optimal error estimates.

2. The Penalty Method. Preliminaries.

Let Ω be a bounded open subset of R^n with smooth boundary, $\partial\Omega$. For simplicity we consider the model problem

$$(2.1) \qquad \Delta u = f \text{ in } \Omega \qquad u = 0 \text{ on } \partial\Omega .$$

Denote by $H^r(\Omega)$ the usual Sobolev space of order r on Ω with norm $||\cdot||_r$. For the space $H^r(\partial\Omega)$ the norm will be denoted by $|\cdot|_r$. For a detailed discussion of these spaces we refer the reader to [14]. For $L_2(\Omega)$ and $L_2(\partial\Omega)$ we will denote the inner products by

$$(u,v) = \int_\Omega uv \, dx$$

and

$$\langle u,v \rangle = \int_{\partial\Omega} uv \, ds \qquad \text{respectively.}$$

We will need the following space of approximants. Let $h \in (0,1)$ and r be an integer. Then V_h^r will denote any finite dimensional subspace of $H^r(\Omega)$ having the following approximation property:

* for any $u \in H^s(\Omega)$, $s \geq 2$, there exists $\bar{u} \in V_h^r$ such that

$$||u - \bar{u}||_0 + h||u - \bar{u}||_1 \leq Ch^{[r,s]}||u||_s$$

where $[r,s] = \min\{r,s\}$, and C is independent of u and h.

Many examples of subspaces satisfying (*) for various choices of r may be found in the literature [3], [5], [9], [16]. A possible choice of V_h^r consists of $(r-1)^{st}$ order splines constructed on a uniform mesh of width h.

The penalty method for the problem (2.1) consists in finding that element $v \in V_h^r$ such that for all $\phi \in V_h^r$

$$(2.2) \qquad (f,\phi) + D(v,\phi) + \gamma h^{-\sigma}\langle v,\phi\rangle = 0.$$

Here $D(\cdot,\cdot)$ denotes the Dirichlet integral, $\gamma > 0$ is a suitably chosen constant, and $\sigma \geq 1$.

It is not difficult to show that (2.2) is equivalent to minimizing, over V_h^r, the functional

$$(2.3) \qquad F(\phi) = D(\phi,\phi) + 2(f,\phi) + \gamma h^{-\sigma}|\phi|_0^2.$$

Furthermore, if F is minimized over $H^1(\Omega)$ by $\tilde{u}$, then $\tilde{u}$ is the weak solution of the problem

$$\text{(2.4)} \qquad \Delta\tilde{u} = f \quad \text{in } \Omega \qquad \tilde{u} + \gamma^{-1}h^{\sigma}\frac{\partial\tilde{u}}{\partial\eta} = 0 \quad \text{on } \partial\Omega$$

where $\frac{\partial}{\partial\eta}$ denotes the exterior normal derivative.

It follows that by solving (2.2) we are approximating the solution of the wrong problem (namely (2.4)) and the error is at best $O(h^{\sigma})$. In fact we have the following error estimate, which is essentially Theorem 3.2 of [2].

Theorem 2.1: Suppose $u \in H^s(\Omega)$, $s \geq 2$, is the solution of (2.1), v is the solution of (2.3), and set $e = u - v$. Then there exists a constant C, independent of h and u, such that

$$||e||_1 \leq Ch^{\nu}||u||_s$$

where $\nu = \min\{[r,s] - \frac{1+\sigma}{2}, \sigma\}$.

If $s \geq r$, the best choice for σ would appear to be $\sigma = (2r-1)/3$, with a resulting loss in accuracy of $O(h^{(r-2)/3})$. Furthermore it is easily shown, under reasonable assumptions on V_h^r, that the condition number of the matrix for the system (2.2) is $O(h^{-1-\sigma})$. Hence there is a tradeoff between accuracy and conditioning in the choice of boundary weight. We should remark that the condition number for the Galerkin method applied to our model problem is $O(h^{\ 2})$.

3. New Error Estimates. Extrapolation

Our new error estimates consist in finding an asymptotic error expansion of the form

$$(3.1)\qquad e = \sum_{j=1}^{k} \gamma^{-j} h^{j\sigma} w_j + O(h^{(k+1)\sigma})$$

where the functions w_j are certain harmonic functions. Such an expansion justifies the application of extrapolation.

Specifically suppose the solution of (2.1) satisfies $u \in H^s(\Omega)$, with $s \geq 2$, and let k be the greatest integer $\leq s - 2$. Let w_j, $1 \leq j \leq k + 1$, be the solution of the problem

$$\Delta w_j = 0 \text{ in } \Omega \qquad w_j = -\frac{\partial w_{j-1}}{\partial \eta} \text{ on } \partial\Omega$$

where $w_0 = u$. Then if $v_{\gamma^{-1}h}$ is the solution of (2.2) we have the following error estimate.

Theorem 3.1: There exists a constant C, independent of h and u, such that

$$(3.2)\qquad ||v_{\gamma^{-1}h} - u - \sum_{j=1}^{k} \gamma^{-j} h^{j\sigma} w_j||_1 \leq Ch^{\lambda} ||u||_s$$

where $\lambda = \min\{[s,r] - \frac{1+\sigma}{2}, (k+1)\sigma\}$.

We note that if $s = r$ and $\sigma = 1$ then

$$||v_{\gamma^{-1}h} - u - \sum_{j=1}^{r-2} \gamma^{-j} h^{j} w_j||_1 \leq Ch^{r-1} ||u||_r .$$

Furthermore it is easily seen that Theorem 3.1 is valid with k replaced by any $\ell = 0,1,\ldots,k$. The choice $\ell = 0$ is just Theorem 2.1.

In order to describe our extrapolation process it is convenient to introduce the solution operators of problems (2.1) and (2.2). Let $T : H^0(\Omega) \to H^2(\Omega)$ be defined by $Tf = u$ where u satisfies (2.1). It is well known that for $f \varepsilon H^s(\Omega)$, $s \geq 0$, $Tf \varepsilon H^{s+2}(\Omega)$. Let $T_{\gamma^{-1}h} : H^0(\Omega) \to V_h^r$ be defined by $T_{\gamma^{-1}h} f = v_{\gamma^{-1}h}$ where $v_{\gamma^{-1}h}$ is the solution of (2.2).

We define the first extrapolation operator by

$$T_h^{(1)} = 2T_{h/2} - T_h$$

and by induction the jth extrapolation operator

$$T_h^{(j)} = \frac{2^j T_{h/2}^{(j-1)} - T_h^{(j)}}{2^j - 1} .$$

Then for $u \varepsilon H^s(\Omega)$, with $s \geq 2$, we have

$$||(T - T_h^{(1)})f||_1 = ||2(T - T_{h/2})f - (T - T_h)f||_1$$

$$= O(h^{\min([s,r] - \frac{1+\sigma}{2}, 2\sigma)})$$

by the triangle inequality and (3.1).

It follows by a straightforward argument using (3.1) that

$$||(T - T_h^{(s-2)})f||_1 \leq Ch^{\lambda}||Tf||_s$$

and hence if $s = r$ and $\sigma = 1$

$$||(T - T_h^{(r-1)})f||_1 = O(h^{r-1}). \tag{3.3}$$

For the specific case of $s = r = 4$ (i.e. cubic splines) and $\sigma = 1$ we have

$$||\tfrac{1}{3}(T_h - 6T_{h/2} + 8T_{h/4})f - Tf||_1 = O(h^3). \tag{3.4}$$

Thus we solve (2.2) with $\sigma = 1$ for $\gamma = 2^0, 2^1, 2^2$ to yield approximate solutions T_hf, $T_{h/2}f$, $T_{h/4}f$ respectively and the linear combination given in (3.4) yields the "optimal" third order occuracy. We should remark that we could also use $\gamma = a^0, a^1, a^2$ for any $a > 0$ and give analogous extrapolation operators satisfying (3.3). In fact Serbin [17] has done some numerical experiments which indicate that better accuracy is achieved for large γ.

In computing $T_h^{(2)}$ in (3.4) we note that if T_hf has been found then $T_{\gamma^{-1}h}f$, for any other γ, may be found without further computation of inner products. Furthermore for $\sigma = 1$ the linear system (2.2) has condition number of order $O(h^{-2})$.

In computing $T_{h/2}f$, $T_{h/4}f$ a possible method would be to use SOR with starting values $T_h f$ and $T_h^{(1)} f$ respectively. See [12] for a discussion of the choice of the SOR parameter in finite element methods.

BIBLIOGRAPHY

[1] J. P. Aubin, Approximation des problèms aux limites non homogenes et régularité de la convergence, Calcolo, 6(1969), pp. 117-139.

[2] I. Babuška, The finite element method with penalty, Math. Comp., 27(1973), pp. 221-228.

[3] __________, Approximations by hill functions, Comment. Math. Univ. Carolinae, 11(1970), pp. 787-811.

[4] A. Berger, R. Scott, G. Strang, Approximate boundary conditions in the finite element method, Symposia Mathematica, Academic Press, New York, 1972, pp. 295-313.

[5] J. H. Bramble and S. Hilbert, Estimation of linear functionals on Sobolev spaces with applications to Fourier transforms and spline interpolation, SIAM Num. Anal., 7(1970), pp. 112-124.

[6] J. H. Bramble and J. Nitsche, A generalized Ritz-least-squares method for Dirichlet problems, SIAM Num. Anal., 10(1973), pp. 81-93.

[7] J. H. Bramble and A. H. Schatz, Raleigh-Ritz-Galerkin methods for Dirichlet's problem using subspaces without boundary conditions, Comm. Pure Appl. Math., 23(1970), pp. 653-675.

[8] ______________________________, Least-squares methods for 2mth order elliptic boundary value problems, Math. Comp., 25(1970), pp. 1-33.

[9] J. H. Bramble, M. Zlamal, Triangular elements in the finite element method, Math. Comp., 24(1970), pp. 809-820.

[10] R. Courant, Variational methods for the solution of problems of equilibrium and vibrations, Bull. Amer. Math. Soc., 49(1942), pp. 1-23.

[11] J. E. Dendy, Penalty Galerkin methods for partial differential equations, Ph.D. thesis, Rice University, 1971.

[12] G. Fix, K. Larsen, On the convergence of SOR iterations for finite element approximations to elliptic boundary value problems, SIAM Num. Anal., 8(1971), pp. 536-547.

[13] J. T. King, New error bounds for the penalty method and extrapolation, Numer. Math., to appear.

[14] J. L. Lions, E. Magenes, Problèmes aux Limites non Homogènes et Applications, Vol. 1, Dunod, Paris, 1968.

[15] J. Nitsche, On Dirichlet problems using subspaces with nearly zero boundary conditions, The Mathematical Foundations of the Finite Element Method with Applications to Partial Differential Equations, Academic Press (A. K. Aziz, editor), New York, 1972, pp. 603-627.

[16] M. H. Schultz, Multivariate spline functions and elliptic problems, SIAM Num. Anal., 6(1969), pp. 523-538.

[17] S. Serbin, A computational investigation of least squares and other projection methods for the approximate solution of boundary value problems, Ph.D. thesis, Cornell Universit[y], 1971.

TRANSONIC DESIGN IN TWO DIMENSIONS*

D. G. Korn

Courant Institute of Mathematical Sciences, New York University

Abstract

This paper contains a general description of the method of complex extension and its applications to a variety of two-dimensional transonic design problems. The geometry of the complex characteristics in the hodograph plane is explored and the selection of initial paths for analytic continuation into the supersonic zone is illustrated. The paper contains a section which explains how to obtain the solution near infinity for the various design problems and another about choosing initial data. An example of a compressor blade that was designed by this method is used to illustrate the practical significance of this method.

* The work presented in this paper is supported by the U. S. Atomic Energy Commission, Contract AT(11-1)-3077 at the AEC Computing and Applied Mathematics Center, Courant Institute of Mathematical Sciences, New York University.

Introduction

This paper is concerned with generating shockless solutions to the two-dimensional compressible fluid dynamics equations by the method of complex extension. We compute a smooth transonic flow and determine the shape or shapes which could generate the computed flow by examining the streamlines. A variety of inverse, or design, problems of physical interest can be solved with this method. These include airfoil design, compressor and turbine blade design, multielement airfoil design, and nozzle design.

We will highlight some of the aspects which are common to all of the preceding design problems. The problem of isolated airfoil design has been described in some detail by Korn [4] for the symmetric case and by Bauer, Garabedian, and Korn [2] for the lifting wing. Both of these references include listings of computer programs which generate solutions to the flow equations using the method of complex extension. A report on the airfoil design problem is in preparation which will include many airfoils of practical significance [1]. A detailed report on the compressor blade design problem is being written.

Complex Extension

Let x and y be Cartesian coordinates and let u and v be the corresponding velocity components. For irrotational flow there exists a velocity potential ϕ such that $\phi_x = u$ and $\phi_y = v$. In terms of this velocity potential the steady state equation of motion for an ideal polytropic gas in two dimensional flow is

$$(1) \qquad (c^2-u^2)\phi_{xx} - 2uv\phi_{xy} + (c^2-v^2)\phi_{yy} = 0$$

where c, the local speed of sound, satisfies Bernoulli's law

$$(2) \qquad c^2 + \frac{\gamma-1}{2} q^2 = \text{const.}; \quad q^2 = u^2 + v^2$$

Notice that the coefficients of the higher order terms do not contain x and y explicitly. Therefore we may use the Legendre transform to introduce the hodograph potential Φ, defined by

$$(3) \qquad \Phi = xu + yv - \phi ; \quad x = \Phi_u , \quad y = \Phi_v$$

From (1) we can derive the second order equation

$$(4) \qquad (c^2-u^2)\Phi_{vv} + 2uv\,\Phi_{uv} + (c^2-v^2)\Phi_{uu} = 0$$

which is linear in the (u,v)-plane, the so-called hodograph plane. We can use equations (1) and (4) to write down the following ordinary differential equations for the characteristics in both the physical plane and the hodograph plane:

$$(5a)\qquad (c^2-u^2)\, dy^2 + 2uv\, dx\, dy + (c^2-v^2)\, dx^2 = 0$$

$$(5b)\qquad (c^2-u^2)\, du^2 - 2uv\, du\, dv + (c^2-v^2)\, dv^2 = 0$$

Solutions to (5a) and (5b) yield the characteristic directions

$$(6)\qquad \frac{du}{dv} = \lambda_{\pm}\ , \qquad \frac{dy}{dx} = -\ \lambda_{\mp}$$

where

$$(7)\qquad \lambda_{\pm} = \frac{uv \pm c\sqrt{q^2-c^2}}{c^2-u^2}$$

Introducing characteristic coordinates ξ and η we arrive at the four canonical equations

$$(8a)\qquad u_{\xi} = \lambda_{+}v_{\xi}\ , \qquad u_{\eta} = \lambda_{-}v_{\eta}$$

$$(8b)\qquad y_{\xi} + \lambda_{-}x_{\xi} = 0\ , \qquad y_{\eta} + \lambda_{+}x_{\eta} = 0$$

which determine u, v, x and y in terms of ξ and η.

It is important to note that $\lambda_{\pm}$ are a complex conjugate pair of roots for subsonic flow, $q < c$. This difficulty is eliminated by extending the independent variables ξ and η into the complex domain. We allow ξ and η to take on complex values and extend all functions of ξ and η into the complex domain by analytic continuation. In this way equations (8a) and (8b) have meaning even for subsonic flow. The only remaining difficulty with equations (8a-8b) once they have been extended is that they become coincident when $q = c$, since $\lambda_+ = \lambda_-$ there. The locus of points where $q = c$ is called the sonic line. Because we have extended our variables into the complex domain, the sonic line is actually a two-dimensional surface inside a four-dimensional space.

It is possible to generate solutions by formulating an initial value problem or a characteristic initial value problem in the complex domain. The characteristic initial value problem has been chosen for design in an unbounded region because it leads to a simple way of generating the appropriate asymptotic behavior of the solution in the physical plane in the neighborhood of infinity. On two initial characteristics $\xi = \xi_0$ and $\eta = \eta_0$ we may choose any four analytic functions to define $v(\xi,\eta_0), v(\xi_0,\eta)$, $x(\xi,\eta_0)$, and $x(\xi_0,\eta)$. The solution to equations (8a) and (8b) with these four initial functions would determine the flow throughout four-space. The corresponding shape would be found by examining streamlines in the real domain. Our method of solution requires that we select a path on the initial characteristic plane $\xi = \xi_0$ starting at η_0 and pick another path on the other characteristic initial plane $\eta = \eta_0$ beginning at ξ_0. These paths lead us to the solution along the two-dimensional surface defined by the intersection of the characteristics from each of these paths. Numerically this can be achieved, as long as λ_+ is not equal to λ_- at any point on this surface by the method of Massau [2]. The region in the four-dimensional domain where the solution has been computed will depend upon the paths which are chosen. The solution at a given point will be independent of the path selected to reach it provided the initial functions are analytic and the resulting solution surface does not wind around the sonic line. By covering the initial characteristics by sets of paths it would be possible to obtain the solution in the four-dimensional domain for any four initial functions.

We are interested in the solution only in the real domain and would like to compute this part of the solution with as few paths as possible. We want each set of paths to yield as many points in the real domain as possible. Therefore, we make use of the many symmetries which exist in the subsonic region. We choose $\xi_0 = \bar{\eta}_0$ to be a point in the subsonic portion of the flow and assign initial data on these initial planes which are complex conjugate at conjugate points. The corresponding solution will be in the real hodograph along the diagonal of each set of conjugate paths. In other words we choose initial functions v and x on one characteristic initial plane and choose a path along it. The other path is obtained by reflection, while we use

$$v(\xi,\eta_0) = \overline{v(\bar{\eta}_0,\bar{\xi})} \tag{9a}$$

$$x(\xi,\eta_0) = \overline{x(\bar{\eta}_0,\bar{\xi})} \tag{9b}$$

for the initial data along these paths. The resulting solution will contain a line of data $\xi = \bar{\eta}$ in the real domain. However, these conjugate paths cannot be used to obtain the solution in the supersonic region because we cannot integrate through the sonic line where $\lambda_+ = \lambda_-$.

Continuation of the Solution into the Supersonic Region

We will now describe a procedure for obtaining the solution on the other side of the sonic line that was first presented by Swenson [5]. For purposes of illustration we choose the Tricomi equation and consider the initial value problem in the complex domain for which only a single initial path need be defined. Figure 1 is a three-dimensional sketch of the continuation of the characteristics of the Tricomi equation as they cross the sonic line $y = 0$. Suppose we would like to find a path in the initial plane $y = 1$ which would yield the solution at some point ξ_1,η_1 in the supersonic region. Each of these characteristics can be traced back to the sonic line, where they form a cusp. It may then be traced back to the initial plane $y = 1$ along either of the two characteristics. To obtain the solution at the point ξ_1,η_1 we must choose a path in

$y = 1$ connecting a point traced back to the initial plane along ξ_1 with a point traced back by η_1. We call the points on the initial plane whose characteristics pass through points on the real sonic line the sonic locus. To obtain the solution at a point in the real supersonic region we must select a path in the initial plane connecting points along the sonic locus, one from the ξ family, the other from the η family which are not complex conjugate. Paths connecting conjugate points along the sonic locus lead to the solution along the sonic line.

It is conjectured, although it has never been proved that there are three branches of the solution at the point ξ_1, η_1 but that only one of these solutions is real. The choices for the path in the initial plane which yield each of these solutions are illustrated in Figure 2. The solid line represents the path which yields a real solution. We of course also require that λ_+ is not equal to λ_- at any point of the resulting solution surface.

The situation for the characteristic initial value problem is similar. We choose a path in the initial plane $\xi = \xi_0$ from η_0 to η_1 and a path in the other initial plane $\eta = \eta_0$ from ξ_0 to ξ_1 in an analogous manner. By extending each initial path so that the extended segment lies along the sonic locus we are able to compute the solution in a region bounded by the sonic line and two characteristics. Figure 3a is a sketch of a typical set of initial paths. The finite difference mesh along these paths used to get the solution in the supersonic region is illustrated in Figure 3b. In general one set of paths is sufficient to obtain the supersonic portion of the flow.

The Asymptotic Solution in the Physical Plane.

For the design problems we have considered we must find the solution in an unbounded region of the physical plane. This will require us to generate solutions which have one or more singularities in the hodograph plane. For the airfoil problems, the appropriate singularity is a pole at a point where the velocity has some prescribed free-stream value. The compressor, turbine, and nozzle flow problems have two logarithmic singularities corresponding to the inlet and exit velocities. In both of these cases the solution for x and y is represented as

(10a) $$x = S * X_1 + X_2$$

(10b) $$y = S * Y_1 + Y_2$$

where S is the appropriate singularity and X_1 and Y_1 are the solution of the homogeneous system (8a) satisfying a characteristic initial value problem. This is similar to the construction of the Riemann function. X_2 and Y_2 are chosen so that x and y are solutions to (8b) and in fact satisfy a set of inhomogeneous differential equations.

Choosing Initial Functions

So far in our discussion we have made no use of the fact that the equations of motion are linear in the hodograph plane. In fact equations (8a) have solutions which can be expressed in terms of elementary functions. Moreover, since any function of a characteristic is itself a characteristic, it is possible to think of the characteristic coordinates as transformed hodograph variables. Choosing the specific transform is equivalent to choosing the analytic function v along the initial characteristics. We choose the characteristic coordinates so that the solution in the characteristic initial plane is single-valued in the region of flow. For the isolated airfoil and compressor blade this is achieved by opening up the typically two-sheeted hodograph plane with a square root transformation. Once we have chosen this transformation the sonic line and therefore the sonic locus is determined.

In practice the initial function $x(\xi_0,\eta)$ is given by a linear combination of simple functions such as powers and logarithms. The weights of these terms are chosen so as to fit a prescribed curve in the subsonic region of the hodograph plane in a least squares sense. This curve represents the surface streamline in the hodograph plane and determines the important physical properties such as velocity and pressure gradients. Not every choice for the initial function will generate a solution which has physical meaning.

In fact, the primary objective of design is to choose an initial function which will lead to improved performance.

Figure 4 is a Calcomp plot of the sonic locus and of the paths of integration in a characteristic initial plane used to design the shape of the compressor blade in Figure 5. The circles in Figure 4 represent points of the prescribed body (i.e. points through which the surface streamline passes). The plus signs show points of the solution lying on the surface streamline. The arrows show the location of the logarithms inside the body and their directions show where the cut was chosen. The asterisks show the location of the inlet and exit velocities on the characteristic initial plane.

Conclusions

The method of complex extension can be successfully applied to a variety of transonic design problems to yield physically realistic solutions. Many solutions are generally required to produce a good design since there is no way to express some of the design constraints in terms of the selection of the initial function. Boundary layer separation is determined only after the solution has been computed.

Many examples of single element airfoils have already been designed by this method and have been compared to experimental data [3]. Work is now under way to design a compressor blade which will be tested in a cascade wind tunnel.

References

[1] F. Bauer, P. Garabedian, A. Jameson, and D. Korn, Handbook of Supercritical Wing Sections, to appear.

[2] F. Bauer, P. Garabedian, and D. Korn, Supercritical Wing Sections, Springer, Berlin, 1972.

[3] J. J. Kacprzynski, L. H. Ohman, P. R. Garabedian and D. G. Korn, Analysis of the flow past a shockless lifting airfoil in design and off-design conditions, N.R.C. of Canada Aeronautical Report LR-554, Ottawa, 1971.

[4] D. G. Korn, Computation of Shock-Free Transonic Flows for Airfoil Design, Report NYU-NYO-125 (1969).

[5] E. V. Swenson, Geometry of the complex characteristics in transonic flow, Comm. Pure Appl. Math., 15, (1968), 175-185.

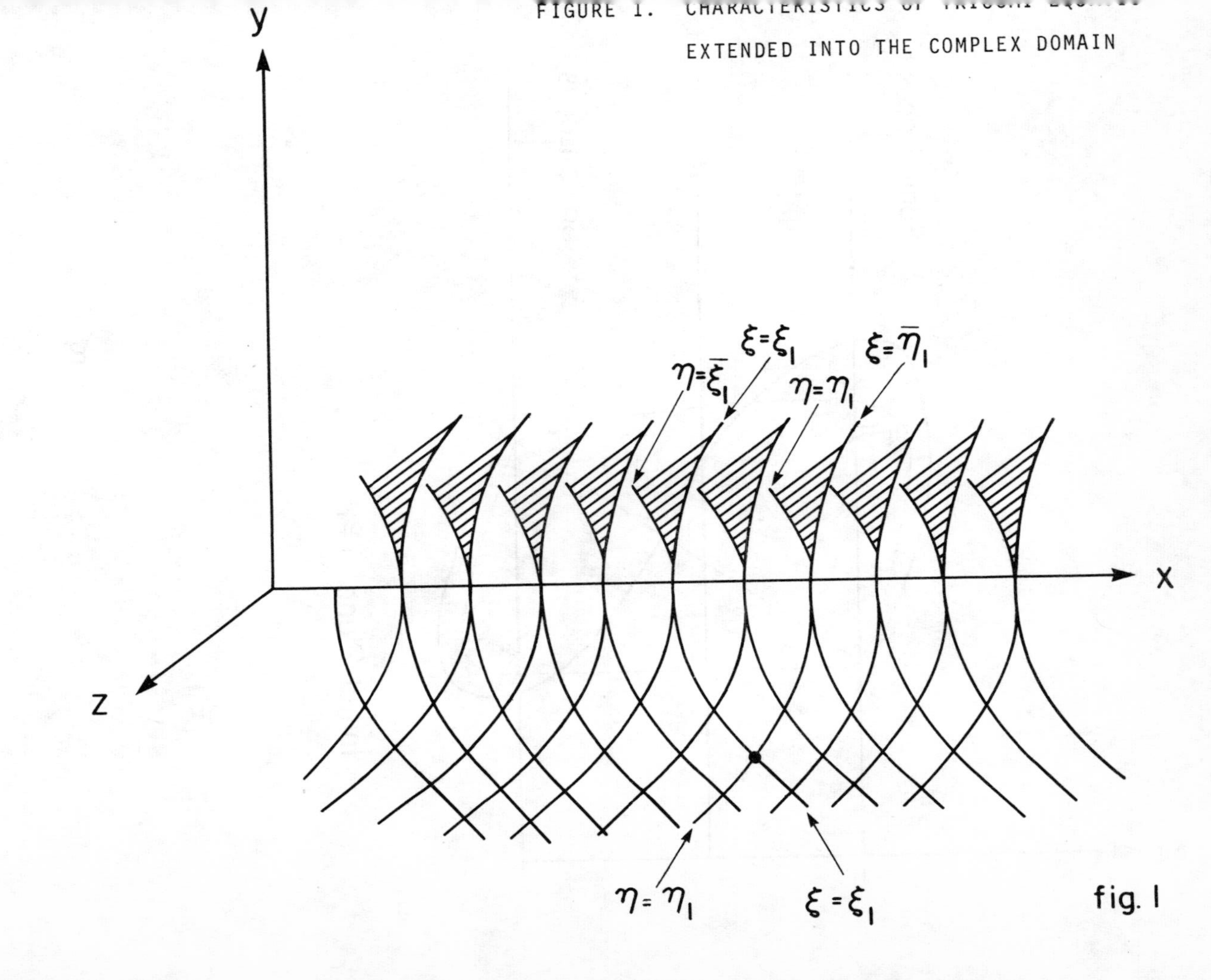

FIGURE 1. CHARACTERISTICS OF TRICOMI EQU
EXTENDED INTO THE COMPLEX DOMAIN

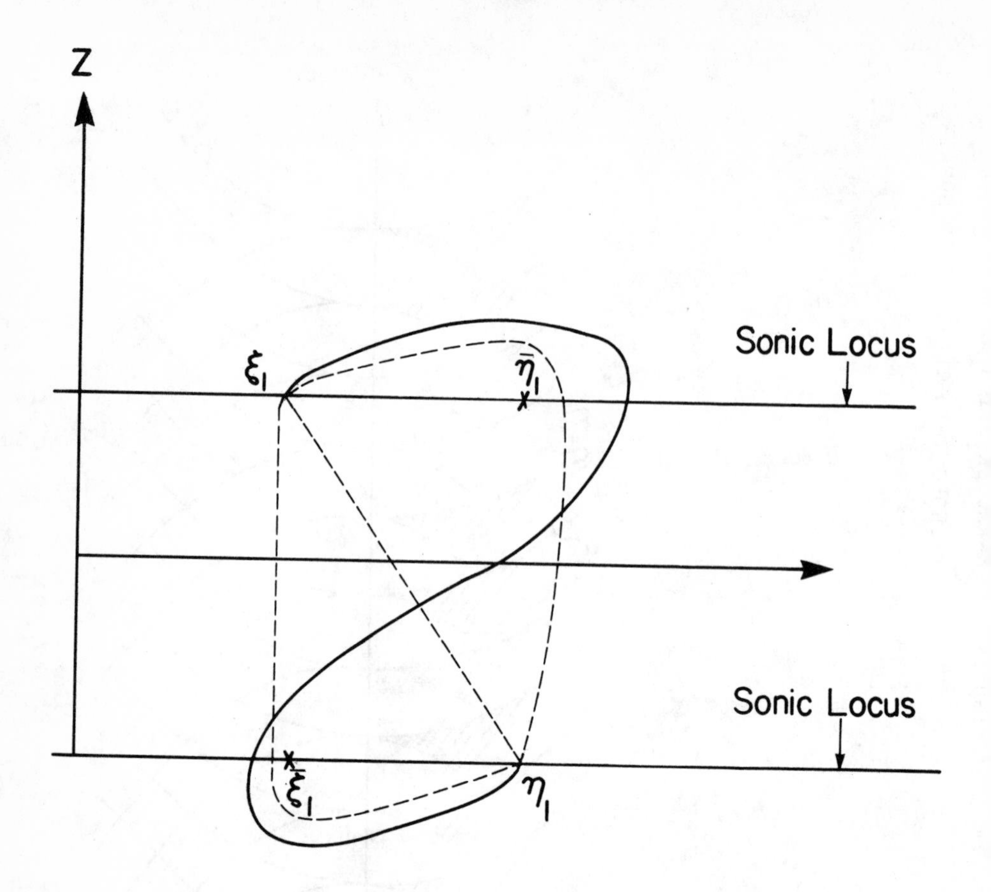

fig. 2

FIGURE 2. CHOOSES FOR THE INITIAL PATH TO GET THE SOLUTION AT THE POINT ξ_1, η_1 IN THE REAL DOMAIN.

Initial paths used to find transonic region.

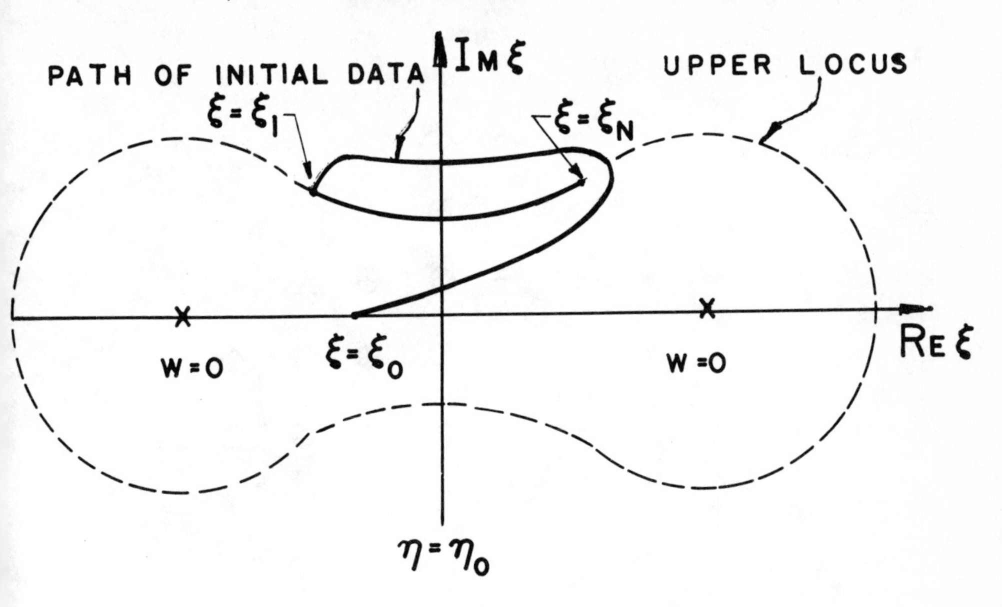

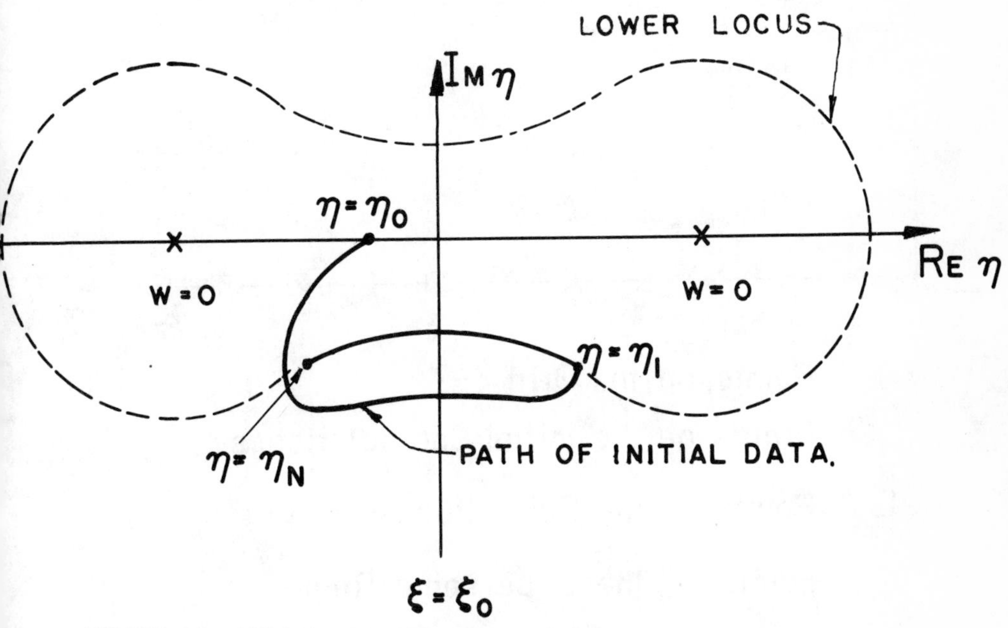

FIGURE 3A. TYPICAL INITIAL PATHS ALONG THE INITIAL CHARACTERISTICS $\eta = \eta_0$ (TOP) AND $\xi = \xi_0$ (BOTTOM) WHICH YIELD THE SOLUTION IN A REGION OF THE SUPERSONIC ZONE.

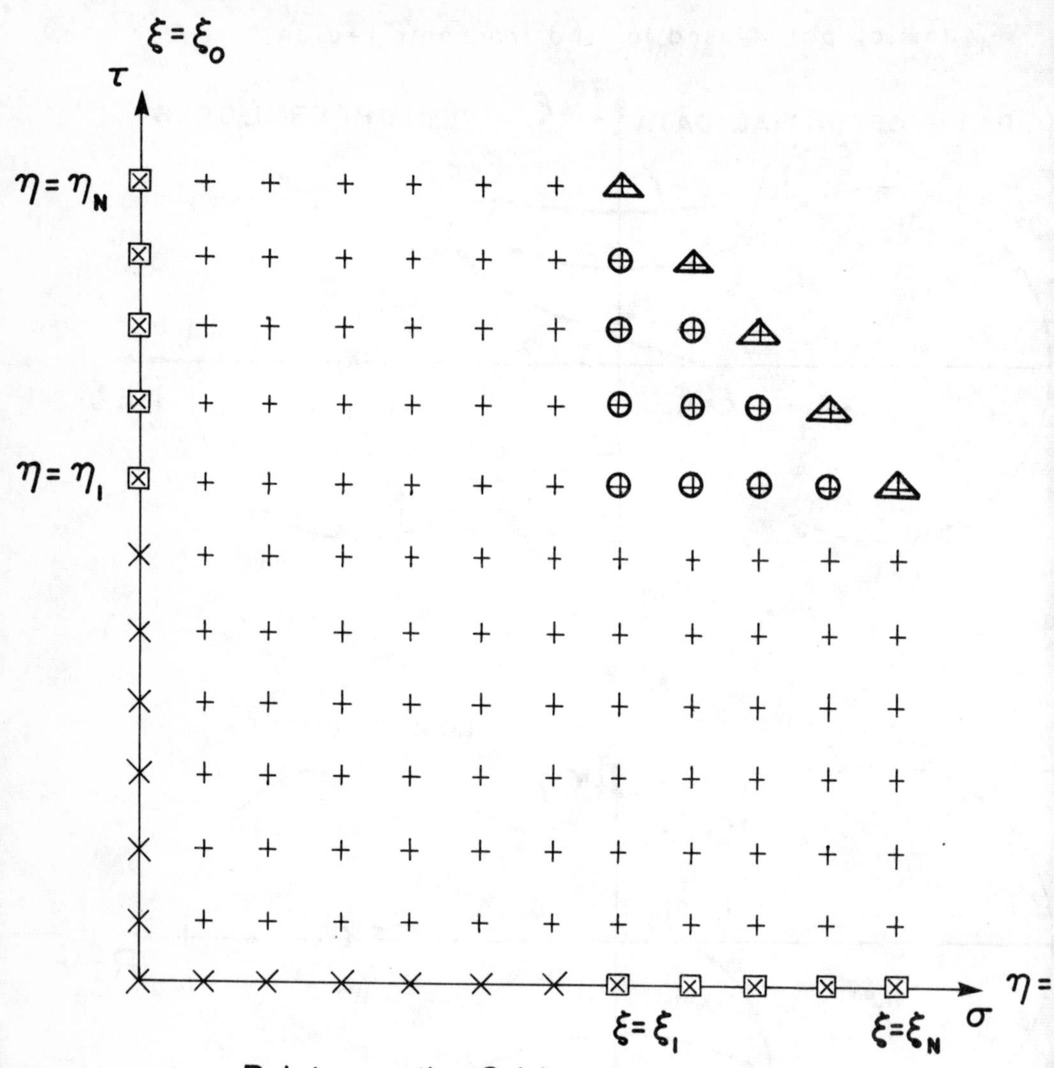

FIGURE 3B. THE CORRESPONDING GRID FOR THE PATHS OF INITIAL DATA IN FIGURE 3A.

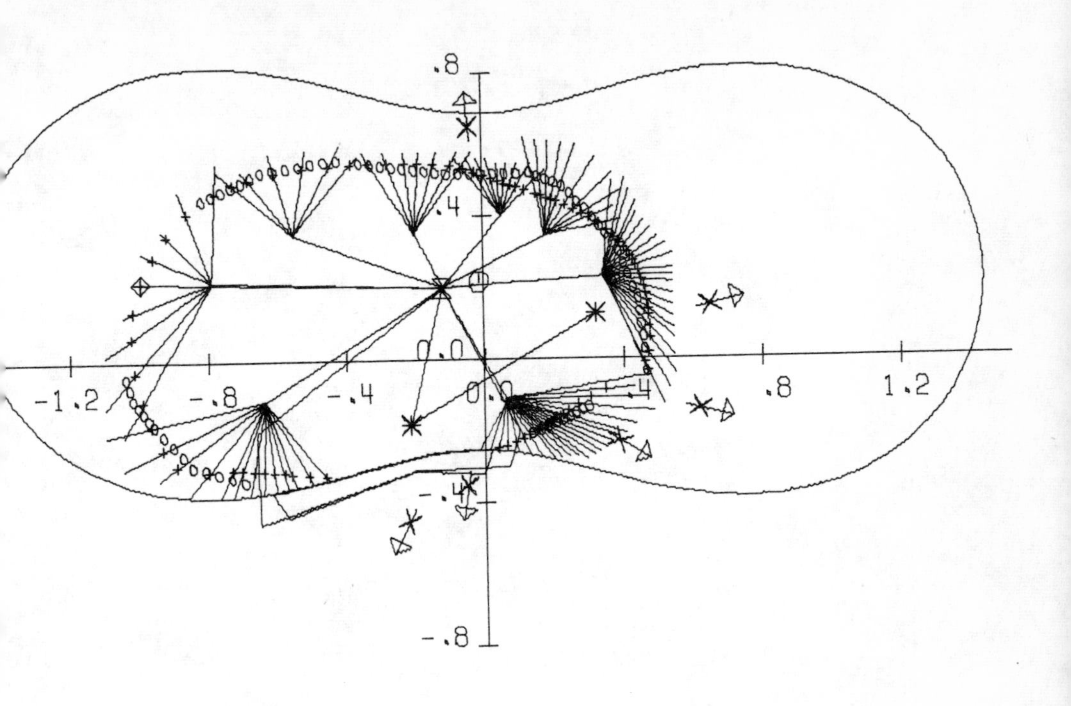

M1= .803   M2= .558

FIGURE 4. INITIAL PLANE FOR A COMPRESSOR BLADE DESIGNED TO TURN THE FLOW 12.01° AT INLET INCIDENCE 48.2° AND INLET MACH NUMBER M1 = .803 FOR A GAP TO CHORD RATIO OF 1.44.

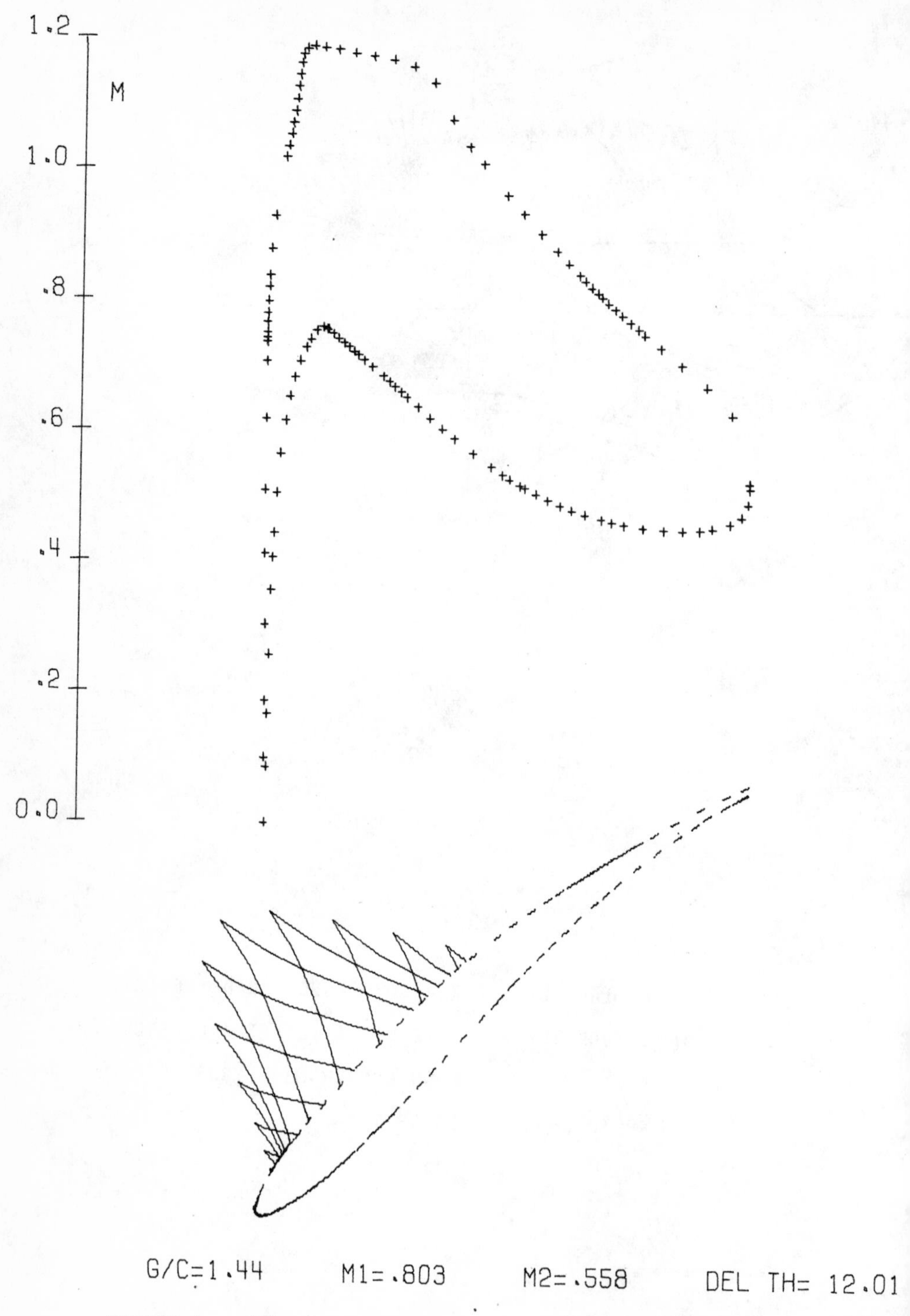

G/C=1.44 M1=.803 M2=.558 DEL TH= 12.01

FIGURE 5. CORRESPONDING COMPRESSOR BLADE FOR INITIAL PLANE OF FIGURE 4.

APPROXIMATE REGULARIZED SOLUTIONS TO IMPROPERLY POSED LINEAR INTEGRAL AND OPERATOR EQUATIONS

M. Zuhair Nashed
School of Mathematics
Georgia Institute of Technology
Atlanta, Georgia 30332

Dedicated to the Memory of My Father

ABSTRACT

Many problems in mathematical analysis lead to what is usually called in the literature improperly posed problems. The theory and numerical methods of investigation and approximation of these problems - which involve additional difficulties that are not encountered in properly posed problems - have been the subject of intensive research during the past two decades, and continue to present many challenging questions. Improperly posed problems have also been the main theme of numerous conferences and addresses.

In the present paper we will report on some recent results for obtaining approximate regularized solutions (and pseudo solutions) of linear operator equations of the first and second kinds. Applications to integral equations will be given.

The underlying philosophy of many approaches to regularization resides in the sense we should understand an approximate solution of an improperly posed problem and in effecting numerically these approximations. We provide computable approximate regularized solutions, as well as convergence rates which are optimal in the context of operator equations considered. We also highlight some aspects of the role of generalized inverses and reproducing kernel spaces in regularization and computational methods for operator and integral equations.

1. Introduction

The notion of a well-posed (correct, properly posed) problem introduced by Hadamard at the beginning of this century plays an important role in the theory and numerical approximations of various linear and nonlinear operator equations arising from problems of mathematical physics, engineering, and analysis.

An operator equation $Ax = y$, where $A : \mathcal{X} \to \mathcal{Y}$ is said to be well-posed if for each $y \varepsilon \mathcal{Y}$ a unique solution exists which depends continuously on the "data"; otherwise the equation is said to be improperly posed (ill-posed). Clearly the notion of well-posedness depends on the class of solutions considered, the admissible data and the "measure" of continuous dependence, i.e. it depends on the spaces $\mathcal{X}$ and $\mathcal{Y}$ and their topologies.

In a narrow sense, continuous dependence on the data is often interpreted as continuous dependence on $y \varepsilon \mathcal{Y}$. However, in a wide sense, the "data" should be interpreted to include y and the operator A. In the case of a differential equation, for example, one may consider as "data" any initial or boundary values, prescribed values of the operator, other coefficients in the equation, and the geometry of the domain of definition of the operator. Similarly for an integral equation, the data may consist of y, the kernel, the domain of the kernel, etc. .

In this wide sense, ill-posed operator equations are harder to analyze. It should be noted that even linear least-squares problems in finite dimensional spaces are ill-posed in this sense since the generalized inverse of a matrix does not depend continuously on perturbations of the matrix. In contrast these problems are well-posed in a narrow sense; however if the condition number of the matrix (i.e., $||A|| \, ||A^{\dagger}||$) is very large, then the problem is numerically ill-posed.

Simple examples of ill-posed problems include integral equations of the first kind, the final value problem for the heat equation, the Dirichlet problem for the wave equation, and the Cauchy problem for Laplace's equation. An integral equation of the second kind with a nonunique solution is also ill-posed in the wide sense (see Section 7).

For many years it was concluded that problems which exhibit discontinuous dependence on the data do not correspond to any real formulations, i.e. they are not problems of mathematical physics or they do not arise in the study of natural phenomena. In other words, there is something wrong with the mathematical model and not with physical problems which it portrays. This attitude has prevailed in several books including Petrovsky's classic book on partial differential equations.

It is now widely recognized that this attitude toward improperly posed problems is erroneous; there are many situations which arise from physical problems and mathematical analysis which require us to consider problems that are improperly posed. Numerous examples are given in [6], [4], [23] and some of the references cited therein. Since this paper is more oriented toward integral equations, we mention that the Fredholm equation of the first kind (which did not receive much attention for several decades after the work of Picard [25]) arises frequently in physical, biological and engineering models: the measurement of physical data by indirect sensing devices using remote sensing experiements, the problem of deducing the structure of a planetary atmosphere from satellite observations, the hapten-binding equation of immunology, the inverse problem of radiography (locating tumors in a body using radiographic techniques), and problems in system identification, antenna theory, polymer chemistry, etc. We shall describe some applications elsewhere [22].

The lack of a continuous dependence of the solution on the data makes <u>direct</u> investigation (and particularly approximation) of ill-posed different, and manifests itself most seriously in the numerical analysis of such problems.

"Regularizations of ill-posed problems" is a phrase that is used for various approaches to circumvent lack of continuous dependence (as well as to bring about existence and uniqueness if necessary). Roughly speaking, this entails an analysis of an ill-posed problem <u>via</u> an analysis of an associated well-posed problem, or a family (usually a sequence, or a filter) of well-posed problems, provided this analysis yields meaningful answers to the given problem.

At the present time there are several methods of investigation and regularization of ill-posed problems. These methods may be classified into the following approaches:

I. Analytic approaches and function-theoretic methods.

1. The notion of a well-posed problem for a given operator depends on the spaces considered. Thus the operator equation $Ax = y$ may be ill-posed relative to the spaces (X, Y), while being well-posed relative to the spaces (X', Y'). To impart continuous dependence in the case when A is considered in the spaces (X, Y), one might consider subsets X', Y' of X, Y respectively, and endow these subsets with topologies which are different from those on X, Y respectively, such that the inverse operator A^{-1}, when considered from Y' to X' is continuous. Note that it may also be necessary to take Y' to be a proper subset of Y to insure that the operator A is onto. Thus one approach to regularization is via a change in the topologies. Here lies one of the challenges of regularization analysis, for the topologies of X' and Y' must still be amenable to the original setting, since it is possible that the topologies which would lead to continuous dependence could also be too restrictive for the problem. The decision on the choice of appropriate topologies cannot be made judiciously from purely mathematical considerations. Physical considerations have the upper hand! A change in topology is related to the mathematical modelling of a physical problem. Physically, this gives a clue as to the type of measurements that are meaningful, and provides a framework where "small"

errors in measurement are tolerable.

2. Continuous dependence on the data can often be restored by imposing global constraints on the class of solutions admitted to consideration. For example, one might ask for nonnegative solutions, or for solutions which satisfy a priori bound, or lie in a compact set. Again these constraints are obtainable from physical considerations. Every regularization method requires for its implementation a numerical or approximation scheme. One difficulty that must be overcome in this approach is the incorporation of the constraint into the computing algorithm.

3. Various function-theoretic methods, logarithmic convexity and related methods have been found useful for the analysis of ill-posed problem. For an excellent and well-motivated description of these methods, see the recent survey paper of Payne [23].

II. Numerical and approximation-theoretic approaches.

The philosophy underlying many methods of regularization lies in deciding in what sense we should understand an approximate solution to an ill-posed problem. Methods along these lines usually involve a change of the concept of a solution, a change of the operator itself, and the introduction of a family of regularization operators. This is tantamount to replacing the original model by a one-parameter family of more tractable models.

As examples of these approaches we mention:

4. The use of regularizing parametric operators, introduced by Tikhonov [27], and further explored by Bakusinskii, Morozov and others; see [6],[7],[13],[4],[19].

5. The method of quasireversibility of Lattes and Lions [8], which is based on the idea of modifying the differential or integrodifferential operator arising in boundary-value problems and unstable control and optimization problems, in order to impart regularity.

6. The replacement of the operator equation by a stable minimization problem depending on a parameter.

III. <u>Probabilistic approaches</u>. The theory of stochastic processes provide another route to regularization, i.e. to obtaining meaningful information from ill-posed problems. Once again we must have some preconceptions about the solution. The idea is to associate random processes with plausible solutions, data sets, errors, etc. See e.g. Lavrentiev [6], Franklin [3], and Morozov [12].

We note that the various approaches to regularization involve one or more of the following intuitive ideas:

(a) a <u>change</u> of the <u>concept</u> of a <u>solution</u>;

(b) a change of the <u>spaces</u> and/or <u>topologies</u>;

(c) a change of the <u>operator</u> itself;

(d) the concept of <u>regularization operators</u>;

(e) probabilistic methods or well-posed <u>stochastic</u> extensions of ill-posed problems.

The various approaches outlined above overlap in many aspects. Usually several of the above ideas manifest themselves in any approach to regularization.

Although the question of well-posedness has been given consideration in the work of Cauchy, Kowaleski, Holmgren, Hadamard, and others, the beginning of work on approximations to solutions of improperly posed operator equations is of more recent interest. There are several surveys of the literature on improperly posed problems. We mention the recent paper by Payne [23] which presents an excellent introduction to the subject of improperly posed partial differential equations, together with interesting historical remarks, and lists 144 references. A brief survey by the author [15], based on a not-too-technical talk lists 200 references related to ill-posed operator equations. (The bibliographies in [23] and [15] have only 17 common entries!). Other related surveys will appear in [24],[22],[21]. Further references are to be found in [4],[6],[8].

This paper consists of an exposition, generally without proofs, of some recent results of the author, as well as joint work with G. Wahba and R. H. Moore, on approximation of least-squares solutions of integral and operator equations of the first and second kinds. Detailed discussion and proofs of results mentioned below will appear elsewhere. No attempt is made to give an unprejudiced survey of the field or a comparison of various methods. We confine our discussion to certain classes of ill-posed linear operator equations with particular reference to integral equations. We provide glimpses into the role that reproducing kernel Hilbert spaces and generalized inverses play in the approximation and regularization of these problems. Our aim in looking at these aspects of ill-posed problems under one roof is to clarify and relate the ideas and techniques involved.

Let $\mathcal{X}$ and $\mathcal{Y}$ be two Hilbert spaces over the same (real or complex) scalars and let A be a linear operator on $\mathcal{D}(A) \subset \mathcal{X}$ into $\mathcal{Y}$. Let $\mathcal{R}(A)$, $\mathcal{N}(A)$ and A^* denote, respectively, the range, null space, and adjoint of A. The orthogonal complement of a subspace S is denoted by $S^{\perp}$; the closure of S is denoted by $\bar{S}$, and the orthogonal projector on a closed subspace $\mathcal{M}$ is denoted by $P_{\mathcal{M}}$.

We consider the linear operator equation

$$Ax = y \tag{1.1}$$

<u>Definition 1.1</u>. An element $u \in X$ is said to be a <u>least squares solution</u> of (1.1) if $\inf\{\|Ax-y\| : x \in \mathcal{X}\} = \|Au-y\|$. A <u>pseudosolution</u> of (1.1) is a least squares solution of minimal norm.

<u>Definition 1.2</u>. The operator equation (1.1) is said to be <u>well-posed</u> (relative to the spaces $\mathcal{X}$ and $\mathcal{Y}$) if for each $y \in \mathcal{Y}$, (1.1) has a unique pseudosolution which depends continuously on y; otherwise the equation is said to be <u>ill-posed</u>.

Obviously (1.1) has a least squares solution for a given $y \in \mathcal{Y}$ if and only if there exists an element $w \in \mathcal{R}(A)$ which is closest to y. From this it follows immediately that (1.1) has a least squares solution if and only if $P_{\overline{\mathcal{R}(A)}} y \in \mathcal{R}(A)$, equivalently $y \in \mathcal{R}(A) + \mathcal{R}(A)^{\perp}$. For such y, it is easy to see that the set S_y of all least squares solutions has a unique element of minimal norm if and only if $P_{\overline{\mathcal{N}(A)}} u \in \mathcal{N}(A)$ for some $u \in S_y$ (in which case this is also true for each $x \in S_y$). The following proposition is an immediate consequence of the preceding observations.

<u>Proposition 1.3</u>. A pseudosolution of (1.1) exists if and only if

$$y \in A[\mathcal{D}(A) \cap \mathcal{N}(A)^{\perp}] + \mathcal{R}(A)^{\perp} . \tag{1.2}$$

In what follows we shall be particularly interested in the cases when A is a <u>closed</u> linear operator on a dense domain $\mathcal{D}(A) \subset X$, <u>or</u> when A is a <u>bounded</u> linear operator on X. In either of these cases condition (1.2) reduces to the condition

$$y \in \mathcal{R}(A) + \mathcal{R}(A)^{\perp} . \tag{1.3}$$

The (linear) map which associates with each y satisfying (1.2) a unique pseudosolution defines the <u>generalized inverse</u> of A, which is denoted by $A^{\dagger}$. (For an exposition on generalized inverses, see [14]). For each $y \in \mathcal{D}(A^{\dagger})$, we have $S_y = A^{\dagger}y + N(A)$.

<u>Proposition 1.4</u>. Let $A : \mathcal{D}(A) \subset \mathcal{X} \to \mathcal{Y}$ be a bounded or a densely defined closed linear operator. Then the following statements are equivalent:

(a) The operator equation (1.1) is well-posed in $(\mathcal{X},\mathcal{Y})$;

(b) A has a closed range in $\mathcal{Y}$;

(c) $A^{\dagger}$ is a bounded operator on $\mathcal{Y}$ into $\mathcal{X}$.

When $\mathcal{R}(A)$ is not closed, $A^{\dagger}$ is an <u>unbounded</u> densely defined operator. The problem of finding least squares solutions of (1.1) is ill-posed in this case relative to the spaces $\mathcal{X},\mathcal{Y}$. An ill-posed problem relative to $(\mathcal{X},\mathcal{Y})$ may be recast in some cases as a well-posed problem relative to new spaces $\mathcal{X}' \subset \mathcal{X}$ and $Y' \subset Y$, with topologies on $\mathcal{X}'$ and $\mathcal{Y}'$ which are different respectively from the topologies on $\mathcal{X}$ and $\mathcal{Y}$. Following Krein it is more appropriate in such case to call the original problem conditionally well-posed. From the point of regularization, the topologies on $\mathcal{X}'$ and $\mathcal{Y}'$ should not be too restrictive and must lend themselves to requirements which are satisfied by a wide class of admissible solutions or pseudosolutions. This is precisely the point which we exploit in connection with the topologies on reproducing kernel Hilbert spaces, which play a central role in several sections of this paper.

2. Some Remarks on Integral Equations of the First Kind, Reproducing Kernel Spaces and Improperly-Posed Problems.

Integral equations of the first kind represent an important class of ill-posed problems. In this section we examine some aspects of the "pathology" associated with these equations, and comment on several approaches to regularization in that context. This also should provide a preview and motivation of some of the schemes of imparting continuous dependence of solutions. We also consider a few examples which may be used to illustrate how some of the methods are applicable.

Example 2.1. Consider the integral equation of the first kind:

$$(2.1) \qquad (Kx)(s) := \int_0^1 k(s,t)x(t)dt = y(s) \qquad 0 \leq s \leq 1$$

where $k(s,t)$ is a (nonseparable) square-integrable kernel on $[0,1] \times [0,1]$. This equation is not well-posed in the space $L_2[0,1]$. It does not have a solution for every $y \in L_2[0,1]$: for if $k(s,t)$ has a certain order of smoothness in s, then there can exist no function $x \in L_2$ satisfying (2.1) for any function $y \in L_2$ that has a lower order of smoothness. Also for (2.1) there is no continuous dependence of the solution on the data y. By the Riemann-Lebesgue Lemma, a change $h(t) = A \sin \omega t$ with a very large frequency ω produces a very small change ε in the data.

This also reflects the fact that the inverse (or the generalized inverse) of the (compact) operator K is unbounded (unless $\mathcal{R}(K)$ is finite dimensional, in which case the integral equation (2.1) has a separable kernel).

To analyze (and later overcome) these difficulties, we review briefly the basic existence theory as given in Picard's "little" classic [25] for operator equations of the first kind:

$$Kx = y \tag{2.2}$$

where $K : \mathcal{H}_1 \to \mathcal{H}_2$ is a compact linear operator, and $\mathcal{H}_1, \mathcal{H}_2$ are Hilbert spaces. The theory is based upon the spectral decomposition of compact, symmetric operators which was developed by E. Schmidt in 1907-1908, and leads to what is called singular-value decomposition of a compact (but not necessarily symmetric) operator. The operators K^*K and KK^* are compact, symmetric, and nonnegative linear operators; hence their spectra are identical and there exist two orthonormal systems of eigenvectors $\{u_n\}$ and $\{v_n\}$ such that

$$KK^* u_n = \lambda_n u_n \quad \text{and} \quad K^*Kv_n = \lambda_n v_n$$

where $\lambda_1 \geq \lambda_2 \geq \cdots \geq \lambda_n \geq \cdots \geq 0$ is a finite or countably infinite set of eigenvalues. Let $\mu_n = \lambda_n^{-1/2}$ for $\lambda_n > 0$. Then

$$u_n = \mu_n K v_n \quad \text{and} \quad v_n = \mu_n K^* u_n .$$

The set $\{u_n, v_n; \mu_n : n = 1,2,\cdots\}$ is called a <u>singular system</u> for K. It follows that the eigenvector set $\{u_n : n = 1,2,\cdots\}$ forms a Schauder basis for $N(K^*)^{\perp} = \overline{R(K)} = \overline{R(KK^*)}$. Equation (2.2) has a solution for a given $y \in H_2$ if and only if

$$(2.3) \qquad y \in N(K^*)^{\perp} \quad \text{and} \quad \sum_{n=1}^{\infty} \mu_n^2 \; |\langle y,u_n\rangle|^2 < \infty$$

(Here $\langle\cdot,\cdot\rangle$ denotes the inner product in H_2). Conditions (2.3) are known as Picard's criteria. This is in contrast to the integral equation of the second kind $Kx - \alpha x = y$, $\alpha \neq 0$, which has a solution for any $y \in N(K^* - \bar{\alpha}I)^{\perp} = R(K - \alpha I)$, where I is the identity operator. The usual Fredholm alternative theorem does not hold for equation (2.2) since $R(K)$ is not closed (unless $R(K)$ is a finite dimensional subspace).

With reference to the integral operator K in (2.1), define

$$H_Q : = \{g \in L_2[0,1] : \sum_n |\langle g,u_n\rangle|^2/\lambda_n < \infty\} ,$$

where if $\lambda_n = 0$, we must have $\langle g,u_n\rangle = 0$, and $\{\lambda_n\}$ are the eigenvalues of K^*K. The vector space H_Q is the set of all functions in $L_2[0,1]$ for which (2.1) has a solution. H_Q is not closed in $L_2[0,1]$. If we define a new inner product in H_Q by

$$\langle f,g\rangle_Q = \sum_n \frac{1}{\lambda_n} \langle f,u_n\rangle \langle g,u_n\rangle ,$$

then H_Q becomes a Hilbert space. Letting

$$Q(s,s') := \int_0^1 k(s,t)\, k(s',t)dt$$

and $Q_s(s') := Q(s,s')$, it follows also that

$$\langle Q_s,z\rangle_Q = z(s), \; z \in H_Q, \qquad 0 \leq s \leq 1 .$$

That is H_Q is a reproducing kernel Hilbert space (RKHS) with reproducing kernel Q . (For details see for instance [18], for properties of RKHS see Aronszajn [2], Shapiro [26] and the recent work of Nashed and Wahba [19],[28]). Furthermore the spaces L_2 and H_Q and their respective norms are related by the following relations:

$$\|z\|_Q = \|Q^{-1/2}z\|_{L_2} \quad , \quad H_Q = Q^{1/2}L_2 , \quad \text{and}$$

$$\|z\|_Q = \inf\{\|p\|_{L_2} : Q^{1/2}p = z\} .$$

Thus if the operator K is regarded as a map $K : L_2 \to H_Q$, then it becomes onto. Hence for any $y \in H_Q$, (2.1) has a unique least squares solution of minimal norm and this solution depends continuously on y (in the topology of H_Q) , i.e. the generalized inverse $K^\dagger : H_Q \to L_2$ is bounded.

In particular if K is one-to-one, then $K^{-1} : H_Q \to L_2$ is bounded.

It should be noted that the introduction of the RKHS H_Q and the casting of Picard's criteria (2.3) in the form $y \varepsilon H_Q$ is not a mere formality. In the above process we have in fact <u>shrunk</u> the image space L_2 to the range of K and endowed the latter space with a <u>new toplogy</u> to guarantee that the range is closed. This is typical of the more general situations where continuous dependence is brought about by a change in the topologies of the spaces.

We may also narrow the domain and endow it with a different topology in order to provide higher orders of smoothness in the solution. For example consider $K : H_P \to H_Q$, where H_P is the space of all absolutely continuous functions with derivatives in $L_2[0,1]$, with the inner product on H_P defined by

$$\langle u,v \rangle = u(0)v(0) + \int_0^1 u'(t)v'(t)dt \; .$$

(H_P is a RKHS with reproducing kernel $P(s,t) = \min(s,t)$) .

Another approach to the "regularization" of (2.1) or (2.2) is to <u>modify</u> the operator, i.e. to replace the operator by a better behaved operator, provided the modified equation can be used to give approximations to the "solution" of the original equation. For example instead of (2.2), we may consider the equation of the second kind:

$$(2.4) \qquad (K^*K + \alpha I)x = K^*y \quad , \quad \alpha > 0$$

The operator $K^*K + \alpha I$ (for $\alpha > 0$) is one-to-one onto and its inverse is continuous. Let $x_\alpha = (K^*K + \alpha I)^{-1}K^*y$. Then as $\alpha \to 0$, $x_\alpha \to K^\dagger y$. However for small α, the equation (2.4) becomes ill-conditioned; thus regularization of (2.2) via (2.4) still leaves us with numerical difficulties in choosing an "optimal" α for a suitable compromise between stability and accuracy.

Example 2.2. The problem of determining the input to a linear time-invariant system when we know the impulse response and the output gives rise to an improperly-posed problem. The output y due to an input x is described by the Volterra equation:

$$y(t) = \int_0^t f(t-s)x(s)ds =: (Ax)(t) .$$

Given a desired output $z(t)$, we wish to find an input $x(t)$ to minimize the least-squares error between the actual output and the desired output over $[0,\tau]$:

$$\min_x \{\int_0^\tau [y(t) - z(t)]^2 dt \ : \ Ax = y\} .$$

Thus the problem is to find $x \in L_2$ which would minimize the L_2-norm of $Ax - y$. The range of A is not closed and the problem is ill-posed. The minimization problem does not have a solution for <u>each</u> $y \varepsilon L_2[0,\tau]$.

The same regularization schemes prescribed for Example 2.1 may be prescribed here to circumvent this impediment.

<u>Example 2.3</u>. Consider the convolution equation

$$\frac{1}{\sqrt{2\pi}} \int_{-\infty}^{\infty} k(s)x(t-s)ds = y(t) ,$$

where $k \varepsilon L_1(-\infty,\infty)$, $y \varepsilon L_2(-\infty,\infty)$, with y being known only through an approximation $\tilde{y}$ such that $||y-\tilde{y}||_{L_2}$ is small. The operator $A : L_2 \to L_2$ defined by

$$(Ax)(t) = \frac{1}{\sqrt{2\pi}} \int_{-\infty}^{\infty} k(s)x(t-s)ds$$

is bounded, but has no continuous inverse on all of $L_2(-\infty,\infty)$. The proof of the last statement uses the isometry and convolution properties of the Fourier and Fourier-Plancherel operators and the Riemann-Lebesgue lemma).

The problem of numerical inversion of the Laplace transform is also ill-posed; some of the techniques presented in this paper can be adapted to this problem to obtain error bounds under suitable assumptions on the function.

3. <u>Generalized Inverses of Linear Operators in Banach and Hilbert Spaces</u>.

We provide in this section a synopsis of results on generalized inverses which will be essential to the main results in the remaining section; the generality will be confined to the context that will be needed. For a unified general approach to generalized inverses see Nashed and Votruba [16],[17].

A <u>projector</u> P on a Banach space X is a continuous linear idempotent (i.e., $P^2 = P$) operator. A subspace M of X is said to have a <u>topological complement</u> if and only if there exists a subspace N such that $X = M \oplus N$ (topological direct sum). In this case M and N are closed; however a closed subspace need not necessarily have a topological complement. A projector P on X induces a decomposition of X into two topological complements PX and $(I-P)X$. Conversely a subspace M of X has a topological complement if and only if there exists a projector P of X onto M.

Let X and Y be Banach spaces and let $A : X \to Y$ be a linear operator with null space $N(A)$ and range $R(A)$, and let $\overline{R(A)}$ denote the closure of $R(A)$. Assume that $N(A)$ has a topological complement M and that $\overline{R(A)}$ has a topological complement S, i.e. there exist projectors U, E with $UX = N(A)$ and $EY = \overline{R(A)}$ so that

$$(3.1) \qquad X = N \oplus M \, , \quad Y = \overline{R(A)} \oplus S \, .$$

Relative to these projectors, A has a unique <u>generalized inverse</u> $A^\dagger$ with domain $D(A^\dagger) = R(A) + S$, $R(A^\dagger) \subseteq M$, and $N(A^\dagger) = S$ such that

$$(3.2)\quad \begin{cases} \text{(a) } A^{\dagger}A = I-U, & \text{(b) } A^{\dagger}AA^{\dagger} = A^{\dagger} \text{ on } \mathcal{D}(A^{\dagger}) \\ \text{(c) } AA^{\dagger} = E \text{ on } \mathcal{D}(A^{\dagger}), \text{ and (hence)} & \text{(d) } AA^{\dagger}A = A . \end{cases}$$

Clearly $A^{\dagger}$ depends on the projectors E, U (or equivalently on the choice of topological complements to $N(A)$ and $\overline{R(A)}$); thus we shall use the notation $A^{\dagger}_{U,E}$. Suppose now that new projectors U', E' are chosen, where $R(U') = N(A)$ and $R(E') = \overline{R(A)}$. Then the restrictions of $A^{\dagger}_{U',E'}$, $A^{\dagger}_{U,E}$ to $R(A)$ are related as follows [16; Theorem 1.3(a)]:

$$(3.3)\quad A^{\dagger}_{U',E'} = (2I-A^{\dagger}_{U,E}A + U')\, A^{\dagger}_{U,E}(2I-AA^{\dagger}_{U,E} + E') .$$

We now consider the important case when X and Y are Hilbert spaces, and A is a <u>bounded</u> or a densely defined <u>closed</u> linear operator. In either case, $N(A)$ is closed. Thus $N(A)$ and $\overline{R(A)}$ have topological complements. In particular we may choose $M = N^{\perp}$ and $S = R(A)^{\perp}$ in (3.1). The generalized inverse corresponding to this choice of M and S (equivalently the choice of orthogonal projectors) is the operator version of the <u>Moore-Penrose</u> inverse. It possesses (and is equivalently characterized by) the following important "<u>best approximation</u>" property. For any $y \in \mathcal{D}(A^{\dagger}) = R(A) + R(A)^{\perp}$, the vector $A^{\dagger}y$ minimizes $||Ax-y||$ over $\mathcal{D}(A)$ and $||A^{\dagger}y||$ is smaller than the norm of any other vector which minimizes $||Ax-y||$.

4. Regularization Via (Stable) Minimization Problems. Regularization Operators.

Let X, Y, Z be Hilbert spaces, let $A : X \to Y$ and $L : X \to Z$ be bounded linear operators. We assume that $R(L)$ is closed in Z but $R(A)$ is not necessarily closed in Y. $N(A)$ and $N(L)$ are both assumed to be nontrivial subspaces. Let

$$S_y = \{u \in X : \|Au-y\| = \inf\|Ax-y\| : x \in X\} .$$

Consider the minimization problem:

(4.1) Find $w \in S_y$ such that $\|Lw\| = \min\{\|Lu\| : u \in S_y\}$.

A necessary and sufficient condition for the existence of a solution to (4.1) for each $y \in D(A^{\dagger})$ is that $N(A) + N(L)$ is closed in X ; for uniqueness it suffices to require $N(A) \cap N(L) = \{0\}$. Assume these conditions are satisfied and let w_y denote the solution of (4.1) for a given y, and let $A_L^{\dagger}$ denote (linear) map induced by $y \to w_y$.

The element w_y is characterized by the conditions: $A^*Aw = A^*y$ and $L^*Lw \in N(A)^{\perp} = \overline{R(A^*)}$. In particular if A has a closed range, then a necessary and sufficient condition for (4.1) to have a solution is the existence of a solution to the system

$$\begin{pmatrix} A^*A & 0 \\ L^*L & -A^* \end{pmatrix} \begin{pmatrix} w \\ v \end{pmatrix} = \begin{pmatrix} A^*y \\ 0 \end{pmatrix} .$$

It should be noted that the subspace

$$(4.2) \qquad \mathcal{M} = \{w \in \mathcal{X} : L^{*}L\,w \perp \mathcal{N}(A)\}$$

is closed in $\mathcal{X}$ and is a topological complement to $\mathcal{N}(A)$. Thus the generalized inverse $A_L^{\dagger}$ can be related easily to the Moore-Penrose inverse $A^{\dagger}$ (when $L = I$) using (3.3); hence

$$(4.3) \qquad A_L^{\dagger} = (2I-A^{\dagger}A-U)A^{\dagger}$$

where U is the projector of $\mathcal{X}$ onto $\mathcal{N}(A)$ along $\mathcal{M}$.

We consider also the minimization problem: Find x_α to minimize the functional

$$(4.4) \qquad J(x;y,\alpha) = \|Ax-y\|_{\mathcal{Y}}^2 + \alpha^2\|Lx\|_{\mathcal{Z}}^2 .$$

This problem is equivalent to a least-squares minimization in a product space. We define a new Hilbert space $\Gamma = \mathcal{Y} \times \mathcal{Z}$ with the usual inner product, and let

$$(4.5) \qquad C_\alpha x = (Ax, \alpha Lx), \ \alpha > 0 .$$

The operator $C_\alpha : X \to \Gamma$ has a generalized inverse defined on all of Γ if $\mathcal{R}(C_\alpha)$ is closed, which would be the case if $\mathcal{R}(L)$ is closed and $\tilde{A} = A|\mathcal{N}(L) \cap \mathcal{N}(A)^{\perp}$ has a closed range. Then $\|C_\alpha^{\dagger}\|^{-1} = \inf\{\|C_\alpha x\| : \|x\| = 1\}$.

Now in the Hilbert space Γ, (4.4) may be rewritten as

$$J(x;y,\alpha) = ||C_\alpha x-b||_\Gamma^2 \text{ , where } b = (y,0) . \tag{4.6}$$

Under the preceding conditions, this minimization problem has a unique solution x_α of minimal-norm, for each $\alpha > 0$: $x_\alpha = C_\alpha^\dagger b$. Furthermore if $N(A) \cap N(L) = \{0\}$, then x_α is the unique minimizer of $J(x;y,\alpha)$.

It follows immediately from (4.4) and (4.6) that x_α satisfies:

$$(A^*A + \alpha^2 L^*L)x_\alpha = C_\alpha^* C_\alpha x_\alpha = A^*y$$

and hence $||(A^*A + \alpha^2 L^*L)^{-1}|| \le ||C_\alpha^\dagger||^2$. Define the <u>resolvent operator</u> by $R_\alpha = (A^*A + \alpha^2 L^*L)^{-1}A^*$. Then $x_\alpha = R_\alpha y$, and $x_\alpha \to A_L^\dagger y$ as $\alpha \to 0$, for $y \in \mathcal{D}(A^\dagger)$.

We now consider the case when $R(A)$ is closed. Then $R_\alpha \to A_L^\dagger$. Let $\tilde{A} = A|M$, where M is defined by (4.2). Then $||Ax|| \geqq ||\tilde{A}^{-1}||\,||x||$ for $x \in M$, and

$$||R_\alpha - A_L^\dagger|| \leq \alpha^2 \, ||\tilde{A}^{-3}|| \, ||L||^2$$

and $||R_\alpha|| \leq ||A_L^\dagger||$.

The resolvent operator R_α may be viewed as a <u>regularization operator</u> for the operator equation

$$Ax = y . \tag{4.7}$$

More generally we consider $A : \mathcal{X} \to \mathcal{Y}$, where A is linear, and $\mathcal{X}$, $\mathcal{Y}$ are normed spaces. Let Λ be a set of real numbers with $0 \in \bar{\Lambda}$.

Following Tikhonov [27], a set of operators $\{T_\alpha : \alpha \in \Lambda\}$ is said to regularize the equation (4.7) if the following conditions hold:

(a) For each $\alpha \in \Lambda$, T_α is a bounded linear operator from Y into X.

(b) For each $x \in X$, $\lim_{\alpha \to 0} \|T_\alpha Ax - x\| = 0$.

Conditions (a) and (b) imply that

$$\lim_{\|Ax-\tilde{y}\| \to 0} \inf_{\alpha \in \Lambda} \|x - T_\alpha \tilde{y}\| = 0, \quad x \in X .$$

If $\{T_\alpha\}$ is a regularization family, then as an approximate solution to (4.7) we may take $\phi_{\alpha,\delta} = T_\alpha y_\delta$. The choice of α is made effectively only if there is definite information about the elements y and $y_\delta - y$.

Regularization operators have been extensively studied, especially by Soviet mathematicians. See for example [6], [7], [11], and [13]. In the next section we consider an effective method for the numerical realization of regularization operators in reproducing kernel Hilbert spaces and obtain rates of convergence for these approximations.

5. Reproducing Kernel Hilbert Space Methods for Simultaneous Regularization and Approximation of Linear Operator Equations.

The objective of this section is to describe an approach to (simultaneous) regularization and approxmation of (ill-posed) linear problems which applies to a large class of operator equations that includes two-point boundary value problems, Fredholm integral equations of the first kind and mixed integrodifferential equations.

We recall that a Hilbert space H of real-valued functions defined on a set S is said to be a <u>reproducing kernel Hilbert</u> space (RKHS) if all the evaluation functionals $f \to f(s)$ for $f \in H$, $s \in S$, are continuous. Then by the Riesz theorem, there exists a unique element in H, call it Q_s, such that $f(s) = \langle Q_s, f \rangle$, $f \in H$. (Here $\langle \cdot,\cdot \rangle$ is the inner product in H). The reproducing kernel (RK) is defined by $Q(s,s') := \langle Q_s, Q_{s'} \rangle$, $s, s' \in S$. Let H_Q denote the RKHS with reproducing kernel Q, and denote the inner product and norm in H_Q by $\langle \cdot,\cdot \rangle_Q$ and $||\cdot||_Q$ respectively.

We take S to be a bounded interval and assume $Q(s, s')$ is continuous on $S \times S$. Then it is easy to show that H_Q is a space of continuous functions. (Note that $L_2[S]$, the space of square integrable real functions on S is <u>not</u> an RKHS since the evaluation functionals are not continuous).

An RKHQ with RK Q induces a symmetric Hilbert-Schmidt operator (also denoted by Q) on $L_2[S]$ into $L_2[S]$ by

$$(5.1) \qquad (Qf)(s) = \int_S Q(s,t)\, f(t)dt\ .$$

The operator Q has an $L_2[S]$ - complete orthonormal system of eigenfunctions $\{\phi_i\}_{i=1}^{\infty}$ and corresponding eigenvalues $\{\lambda_i\}_{i=1}^{\infty}$ with $\lambda_i \geq 0$ and $\sum_{i=1}^{\infty} \lambda_i < \infty$. We also have the following uniformly convergent Fourier expansions:

$$Q(s,s') = \sum_{i=1}^{\infty} \lambda_i\ \phi_i(s)\ \phi_i(s')\ ,$$

$$Qf = \sum_{i=1}^{\infty} \lambda_i\ (f,\phi_i)_{L_2}\ \phi_i\ .$$

Furthermore

$$H_Q = \{f \in L_2[S] : \sum_{i=1}^{\infty} \lambda_i^{-1}(f,\phi_i)_{L_2}^2 < \infty\}\ ,$$

where the notational convention $\frac{0}{0} = 0$ is being adopted, and

$$\langle f\ ,\ g\rangle_Q = \sum_{i=1}^{\infty} \lambda_i^{-1}(f,\phi_i)_{L_2}\ (g,\phi_i)_{L_2}\ .$$

The symmetric square root of the operator Q is given by

$$Q^{1/2}f = \sum_{i=1}^{\infty} \sqrt{\lambda_i}\ (f,\phi_i)_{L_2}\ \phi_i\ ,$$

and since $N(Q) = N(Q^{1/2})$, we get

$$H_Q = Q^{1/2}L_2[S] = Q^{1/2}(L_2[S] \ominus N(Q))\ .$$

$(Q^{1/2})^{+}$ has the representation

$$(Q^{1/2})^{\dagger}f = \sum_{i=1}^{\infty} (\sqrt{\lambda_i})^{\dagger}\ (f,\phi_i)_{L_2}\ \phi_i$$

on $H_Q + H_Q^{\perp}$ ($\perp$ is in L_2) , when for α a real number, $\alpha^{\dagger} = \alpha^{-1}$ if $\alpha \neq 0$ and $0^{\dagger} = 0$. Similarly

$$Q^{\dagger}f = \sum_{i=1}^{\infty} \lambda_i^{\dagger}\ (f,\phi_i)_{L_2}\ \phi_i$$

on its domain. We adopt the notational conventions

$$Q^{-1/2} := (Q^{1/2})^{\dagger} \quad \text{and} \quad Q^{-1} := Q^{\dagger} .$$

We have the relations for $f , g \in H_Q$:

$$||f||_Q = \inf\{||p||_{L_2} : p \in L_2[S] , Q^{1/2}p = f\}$$

$$<f , g>_Q = (Q^{-1/2}f , Q^{-1/2}g)_{L_2} .$$

Let H_Q and H_P be RKHS with norms $||\cdot||_Q$ and $||\cdot||_P$ respectively. Let A be a linear operator with a dense domain in $X = L_2[S]$ into $Y = L_2[T]$, where S , T are bounded intervals. We assume throughout that H_Q is chosen so that $A[H_Q] = H_R$, where H_R is some RKHS contained (as a set) in $L_2[T]$, and H_P is a subset of $L_2[T]$.

<u>Definition 5.1</u>. A <u>regularized</u> <u>pseudosolution</u> (in RKHS) of the equation

(5.2) $\qquad Af = g$

is a solution to the variational problem: Find $f_\lambda \in H_Q$ to minimize

(5.3) $$\phi(f;g,\lambda) = ||Af - g||_P^2 + \lambda\, ||f||_Q^2 , \quad \lambda > 0$$

($\phi(f;g,\lambda)$ will be assigned $+\infty$ if $Af - g \notin H_P$) .

For $\lambda > 0$, let $H_{\lambda P}$ be the RKHS with kernel $\lambda P(t , t')$, where $P(t,t')$ is the (continuous) RK of H_P . Then $H_P = H_{\lambda P}$ and $||\cdot||_P^2 = \lambda\, ||\cdot||_{\lambda P}^2$. Let $R(\lambda) = R + \lambda P$, where the operators R and P are induced as in (5.1), and let $H_{R(\lambda)}$ be the RKHS with kernel $R(\lambda; t , t')$. Then (see Aronszajn [2; p. 352]) $H_{R(\lambda)}$ is the Hilbert space of all functions of the form $g = g_0 + \xi$, where $g_0 \in H_R$

and $\xi \in H_P$. The norm on $H_{R(\lambda)}$ is given by

$$||g||^2_{R(\lambda)} = \min\{||g_0||^2_R + ||\xi||^2_{\lambda P} : g_0 \in H_P , g_0 + \xi = g\}$$

Following Aronszajn [2], we note that the decomposition $g = g_0 + \xi$ is not unique unless $H_P \cap H_R = \{0\}$. However, the g_0 and ξ attaining the minimum in the above expression are easily shown to be unique by the strict convexity of the norm.

Theorem 5.2. Suppose $\mathcal{D}(A^*)$, the domain of the adjoint (in L_2) of A, is dense in Y, $H_Q \subset \mathcal{D}(A)$, and A and H_Q are such that the linear functionals E_t defined by $E_t f = (Af)(t)$ are continuous in H_Q. Suppose H_Q, H_R and $H_P \subset Y$ all have continuous kernels. Then for $g \in H_{R(\lambda)}$, the unique minimizing element $f_\lambda \in H_Q$ of the functional $\phi(f; g, \lambda)$, defined by (5.3) is given by

$$f_\lambda(s) = \langle AQ_s , g\rangle_{R(\lambda)} = (QA^*(AQA^* + \lambda P)^{\dagger} g(s) .$$

Theorem 5.3. In addition to the assumptions of Theorem 5.2 suppose that $H_P \cap H_R = \{0\}$. Then the minimizing element f_λ of (5.3) is the solution to the least-square problem: Find $f \in \Omega$ to minimize $||f||_Q$, where

$$\Omega = \{f \in H_Q : ||Af - g||_{R(\lambda)} = \inf_{h \in H_Q} ||Ah - g||_{R(\lambda)}\} .$$

The proofs of Theorems 5.2 and 5.3 will appear in [19]. We note that if $H_R \cap H_P = \{0\}$, then H_R is a closed subspace of $H_{R+\lambda P}$; in this case Theorem 5.3 says that regularization operator is a generalized inverse in an appropriate RKHS. However, the topology in H_R is not, in general, the restriction of the topology of $H_{R+\lambda p}$, with the notable exception of the case $H_R \cap H_p = \{0\}$.

We now describe a procedure for obtaining computable approximations $\{f_{\lambda,n}\}$ to f_λ using moment discretization, which are uniformly pointwise convergent. Let $T_n = \{t_1, t_2, \ldots, t_n\}$, where $t_i \in T$, $t_1 < t_2 < \ldots < t_n$. For a generic function h on T, let $h_n := (h(t_1), \ldots, h(t_n))$. Let P_n be the n x n matrix whose ijth entry is $P(t_i, t_j)$, and define

$$||h_n||_{P_n} = \min\{||e|| : e \in R^n, P_n^{1/2} e = h_n\}.$$

Let $f_{\lambda,n}$ be the minimizing element in H_Q of the functional

(5.4) $$J_n(f) = ||(Af)_n - g_n||^2_{P_n} + \lambda ||f||^2_Q, \quad \lambda > 0.$$

If the matrix $R_n(\lambda) = R_n + \lambda P_n$ is nonsingular, an explicit expression for $f_{\lambda,n}$ is given by (see [20])

(5.5) $$f_{\lambda,n}(s) = (\eta_{t_1}(s), \ldots, \eta_{t_n}(s)) R_n^{-1}(g(t_1), \ldots, g(t_n)),$$

where η_t is defined by

(5.6) $$E_t f := (Af)(t) = \langle \eta_t, f \rangle_Q.$$

Such a representation exists by Riesz's theorem since E_f by assumption is a continuous functional in H_Q). Define η^*_s by $\eta^*_s(t) = \eta_t(s)$, and let $P_{T_n}(\lambda)$ be the orthogonal projector in $H_{R(\lambda)}$ onto the subspace spanned by $\{R_t(\lambda): t = t_1,\dots t_n\}$. Here $R_t(\lambda)$ is the representer of the evaluation functional at t in $H_{R(\lambda)}$, i.e. $R_t(\lambda)(t') = R(\lambda;t\ ,\ t')$. It is easy to show that

$$f_{\lambda,n}(s) = \langle P_{T_n}(\lambda)\ \eta_s^*\ ,\ P_{T_n}(\lambda)g\rangle_{R(\lambda)}\ ,$$

whereas

$$f_\lambda(s) = \langle \eta_s^*\ ,\ g\rangle_{R(\lambda)}\ . \tag{5.5}$$

<u>Theorem 5.4</u>. [20]. Let A be a linear operator from H_Q into H_R which satisfies the assumptions of Theorem 5.2. For $g \in H_{R(\lambda)}$, let $f_\lambda(s)$ and $f_{\lambda,n}(s)$ be given by (5.5) and (5.4) respectively. Suppose that $R(\lambda\ ;\ t\ ,\ t') = R(t\ ,\ t') + \lambda P(t\ ,\ t')$ satisfies for some positive integer conditions (i) and (ii) of Theorem 6.1 of Section 6 below (with $Q(t\ ,\ t')$ replaced by $R(\lambda\ ;\ t\ ,\ t')$ and $[0,1]$ replaced by T).

Suppose further that $g \in H_{R(\lambda)}$ satisfies

(a) $g = (R + \lambda P)\rho$ for some bounded ρ , or

(b) $\eta_s^* = AQ(\cdot\ ,\ s) = (R + \lambda P)\rho_s$ for some bounded ρ_s .

Then

$$|f_\lambda(s) - f_{\lambda,n}(s)| = O(\Delta_n^m)\ ,$$

where $\quad \Delta_n := \sup_{t\in T}(\inf_{t_i\in T_n} |t - t_i|)$.

If g satisfies both (a) and (b) , then

$$|f_\lambda(s) - f_{\lambda,n}(s)| = O(\Delta_n^{2m})\ .$$

In the case $H_P \cap H_R = \{0\}$, $f_{\lambda,n}$ converges uniformly to $A^\dagger g$

6. Convergence Rates of Approximate Least Squares Solutions of Linear Integral Equations of the First Kind Using Moment Discretization.

We consider the numerical solution of the integral equation of the first kind

$$(6.1) \qquad Kx := \int_0^1 K(t,s)x(s)ds = y$$

via moment discretization using n values $y(t_i)$ of $y(t)$. We consider first the case when $y \in R(K)$, but K is not necessarily one-to-one. Suppose that $y(t)$ is known on the set $T_n = \{t_1,\cdots,t_n\} \subset [0,1]$ and consider the set of linear equations obtained by moment discretization on T_n of (6.1):

$$(6.2) \qquad \int_0^1 K(t_i,s)x(s)ds = y(t_i), \quad i = 1,\cdots,n .$$

Let x_n be the element of minimal $L_2[0,1]$ - norm which minimizes the sum of the squares of the residuals in (6.2). It is easy to show that x_n is given explicitly by

$$(6.3) \qquad x_n(\cdot) = (y_1,\cdots,y_n)\, Q_n^{\dagger}\, (k_{t_1},\cdots,k_{t_n})' ,$$

where the prime denote transpose, Q_n is the $n \times n$ matrix whose ij^{th} element is given by

$$Q(t_i,t_j) = \int_0^1 K(t_i,s)K(t_j,s)ds$$

and $k_t := K(t,\cdot)$. We shall assume for simplicity that the set $\{k_t ; t \in T_n\}$ is linearly independent for every finite n. Under this assumption, $Q_n^\dagger$ becomes Q_n^{-1}. Let

$$(6.4) \qquad \Delta_n := \sup_{t\in[0,1]} \left(\inf_{t_i \in T_n} |t-t_i| \right) .$$

Note that if $\lim_{n\to\infty} \Delta_n = 0$, then $\bigcup_n T_n$ is dense in $[0,1]$. The following theorem is established in [18].

Theorem 6.1. (a) If $Q(t,t')$ is continuous and if $y \in R(K)$, then

$$\lim_{\Delta_n \to 0} ||x_n - K^\dagger y|| = 0$$

where Δ_n is defined by (6.4).

(b) Let Q is the integral operator whose kernel is $Q(t,t')$. If $y \in R(K)$, $K^\dagger y \in K^* L_2$, or equivalently $y \in Q$ and $Q(t,t')$ satisfies

(i) $\frac{\partial^\ell}{\partial t^\ell} Q(t,t')$ exists and is continuous on $[0,1] \times [0,1$ for $t \neq t'$, $\ell = 0,1,2,\cdots,2m$; $\frac{\partial^\ell}{\partial t^\ell} Q(t,t')$ exists and is continuous on $[0,1] \times [0,1]$ for $\ell = 0,1,2,\cdots,2m-2$;

and

(ii) $\lim_{t\nearrow t'} \frac{\partial^{2m-1}}{\partial t^{2m-1}} Q(t,t')$ and $\lim_{t\searrow t'} \frac{\partial^{2m-1}}{\partial t^{2m-1}} Q(t,t')$ exist and are bounded for all $t' \in [0,1]$

then

$$\|x_n - K^{\dagger}y\|^2 = O(\Delta_n^m) .$$

(c) If y does not coincide with some element $y_0 \in R(K)$ on $\bigcup_n T_n$, then $\|x_n\| \to \infty$.

In the case $y \notin R(K)$, similar convergence results to $K^{\dagger}y$ for $y \in R(K) + R(K)^{\perp}$ are also derived in [18], as well as generalization to a class linear operator equations on Hilbert spaces. Theorem 6.1 does not provide a regularization method since y can fail to be in $R(K)$ both conceptually and numerically. However the techniques and results apply if instead of the integral equation (6.1), we consider the problem of minimizing the functional:

$$(6.5) \qquad J(f;g,\alpha,L_m) = \|Kf - y\|^2 + \alpha^2 \|L_m f\|^2$$

where L_m is usually taken to be an mth order linear differential operator (compare with sections 4 and 5); here all the norms are L_2-norms. Common choices for L_m include $L_0 f = f$, $L_1 f = f'$, $L_2 f = f''$. Moment discretizations, applied to (6.5), provide a rapidly convergent numerical filter for the components of the solutions which correspond to very small eigenvalues of KK^* . The convergence of approximations to smooth solutions of (6.1) is subsumed by the RKHS theory (see [18; pp. 75-76]).

7. Pointwise Convergent Approximations to Least Squares Solutions of Linear Integral Equations of the Second Kind

Let X be a (real or complex) Banach space and let $[X]$ denote the space of continuous linear operators $A : X \to X$ equipped with the uniform norm $\|A\| := \sup\{\|Ax\| : x \in B\}$, where $B = \{x \in X : \|x\| \leq 1\}$. All operators considered in this section are in $[X]$.

Let A be an operator with a closed range. Then $A^\dagger \in [X]$. However the map $A \to A^\dagger$ of $[X]$ into $[X]$ is not continous, even in the case when X is a finite dimensional space. (This is in contrast to the case of invertible operators, where the map $A \to A^{-1}$ is continous.) Thus if A_n is a sequence in $[X]$ which converges to A, it is not necessarily true that $A_n^\dagger$ converges to $A^\dagger$. Approximations (e.g. by numerical quadratures) to least-squares of an integral equation of the second kind give rise to difficulties which are not present in the case when the equation has a unique solution (for a thorough treatment of the latter case see the recent monograph of Anselone [1]). In other words, while the operator equation $Ax = y$, where A has a closed range, is well-posed in the space $[X]$, (i.e. it has a unique pseudosolution (for each y) which depends continuously on y), the pseudosolution does not depend continuously on A, even when the perturbated operators have closed range.

A general setting for approximating generalized inverses of linear operators on Banach spaces was presented in [9] and further developed in [10]. In this section we highlight the application of the theory in [10] to the approximation of least squares solutions of integral equations of the second kind in the space $X = C = C[a,b]$ equipped with the uniform norm. Consider

(7.1) $$Ax := (I-K)x = y \quad ,$$

where $N(A) \neq \{0\}$ and where

(7.2) $$(Kx)(s) = \int_a^b k(s,t)\, x(t)\, dt \; .$$

The integral operator in (7.2) will be regarded as both an operator K on C and an operator $\tilde{K}$ on $L_2 = L_2[a,b]$; thus $K = \tilde{K}\big|_C$. Besides (7.2) consider the equation in L_2:

(7.1)~ $$\tilde{A}x := (I-\tilde{K})x = y \; .$$

Relative to <u>orthogonal</u> projectors (in L_2) we consider the generalized inverse $\tilde{A}^{\dagger} = (I-\tilde{K})^{\dagger}$. From the assumptions below it will follow that $\tilde{A}^{\dagger}\big|_C$, the restriction to C of $\tilde{A}^{\dagger}$, is in fact the generalized inverse $A^{\dagger} = (I-K)^{\dagger}$ of A relative to the (restricted) orthogonal projectors. This observation enables us to exploit the geometry of Hilbert spaces and in particular the least-squares property of the generalized inverse in L_2 , while treating the equation (7.1) in the space C .

The basic assumptions are:

(A1) $K \in [C]$

(A2) $\tilde{K} \in [L_2]$ (and hence $\tilde{K}^* \in [L_2]$) .

(A3) $\tilde{K}$ and $\tilde{K}^*$ map L_2 into C .

(A4) For some $N > 0$, $\tilde{K}^N$ is a compact operator on L_2 ; consequently (A4a) $(I-\tilde{K})L_2$ is closed;
(A4b) $\dim N(I-\tilde{K})$ and $\operatorname{codim}(I-\tilde{K})L_2 = \dim((I-\tilde{K})L_2)^{\perp}$ are finite.

(As usual $S^{\perp}$ denotes the subspace orthogonal to S in L_2.)

These assumptions are satisfied if $k(s,t)$ is continuous, or mildly singular, for example $k(s,t) = \ln|s-t|$ or $|s-t|^r$ for $r > -1$.

Let $N = N(\tilde{A})$, $N^* = N(\tilde{A}^*) = N(I-\tilde{K}^*)$, $\tilde{R} = \tilde{A}L_2$, $R = AC$. By (A3)

$$N \subset C \;, \quad N^* \subset C \; . \tag{7.3}$$

$$\tilde{A}x = (I-\tilde{K})x \in C \Longleftrightarrow x \in C \;, \text{ that is } \tilde{R} \cap C = R \; . \tag{7.4}$$

By (A2) and (A4a) N and $\tilde{R}$ are closed, and $\tilde{R}^{\perp} = N^*$. Hence

$$L_2 = N \oplus N^{\perp} = \tilde{R} \oplus N^* \; . \tag{7.5}$$

Then (7.3)-(7.5) yield the <u>orthogonal</u> decomposition of C :

$$C = N \oplus (N^{\perp} \cap C) = R \oplus N^* \; . \tag{7.6}$$

Let $\tilde{E}$, $\tilde{U}$, $\tilde{P} = I-\tilde{U}$ be the (continuous) orthogonal projectors of L_2 onto $\tilde{R}$, N, $N^{\perp}$, respectively. Then by the above

$$(7.7) \qquad \tilde{E}\, x \in C \Leftrightarrow x \in C$$

$$(7.8) \qquad \tilde{P}\, x \in C \Leftrightarrow x \in C\ .$$

Let $\tilde{A}^{\dagger} = (I-\tilde{K})^{\dagger}$ be the generalized inverse of $\tilde{A}$ relative to the projectors $\tilde{E}$ and $\tilde{P}$. Then for any $y \in L_2$, $A^{\dagger}y$ is the least squares solution of minimal L_2-norm of (7.1)´, and the set S of all least squares solutions of (7.1)´ is given by $S = \tilde{A}^{\dagger}y + N$. Let $\tilde{K}$ satisfy (A2), (A3) and (A4a). Then

$$(7.9a) \qquad \tilde{A}^{\dagger}y = (I-\tilde{K})^{\dagger}y \in C \Leftrightarrow y \in C\ ,$$

$$(7.9b) \qquad y \in C \Longrightarrow S \subset C\ ;\ y \notin C \Longrightarrow S \cap C = \phi\ .$$

So far we have been developing results in L_2, that is for the equation (7.1)´. To apply the above to $A = I-K$ in the space C, note that $A = \tilde{A}\,\big|_C$, let $A^{\#} := \tilde{A}^{\dagger}\,\big|_C$ and let $P = \tilde{P}\,\big|_C$, $E = \tilde{E}\,\big|_C$ be the restricted projectors. In view of (7.4) and (7.9a), the restrictions of the Moore-Penrose relations for the generalized inverse $\tilde{A}^{\dagger}$ to C imply

$$(7.10) \begin{cases} (a)\quad A^{\#}A = P\ , & (b)\quad A^{\#}AA^{\#} = A^{\#} \\ (c)\quad AA^{\#} = E & (d)\quad AA^{\#}A = A\ . \end{cases}$$

Furthermore $A^{\#}$ is in fact $A^{\dagger}$. More precisely [10;Prop. 4.3] we have the following result which does not depend on (A4b).

<u>Lemma 7.1</u>. Let $K, \tilde{K}$ satisfy (A1), (A2), (A3), (A4a). Then $A^{\#}$ is the generalized inverse $A^{\dagger}$ of A relative to P and E.

We are now in a position to consider <u>convergence properties</u> of <u>quadrature approximations</u> for (7.1). Let the integral operator K in (7.2) be approximated (in C) by K_n defined by a quadrature rule with nodes t_{n_j} and weights w_{n_j}:

$$(7.11) \qquad (K_n x)(s) = \sum_{j=1}^{n} k(s,t_{n_j})x(t_{n_j})w_{n_j}$$

and suppose that

$$(7.12) \qquad K \text{ is compact and } \|K_n x - Kx\| \to 0,\ x \in C .$$

<u>Theorem 7.2</u>. Let K defined by (7.2) and K_n by (7.11), $n = 1,2,\cdots$, satisfy (7.12). Let the assumptions (A1) - (A4) above hold. Let P and E be the orthogonal projectors onto $M = PC = N^{\perp} \cap C$ and onto $AC = EC$, respectively, and let $A^{\dagger} = (I-K)^{\dagger}$ be the corresponding generalized inverse, as above. Let

$$H_n := (I-K)^{\dagger}(K_n-K)(I-E+EK_n), \text{ and } \theta_n = \|H_n\| .$$

Then for n sufficiently large, $\theta_n < 1$ and there exists an approximation A_n^{ϕ} (to $A^{\dagger}$) defined by

$$A_n^{\dagger} = (I-H_n)^{-1} \, C_n \quad ,$$

where

$$C_n = (I-K)^{\dagger}[I + (K_n-K)E] \quad ,$$

and $$\|A_n^{\phi} - C_n\| \leqq \frac{\theta_n}{1 - \theta_n} \|C_n\| \; .$$

The continuous linear operator A_n^{ϕ} from C onto M satisfies the following properties:

(a) $A_n^{\phi} \, A_n|_M = I_M$ so $A_n^{\phi} \, A_n \, A_n^{\phi} = A_n^{\phi}$ and $A_n \, A_n^{\phi} \, A_n|_M = A_n|_M$;

(b) $F_n \, := A_n \, A_n^{\phi}$, $Q_n \, := A_n^{\phi} \, A_n$ and $V_n \, := I-Q_n$ are projectors;

(c) $Q_n C = M$, $F_n C = A_n M$

(d) $A_n|_M$ is one-to-one, $(A_n|_M)^{-1} = A_n^{\phi}|_{F_n C}$.

(e) $N(A_n) \qquad N(F_n A_n) = V_n C$;

(f) $\dim N(A_n) \leqq \dim N(A)$;

(g) $A_n^{\phi} \, (I-E) = 0$ and $N(A^{\dagger}) = (I-E)C \qquad (I-F)C = N(A_n^{\phi})$;

(h) $\|A_n^{\phi}x - A^{\dagger}x\| \leqq \frac{\|A^{\dagger}\|}{1-\theta} \, (\theta\|x\| + \|(K-K_n)Ex\|) \to 0$, $n \to \infty$, $x \in C$

(i) $\|F_n x - Ex\| \leqq \|A_n\| \; \|A_n^\phi x - A^\dagger x\| + \|(K_n - K)A^\dagger x\| \to 0$,

$n \to \infty$, $x \in \mathcal{C}$.

(j) $(I-F_n)\mathcal{C} = (I-E)\mathcal{C}$.

Furthermore we have the following <u>asymptotic</u> <u>orthogality</u> <u>relations</u> as $n \to \infty$, for $x, x' \in \mathcal{C}$:

(k) $|\langle (I-F_n)x, F_n x' \rangle| \to 0$;

(ℓ) $|\langle Q_n x , (I-Q_n)x' \rangle| \to 0$.

By Lemma 7.1, $A^\dagger y = A^\# y = \tilde{A}^\dagger y$ is the least squares solution of minimal L_2-norm of $Ax = y$, $y \in \mathcal{C}$. Since E is an orthogonal projector, relation (j) of Theorem 7.2 implies that $(I-F_n)\mathcal{C}$ is orthogonal to $E\mathcal{C} = A\mathcal{C}$. Thus for $x_n = A_n^\phi y$, the residual $y - A_n x_n = y - A_n A_n^\phi y = (I-F_n)y$ is orthogonal to $E\mathcal{C}$. Also by (k), $(I-F_n)\mathcal{C}$ is asymptotically orthogonal to $F_n\mathcal{C}$. Thus x_n is asymptotically a least squares solution of $A_n x = y$. Furthermore x_n is orthogonal to N since $x_n = A_n^\phi y \in Q_n\mathcal{C} = M = N^\perp \cap \mathcal{C}$. Since by (ℓ) $Q_n\mathcal{C}$ is asymptotically orthogonal to $(I-Q_n)\mathcal{C}$, x_n is asymptotically an element of minimal L_2-norm among the solutions of $A_n x = F_n y$.

REFERENCES

[1] P. M. Anselone, *Collectively Compact Operator Approximation Theory*, Prentice-Hall, Englewood Cliffs, N. J., 1971.

[2] N. Aronszajn, Theory of reproducing kernels, *Trans. Amer. Math. Soc.* 68 (1950), pp. 337-404.

[3] J. N. Franklin, Well-posed stochastic extensions of ill-posed linear problems, *J. Math. Anal. Appl.* 31 (1970), pp. 682-716.

[4] J. W. Hilgers, Non-iterative methods for solving operator equations of the first kind, MRC Tech. Summary Report #1413, University of Wisconsin-Madison, January 1974.

[5] W. J. Kammerer and M. Z. Nashed, Iterative methods for best approximate solutions of linear integral equations of the first and second kinds, *J. Math. Anal. Appl.* 40 (1972), pp. 547-573.

[6] M. M. Lavrentiev, *Some Improperly Posed Problems of Mathematical Physics*, Izdat. Sibirsk. Otdel. Akad. Nauk SSSR, Novosibirsk, 1962; English Transl., Springer-Verlag Tracts in Natural Philosophy, Vol. II, Springer-Verlag, Berlin, 1967.

[7] M. M. Lavrentiev, Numerical solution of conditionally properly posed problems, in *Numerical Solution of Partial Differential Equations*-II SYNSPADE (1970), B. Hubbard, ed., Academic Press, New York-London, 1971, pp. 417-432.

[8] R. Lattes and J. L. Lions, *Theory and Applications of the Method of Quasi-Reversibility*, American Elsevier Publishing Co., New York, 1969.

[9] R. H. Moore and M. Z. Nashed, Approximation of generalized inverses of linear operators in Banach spaces, in Approximation Theory, G. G. Lorentz, ed., Academic Press, New York, 1973, pp. 425-429.

[10] R. H. Moore and M. Z. Nashed, Approximations to generalized inverses of linear operators, SIAM J. Appl. Math. 26 (1974).

[11] V. A. Morozov, Convergence of an approximate method of solving operator equations of the first kind, Zh. vȳchisl. Mat. mat. Fiz. 13 (1973), pp. 3-17.

[12] V. A. Morozov, Optimal regularization of operator equations, ibid. 10 (1970), pp. 818-829.

[13] V. A. Morozov, Error estimates for the solution of an incorrectly posed problem involving unbounded linear operators, ibid. 10 (1970), pp. 1081-1091.

[14] M. Z. Nashed, Generalized inverses, normal solvability, and iteration for singular operator equations, in Nonlinear Functional Analysis and Applications, L. B. Rall, ed., Academic Press, New York, 1971, pp. 311-359.

[15] M. Z. Nashed, Some aspects of regularization and approximation of solutions of ill-posed operator equations, Proceedings of the 1972 Army Numerical Analysis Conference, pp. 163-181.

[16] M. Z. Nashed and G. F. Votruba, A unified approach to generalized inverses of linear operators: I. Algebraic, topological, and projectional properties, Bull. Amer. Math. Soc. 80 (1974).

[17] M. Z. Nashed and G. F. Votruba, A unified approach to generalized inverses of linear operators: II. Extremal and proximal properties, Bull. Amer. Math. Soc. 80 (1974).

[18] M. Z. Nashed and Grace Wahba, Convergence rates of approximate least squares solutions of linear integral and operator equations of the first kind, Math. Comp. 28 (1974), pp. 69-80.

[19] M. Z. Nashed and Grace Wahba, Generalized inverses in reproducing kernel spaces: An approach to regularization of linear operator equations, SIAM J. Math. Anal., 5 (1974), to appear.

[20] M. Z. Nashed and Grace Wahba, Approximate regularized pseudosolutions of linear operator equations when the data-vector is not in the range of the operator, to appear.

[21] M. Z. Nashed, ed., Generalized Inverses and Applications, Academic Press, New York, 1975.

[22] M. Z. Nashed, Regularization and approximations of ill-posed linear operator equations (An expanded version of an invited address delivered to the 710th meeting of the American Mathematical Society, November 16, 1973), Bull. Amer. Math. Soc., to be submitted.

[23] L. E. Payne, Some general remarks on improperly posed problems for partial differential equations, in Symposium on Non-Well-Posed Problems and Logarithmic Convexity, Lecture Notes in Mathematics Vol. 316, Springer-Verlag, Berlin, 1973, pp. 1-30.

[24] L. E. Payne, Improperly Posed Problems in Partial Differential Equations (An expanded version of a series of lectures given at an NSF Regional Conference, The University of New Mexico, May 20-24, 1974), to be published by SIAM in the CBSM Regional Conference Series in Applied Mathematics.

[25] E. Picard, Sur un théorme générale relatif aux equations integrales de premiére especes et sur quel ques problemes de physique mathematiques, _R. C. Mat. Palermo_ 29 (1970), pp. 615-619.

[26] H. L. Shapiro, _Topics in Approximation Theory_, Lecture Notes in Mathematics, Springer-Verlag, Berlin, 1970.

[27] A. N. Tikhonov, On the solution of incorrectly formulated problems and the regularization method, _Dokl. Akad. Nauk_ SSSR 151 (1963), pp. 501-504; _Soviet Math._ 4 (1963), pp. 1035-1038.

[28] G. Wahba, Convergence rates for certain approximate solutions to Fredholm integral equations of the first kind, _J. Approximation Theory_ 7 (1973), pp. 167-185.

A Majorization Technique for Hyperbolic Equations

EDWARD NEWBERGER[†]

Department of Mathematics
SUNY, College at Buffalo
Buffalo, New York 14222, U.S.A.

Classes of infinitely differentiable functions in the sense of Mandelbrojt are introduced. These classes, called asymptotic Gevrey classes, give finer asymptotic estimates for infinitely differentiable functions than Gevrey classes, the asymptoticity being with respect to increasing order of differentiation. The Cauchy problem for a family of hyperbolic operators is considered, with data in these asymptotic Gevrey classes. A technique of double majorization, originally introduced by J. Leray and Y. Ohya is used. The first majorization scheme is based on Gårding's inequality for hyperbolic operators. The second scheme, which majorizes the first majorized problem, reduces the problem to an analytic one.

1. INTRODUCTION

This paper is an application of a majorization technique for hyperbolic equations and systems which was introduced by J. Leray and Y. Ohya [5].

We will work throughout the paper with a "special" equation; however, this equation exhibits all of the features of the general problem — with considerable reduction in notational and technical difficulties.

[†]The initial formulation of many of the ideas in this paper was supported by the Air Force Office of Scientific Research through AFOSR - Grant 1206-67.

Let t, x be coordinates on R_2. Let T be a positive real number. Let X denote the strip in R_2 given by $0 \leq t \leq T$.

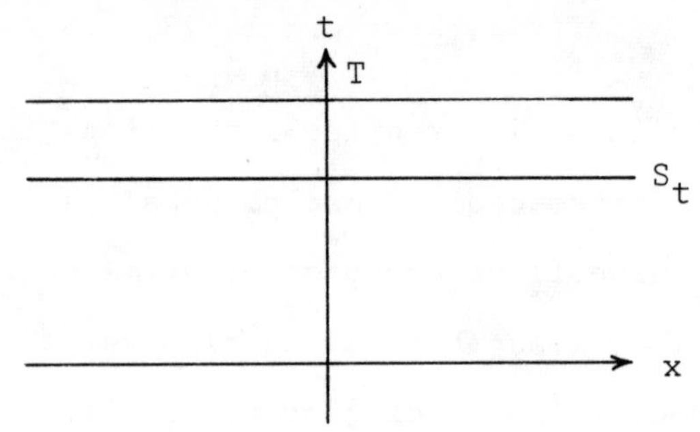

S_t denotes an arbitrary (horizontal) line in the strip X. Consider the Cauchy problem

$$\begin{aligned} &(D_t^p - bD_x^q)u = v \\ &u|S_0 = D_t u|S_0 = \cdots = D_t^{p-1}u|S_0 = 0 \end{aligned} \tag{1.1}$$

where $p > q > 0$; $b = b(t,x)$ and $v = v(t,x)$ are complex-valued measurable functions defined almost everywhere on X. Let n be an integer $\geq p$ and let α be a real number such that $1 < \alpha < \frac{p}{q}$.

We now define the norms $|\ |_{2,n}$ and $|\ |_{\infty,n}$.

$$|v|_{2,n} = \max_{\mu+\nu \leq n} \ \operatorname*{ess\,sup}_{t \in [0,T]} \left(\int_{S_t} |D_t^\mu D_x^\nu v|^2 \, dx \right)^{\frac{1}{2}}$$

$$|b|_{\infty,n} = \max_{\mu+\nu \leq n} \ \operatorname*{ess\,sup}_{\substack{S_t \\ t \in [0,T]}} |D_t^\mu D_x^\nu b| \ .$$

Then the existence theorem of Leray and Ohya [5, Section 23] for (1.1) states that if (1.2) and (1.3) hold, then so does (1.4):

$$\overline{\lim_{s\to\infty}} \frac{1}{s^{\alpha}} |D_x^s v|_{2,n}^{\frac{1}{s}} < \infty \tag{1.2}$$

$$\overline{\lim_{s\to\infty}} \frac{1}{s^{\alpha}} |D_x^s b|_{\infty,n}^{\frac{1}{s}} < \infty \tag{1.3}$$

$$\overline{\lim_{s\to\infty}} \frac{1}{s^{\alpha}} |D_x^s u|_{2,p+n}^{\frac{1}{s}} < \infty . \tag{1.4}$$

Of course, we are assuming that v and b have the correct regularity properties so that the norms $|D_x^s v|_{2,n}$, $|D_x^s b|_{\infty,n}$ are well defined for all integers $s \geq 0$.

Suppose however that v and b, instead of satisfying (1.2) and (1.3) respectively, satisfy instead

$$\overline{\lim_{s\to\infty}} \left(\frac{\log s}{s^{\alpha}}\right) |D_x^s v|_{2,n}^{\frac{1}{s}} < \infty \tag{1.5}$$

$$\overline{\lim_{s\to\infty}} \left(\frac{\log s}{s^{\alpha}}\right) |D_x^s b|_{\infty,n}^{\frac{1}{s}} < \infty \tag{1.6}$$

Of course, (1.5) and (1.6) are stronger than (1.2) and (1.3) respectively, so conclusion (1.4) still follows. But it would be desirable to prove that

$$\overline{\lim_{s\to\infty}} \left(\frac{\log s}{s^{\alpha}}\right) |D_x^s u|_{2,p+n}^{\frac{1}{s}} < \infty .$$

This is an example of the type of result that we are after.[1]

2. INTERPOLATIVE INEQUALITIES

LEMMA 2.1. Let $f \in C^2(R)$ and let f, $f^{(1)}$ and $f^{(2)} \in L_2(R)$. Then

$$\int_R |f^{(1)}|^2 \, dx \le \left(\int_R |f|^2 \, dx\right)^{\frac{1}{2}} \left(\int_R |f^{(2)}|^2 \, dx\right)^{\frac{1}{2}} .$$

Proof. Let a and b be finite real numbers such that $a < b$. Since $f \in C^2(R)$, it is permissible to integrate $\int_a^b f^{(1)}\overline{f}^{(1)} \, dx$ by parts. We have

$$\int_a^b f^{(1)}\overline{f}^{(1)} \, dx = E(f,a,b) - \int_a^b \overline{f}f^{(2)} \, dx \tag{2.1}$$

where $E(f,a,b) = f^{(1)}(b)\overline{f}(b) - f^{(1)}(a)\overline{f}(a)$. We will prove that

$$\lim_{\substack{a\to-\infty \\ b\to\infty}} E(f,a,b) = 0 . \tag{2.2}$$

However, first suppose that (2.2) is proved. Then letting $a \to -\infty$ and $b \to \infty$ in (2.1), we get $\int_R |f^{(1)}|^2 \, dx = -\int_R \overline{f}f^{(2)} \, dx$ which is $\le \left(\int_R |f|^2 \, dx\right)^{\frac{1}{2}} \cdot \left(\int_R |f^{(2)}|^2 \, dx\right)^{\frac{1}{2}}$ by the Schwarz inequality.

Proof of (2.2). We first prove that

$$f(x) \to 0 \quad \text{as} \quad x \to \infty . \tag{2.3}$$

[1] That an extension of the results of [5] to non-quasi-analytic classes other than Gevrey classes might be possible was suggested to me by Professor Leray.

Let N be a positive integer. Since $f \in L_2(R)$, it follows that

$$\left(\int_N^{N+1} |f|^2 \, dx \right)^{\frac{1}{2}} \to 0 \quad \text{as} \quad N \to \infty . \tag{2.4}$$

Similarly, since $f^{(1)} \in L_2(R)$, it follows that

$$\left(\int_N^{N+1} |f^{(1)}|^2 \, dx \right)^{\frac{1}{2}} \to 0 \quad \text{as} \quad N \to \infty . \tag{2.5}$$

By Sobolev's inequality,

$$\max_{x \in [N,N+1]} |f| \le C \max_{i=0,1} \left(\int_N^{N+1} |f^{(i)}|^2 \, dx \right)^{\frac{1}{2}} \tag{2.6}$$

where C is an absolute constant. (2.4), (2.5) and (2.6) imply (2.3).

In the same way as (2.3) was proved, it also follows that $f(x) \to 0$ as $x \to -\infty$ and that $f^{(1)}(x) \to 0$ as $x \to \infty$ and as $x \to -\infty$. This together with (2.3) implies (2.2).

LEMMA 2.2. Let $k > 1$. Let $b_0, b_1, \ldots, b_k$ be positive real numbers satisfying

$$b_s^2 \le b_{s-1} b_{s+1} , \quad 0 < s < k . \tag{2.7}$$

Then for all positive integers j such that $j < k$, we have

$$b_j \le b_0^{1 - \frac{j}{k}} \cdot b_k^{\frac{j}{k}} \tag{2.8}$$

Proof. From (2.7), we have

$$\frac{b_1}{b_0} \le \frac{b_2}{b_1} \le \cdots \le \frac{b_k}{b_{k-1}} . \tag{2.9}$$

From (2.9), we have

$$\frac{b_k}{b_j} = \frac{b_{j+1}}{b_j}\frac{b_{j+2}}{b_{j+1}} \cdots \frac{b_k}{b_{k-1}} \ge \left(\frac{b_{j+1}}{b_j}\right)^{k-j}$$

and

$$\frac{b_j}{b_0} = \frac{b_1}{b_0}\frac{b_2}{b_1} \cdots \frac{b_j}{b_{j-1}} \le \left(\frac{b_{j+1}}{b_j}\right)^{j} .$$

It follows that $\displaystyle \frac{b_j}{b_0} \le \left(\frac{b_k}{b_j}\right)^{\frac{j}{k-j}}$.

This implies (2.8).

PROPOSITION 2.1. Let $f \in C^k(R)$, $k > 1$. Let $f^{(s)} \in L_2(R)$ for all s such that $s \le k$. Then for all positive integers j such that $j < k$, we have

$$\|f^{(j)}\|_2 \le \|f\|_2^{1-\frac{j}{k}} \cdot \|f^{(k)}\|_2^{\frac{j}{k}} . \tag{2.10}$$

Proof.[2] Let $b_s = \|f^{(s)}\|_2$, $0 \le s \le k$. Since 0 is the only constant in $L_2(R)$, it follows that either $b_0 = b_1 = \cdots = b_k = 0$, in which case (2.10) holds trivially, or $b_s > 0$ for all $s = 0,1,\ldots,k$, which we now assume. It follows that if s is a positive integer less than k, then from

---

[2]The interpolative inequality (2.10) is well known [9, Lecture II], however with a multiplicative constant C on the right hand side that may depend on j and k. Here $C = 1$.

lemma 2.1

$$\int_R |f^{(s)}|^2\,dx \le \left(\int_R |f^{(s-1)}|^2\,dx\right)^{\frac{1}{2}} \cdot \left(\int_R |f^{(s+1)}|^2\,dx\right)^{\frac{1}{2}},$$

i.e., that (2.7) holds. Therefore, by lemma 2.2, proposition 2.1 follows.

COROLLARY 2.1. Let $p \in \{2,\infty\}$. Let $f \in C^k(R)$, $k > 1$. Let $f^{(s)} \in L_2(R)$ for all s such that $s \le k$. Then for all positive integers j such that $j < k$, we have

$$\|f^{(j)}\|_p \le C(p)\|f\|_p^{1-\frac{j}{k}} \cdot \|f^{(k)}\|_p^{\frac{j}{k}} .$$

$C(2) = 1$ and $C(\infty) = \frac{\pi}{2}$.

The case $p = \infty$ is proved in Kolmogoroff [4].

3. MANDELBROJT CLASSES.

Let $p \in \{2,\infty\}$. Let M_s, $s \ge 0$ be a sequence of positive real numbers. Let $f^{(s)} \in L_p(R)$ for all non-negative integers s. We say that $f \in \mathscr{C}(p,M_s)$ if

$$\overline{\lim_{s\to\infty}} \left(\frac{\|f^{(s)}\|_p}{M_s}\right)^{\frac{1}{s}} < \infty .$$

We also say that the sequence M_s, $s \ge 0$ determines the class $\mathscr{C}(p,M_s)$.

A standard problem in classes of this type is: when do two sequences M_s, M_s', $s \ge 0$ of positive real numbers determine the same class. It is to this end that propositions 3.1 and 3.2 are directed.

PROPOSITION 3.1. Let $p \in \{2,\infty\}$. Let M_s, $s \geq 0$ be a sequence of positive real numbers satisfying $M_s^{\frac{1}{s}} \to \infty$ as $s \to \infty$. Then there exists a logarithmically convex sequence M'_s, $s \geq 0$ of positive real numbers[3] such that

$$\mathscr{C}(p,M_s) = \mathscr{C}(p,M'_s) .$$

The sequence M'_s, $s \geq 0$ is independent of the choice of $p \in \{2,\infty\}$. Furthermore the sequence M'_s, $s \geq 0$ satisfies

$$M'_0 = M'_1 = 1, \quad {M'_s}^{\frac{1}{s}} \to \infty \quad \text{as} \quad s \to \infty .$$

This proposition makes essential use of the interpolative inequalities of corollary 2.1; the proof is based on Newton's polygon method and the argument is standard (see [8]).

REMARK 3.1. If N_s, $s \geq 0$ is a logarithmically convex sequence of positive real numbers and if $N_0 = 1$, then

$$N_r N_s \leq N_{r+s} \quad \text{for all integers} \quad r,s \geq 0 .$$

Proof. We have $N_s^2 \leq N_{s-1}N_{s+1}$ for all $s \geq 1$; this implies that the sequence $\frac{N_{s+1}}{N_s}$, $s \geq 0$ is non-decreasing. It follows that the sequence $\frac{N_{s+r}}{N_s}$, $s \geq 0$ is non-decreasing for each fixed integer $r \geq 1$. Therefore

$$\frac{N_{s+r}}{N_s} \geq \frac{N_r}{N_0} = N_r .$$

---

[3] This means that the sequence $\log M'_s$, $s \geq 0$ is convex.

LEMMA 3.1. Let M_s, $s \geq 0$ be a sequence of positive real numbers satisfying $M_s^{\frac{1}{s}} \to \infty$ as $s \to \infty$. Then $\mathscr{C}(\infty, M_s)$ is an algebra.

Proof. By proposition 3.1, we may regard the sequence M_s, $s \geq 0$ as logarithmically convex and such that $M_0 = 1$. Therefore, by remark 3.1,

$$M_r M_s \leq M_{r+s} \quad \text{for all} \quad r,s \geq 0 . \tag{3.1}$$

Let $f,g \in \mathscr{C}(\infty, M_s)$. From Leibnitz's formula, we have

$$\|(fg)^{(s)}\|_\infty \leq \sum_{i+j=s} \frac{s!}{i!j!} \|f^{(i)}\|_\infty \cdot \|g^{(j)}\|_\infty \tag{3.2}$$

(3.1), (3.2) and the equality $\sum_{i+j=s} \frac{s!}{i!j!} = 2^s$ imply that $fg \in \mathscr{C}(\infty, M_s)$.

PROPOSITION 3.2. Let M_s, $s \geq 0$ be a logarithmically convex sequence of positive real numbers satisfying $M_s^{\frac{1}{s}} \to \infty$ as $s \to \infty$. Let M'_s, $s \geq 0$ be another sequence of positive real numbers.

(A) If $\mathscr{C}(\infty, M_s) \subset \mathscr{C}(\infty, M'_s)$, then $\overline{\lim_{s\to\infty}} \left(\frac{M_s}{M'_s}\right)^{\frac{1}{s}} < \infty$.

(B) Let $\mathscr{C}(2, M'_s)$ be differentiable, i.e., suppose $f^{(1)} \in \mathscr{C}(2, M'_s)$ whenever $f \in (2, M'_s)$. Suppose $\mathscr{C}(2, M_s)$ is non-quasi-analytic, i.e., $\mathscr{C}(2, M_s)$ contains a non-zero function φ with compact support. If $\mathscr{C}(2, M_s) \subset \mathscr{C}(2, M'_s)$, then $\overline{\lim_{s\to\infty}} \left(\frac{M_s}{M'_s}\right)^{\frac{1}{s}} < \infty$.

Discussion of (A). By proposition 3.1, it is no loss of generality to assume that the sequence M_s, $s \geq 0$ is logarithmically convex. In [8], it is proved that if

$$\overline{\lim_{s\to\infty}} \left(\frac{M_s}{M'_s}\right)^{\frac{1}{s}} = \infty ,$$

then there exists a periodic function f such that

$$f \in \mathscr{C}(\infty, M_s) \quad \text{but} \notin \mathscr{C}(\infty, M'_s) . \tag{3.3}$$

We can assume that f has period 2.

Proof of (B). Suppose that

$$\overline{\lim_{s\to\infty}} \left(\frac{M_s}{M'_s}\right)^{\frac{1}{s}} = \infty .$$

We can assume that

$$\operatorname{supp} \varphi \subset [-1,1], \quad \varphi \geq 0, \quad \int_R \varphi \, dx = 1 . \tag{3.4}$$

Define $F(x) = f(x)\psi(x)$ where

$$\psi(x) = \int_{-2}^{2} \varphi(x-y) \, dy . \tag{3.5}$$

Proof that $F \in \mathscr{C}(2, M_s)$. It follows from (3.4) and (3.5) that

$$\begin{aligned} \psi(x) &= 1 \quad \text{if} \quad x \in [-1,1] \\ &= 0 \quad \text{if} \quad x \notin [-3,3] . \end{aligned}$$

From (3.5), we have

$$\psi^{(s)}(x) = \int_{x-2}^{x+2} \varphi^{(s)}(y)\, dy ,$$

so

$$|\psi^{(s)}(x)| \le 2\left(\int_{x-2}^{x+2} |\varphi^{(s)}(y)|^2\, dy\right)^{\frac{1}{2}}$$

$$\le 2\|\varphi^{(s)}\|_2 .$$

Since $\varphi \in \mathscr{C}(2,M_s)$, it follows that $\psi \in \mathscr{C}(\infty,M_s)$. Therefore $F = f\psi \in \mathscr{C}(\infty,M_s)$ since $\mathscr{C}(\infty,M_s)$ is an algebra by lemma 3.1. Now F has compact support since ψ does; therefore $F \in \mathscr{C}(2,M_s)$.

Proof that $F \notin \mathscr{C}(2,M_s')$. Suppose that $F \in \mathscr{C}(2,M_s')$. Then so does $F^{(1)}$. By Sobolev's inequality applied to $F^{(s)}$, we have

$$|F^{(s)}(x)| \le C \max(\|F^{(s)}\|_2, \|F^{(s+1)}\|_2) \tag{3.6}$$

for all $x \in R$ where C is an absolute constant. Since (3.6) holds for all $x \in R$, it holds in particular for all $x \in [-1,1]$. But $F = f$ for $x \in [-1,1]$; and since f is periodic of period 2, it follows from the inequality (3.6) that

$$|f^{(s)}(x)| \le C \max(\|F^{(s)}\|_2, \|F^{(s+1)}\|_2) \tag{3.7}$$

for all $x \in R$. From (3.7) together with the fact that F and $F^{(1)} \in \mathscr{C}(2,M_s')$, we have that $f \in \mathscr{C}(\infty,M_s')$. But this contradicts (3.3).

4. ASYMPTOTIC GEVREY CLASSES

Let $p \in \{2,\infty\}$. Let α be a real number > 1. Let $f^{(s)} \in L_p(R)$ for all non-negative integers s. We say that $f \in \gamma_p^{(\alpha)}(R)$ if

$$\overline{\lim_{s\to\infty}} \frac{1}{s^\alpha} \left\| f^{(s)} \right\|_p^{\frac{1}{s}} < \infty .$$

$\gamma_p^{(\alpha)}(R)$ is called a Gevrey class α and it is clear that $\gamma_p^{(\alpha)}(R) = \mathscr{C}(p, M_s)$ where

$$M_s = M_s(\alpha) = s^{\alpha s} \quad \text{if} \quad s \geq 1$$
$$= 1 \quad \text{if} \quad s = 0 .$$

Properties of $\gamma_p^{(\alpha)}(R)$.

(A) $\gamma_p^{(\alpha)}(R)$ is differentiable. Also, $\gamma_p^{(\alpha)}(R)$ is non-quasi-analytic, i.e , it contains a non-zero function with compact support. Such a function φ (see [3]) is constructed as follows:

given α, let $a > 0$ be such that $\alpha > 1 + \frac{1}{a}$. Let $\varphi(x) = \psi(1+x)\psi(1-x)$ where

$$\psi(x) = e^{-2x^{-a}} \quad \text{if} \quad x > 0$$
$$= 0 \quad \text{if} \quad x \leq 0 .$$

(B) The class $\gamma_p^{(\alpha)}(R)$ is strictly increasing with α, i.e., if $1 < \alpha' < \alpha''$, then

$$\gamma_p^{(\alpha')}(R) \underset{\neq}{\subset} \gamma_p^{(\alpha'')}(R) ;$$

this follows from proposition 3.2.

In view of (B), we make the following

DEFINITION 4.1. Let $p \in \{2,\infty\}$. Let M_s, $s \geq 0$ be a sequence of positive real numbers. Let α be a fixed real number > 1. We say that the class $\mathscr{C}(p,M_s)$ is an asymptotic Gevrey class α if (4.1) and (4.2) hold:

$$\gamma_p^{(\alpha')}(R) \subset \mathscr{C}(p,M_s) \subset \gamma_p^{(\alpha'')}(R) \tag{4.1}$$

for all real numbers α', α'' such that $1 < \alpha' < \alpha < \alpha''$, i.e., $\mathscr{C}(p,M_s)$ contains every Gevrey class contained in the Gevrey class α; and $\mathscr{C}(p,M_s)$ is contained in every Gevrey class containing the Gevrey class α.

$$\mathscr{C}(p,M_s) \neq \gamma_p^{(\alpha)}(R) , \tag{4.2}$$

i.e., $\mathscr{C}(p,M_s)$ is not the Gevrey class α.

It is clear that such a class is non-quasi-analytic.

REMARK 4.1. Let $p \in \{2,\infty\}$. Let M_s, $s \geq 0$ be a sequence of positive real numbers such that

$$\varliminf_{s\to\infty} M_s^{\frac{1}{s}} < \infty . \tag{4.3}$$

Then $\mathscr{C}(p,M_s)$ is not an asymptotic Gevrey class α for any $\alpha > 1$.

Proof. From (4.3), there is a strictly increasing sequence s_i, $i \geq 0$ with $s_0 = 0$ and a constant $c > 0$ such that

$$M_{s_i} \leq c^{s_i} \quad \text{for all } i \geq 0 .$$

From corollary 2.1, it easily follows that $\mathscr{C}(p,M_s) \subset \mathscr{C}(p,N_s)$ where $N_s = 1$ for all $s \geq 0$. But $\mathscr{C}(p,N_s)$ contains no non-zero function with compact support. Therefore $\mathscr{C}(p,M_s)$ cannot be an asymptotic Gevrey class α for any $\alpha > 1$.

THEOREM 4.1. Let $p \in \{2,\infty\}$. Let M_s, $s \geq 0$ be a logarithmically convex sequence of positive real numbers satisfying

$$M_s^{\frac{1}{s}} \to \infty \quad \text{as} \quad s \to \infty\,; \quad M_0 = M_1 = 1\,.$$

In the case $p = 2$, assume that the class $\mathscr{C}(2,M_s)$ is differentiable. Let α be a fixed real number > 1. Define a sequence $g(s)$, $s > 1$ of real numbers by

$$M_s = s^{s[\alpha+g(s)]} \qquad \text{for all} \quad s > 1\,.$$

Then $\mathscr{C}(p,M_s)$ is an asymptotic Gevrey class α if and only if (4.4) and (4.5) hold:

$$g(s) \to 0 \quad \text{as} \quad s \to \infty \tag{4.4}$$

$$\overline{\lim_{s\to\infty}}\, |g(s)| \log s = \infty\,. \tag{4.5}$$

Proof. From proposition 3.2, the following four equivalences are immediate:

Let α' be such that $1 < \alpha' < \alpha$. Then

(1) $\gamma_p^{(\alpha')}(R) \subset \mathscr{C}(p,M_s)$

$\Leftrightarrow$

(1') $E_1(\alpha') = \overline{\lim_{s\to\infty}} \dfrac{1}{s^{\alpha-\alpha'+g(s)}} < \infty$.

Let $\alpha'' > \alpha$. Then

(2) $\mathscr{C}(p,M_s) \subset \gamma_p^{(\alpha'')}(R)$

$\Leftrightarrow$

(2') $E_2(\alpha'') = \overline{\lim\limits_{s\to\infty}}\, s^{g(s)-(\alpha''-\alpha)} < \infty$

(3) $\gamma_p^{(\alpha)}(R) \subset \mathscr{C}(p,M_s) \Leftrightarrow \overline{\lim\limits_{s\to\infty}} \frac{1}{s^{g(s)}} < \infty$

(4) $\mathscr{C}(p,M_s) \subset \gamma_p^{(\alpha)}(R) \Leftrightarrow \overline{\lim\limits_{s\to\infty}}\, s^{g(s)} < \infty$.

From these four equivalences, we now prove theorem 4.1.

Proof that (4.2) $\Leftrightarrow$ (4.5).

From (3) and (4), $\mathscr{C}(p,M_s) = \gamma_p^{(\alpha)}(R) \Leftrightarrow$

$\overline{\lim\limits_{s\to\infty}} -g(s) \log s < \infty$ and $\overline{\lim\limits_{s\to\infty}}\, g(s) \log s < \infty \Leftrightarrow$

$\overline{\lim\limits_{s\to\infty}} |g(s)| \log s < \infty$.

Proof that (4.4) $\Rightarrow$ (4.1).

(4.4) $\Rightarrow E_1(\alpha') = E_2(\alpha'') = 0$ for all α', α'' such that

$$1 < \alpha' < \alpha < \alpha'' .$$

Proof that (4.1) $\Rightarrow$ (4.4).

The left inclusion relation in (4.1) implies that $E_1(\alpha') < \infty$ for all α' such that $1 < \alpha' < \alpha$. Let α' be fixed and let $\varepsilon = \alpha - \alpha'$. Then

$$\overline{\lim\limits_{s\to\infty}} \frac{1}{s^{\varepsilon+g(s)}} < \infty . \tag{4.6}$$

From (4.6), there is a constant $C(\varepsilon) > 1$ such that

$$\frac{1}{s^{\varepsilon+g(s)}} \le C(\varepsilon) \qquad \text{for all} \quad s > 1 . \tag{4.7}$$

Let $C_1(\varepsilon) = \log C(\varepsilon)$. From (4.7), it follows that

$$g(s) \ge -\varepsilon - \frac{C_1(\varepsilon)}{\log s} \qquad \text{for all} \quad s > 1 . \tag{4.8}$$

(4.8) implies that $\underline{\lim}_{s\to\infty} g(s) \ge -\varepsilon$.

Since $\varepsilon = \alpha - \alpha'$ can be an arbitrarily small positive number, it follows that

$$\underline{\lim}_{s\to\infty} g(s) \ge 0 . \tag{4.9}$$

The right inclusion relation in (4.1) implies that $E_2(\alpha'') < \infty$ for all α'' such that $\alpha'' > \alpha$. Let α'' be fixed and let $\eta = \alpha'' - \alpha$. Then

$$\overline{\lim}_{s\to\infty} s^{g(s)-\eta} < \infty . \tag{4.10}$$

From (4.10), there is a constant $C(\eta) > 1$ such that

$$s^{g(s)-\eta} \le C(\eta) \qquad \text{for all} \quad s > 1 . \tag{4.11}$$

Let $C_1(\eta) = \log C(\eta)$. From (4.11), it follows that

$$g(s) \le \eta + \frac{C_1(\eta)}{\log s} \qquad \text{for all} \quad s > 1 . \tag{4.12}$$

(4.12) implies that $\overline{\lim}_{s\to\infty} g(s) \le \eta$. Since $\eta = \alpha'' - \alpha$ can be an arbitrarily small positive number, it follows that

$$\overline{\lim}_{s\to\infty} g(s) \le 0 . \tag{4.13}$$

(4.13) and (4.9) imply (4.4).

Notation. Suppose $\mathscr{C}(2,M_s)$ is differentiable. In accordance with theorem 4.1 (and remark 4.1), if $\mathscr{C}(p,M_s)$ is an asymptotic Gevrey class α, we may denote it by $\gamma_p^{(\alpha+g)}(R)$.

5. CLASSES IN WHICH THE DATA LIE.

With the discussion in the Introduction (section 1) as motivation, we now wish to extend the development of sections 2, 3 and 4. More precisely, we will merely indicate how the extension proceeds — as the only difficulties of the extension are notational.

Let $f = f(t,x)$ be a measurable function defined almost everywhere on the strip X (see section 1). Let n be a non-negative integer.

We say that f has the regularity property $R_{2,n}$ if the following norm is well-defined:

$$|f|_{2,n} = \max_{\mu+\nu\leq n} \operatorname*{ess\,sup}_{t\in[0,T]} \left(\int_{S_t} |D_t^\mu D_x^\nu f|^2 \, dx \right)^{\frac{1}{2}} .$$

This means that the restriction of each derivative of f of total order $\leq n$ to an arbitrary (horizontal) line S_t in the strip X is in L_2 for almost all $t \in [0,T]$; and furthermore, the L_2 norm of $D_t^\mu D_x^\nu f|S_t$, considered as a function of $t \in [0,T]$, is in $L_\infty[0,T]$.

Similarly, we say that f has the regularity property $R_{\infty,n}$ if the following norm is well-defined:

$$|f|_{\infty,n} = \max_{\mu+\nu\leq n} \operatorname*{ess\,sup}_{\substack{S_t \\ t\in[0,T]}} |D_t^\mu D_x^\nu f| .$$

The explicit statement of the regularity property $R_{\infty,n}$ is directly analogous to that for $R_{2,n}$, with the L_2 norm now being replaced by the corresponding L_∞ norm.

Let $p \in \{2,\infty\}$, $n \geq 0$. Let M_s, $s \geq 0$ be a sequence of positive real numbers. Let f be a measurable function defined almost everywhere on X. Suppose that all spatial derivatives (i.e., derivatives with respect to x) of f have the regularity property $R_{p,n}$.

Then we are interested in the class determined by the condition

$$\overline{\lim_{s\to\infty}} \left(\frac{|D_x^s f|_{p,n}}{M_s} \right)^{\frac{1}{s}} < \infty . \tag{5.1}$$

If
$$M_s = s^{\alpha s}, \quad s \geq 1, \quad \alpha \text{ real}, > 1$$
$$= 1 , \quad s = 0$$

the corresponding class is called a Gevrey class α and is denoted by $\gamma_p^{n,(\alpha)}(X)$.

In analogy with definition 4.1, we have

DEFINITION 5.1. Let $p \in \{2,\infty\}$, $n \geq 0$. Let M_s, $s \geq 0$ be a sequence of positive real numbers. Let α be a fixed real number > 1. Let $\mathscr{C} = \mathscr{C}(p,n,M_s,X)$ denote the class determined by (5.1). We say that $\mathscr{C}$ is an asymptotic Gevrey class α if (5.2) and (5.3) hold:

$$\gamma_p^{n,(\alpha')}(X) \subset \mathscr{C} \subset \gamma_p^{n,(\alpha'')}(X) \tag{5.2}$$

for all real numbers α', α'' such that $1 < \alpha' < \alpha < \alpha''$.

$$\mathscr{C} \neq \gamma_p^{n,(\alpha)}(X) \ . \tag{5.3}$$

In analogy with theorem 4.1, we have

THEOREM 5.1. Let $p \in \{2,\infty\}$, $n \geq 0$. Let M_s, $s \geq 0$ be a logarithmically convex sequence of positive real numbers satisfying

$$M_s^{\frac{1}{s}} \to \infty \quad \text{as} \quad s \to \infty ; \quad M_0 = M_1 = 1 \ .$$

Let $\mathscr{C} = \mathscr{C}(p,n,M_s,X)$ denote the class determined by (5.1). In the case $p = 2$, assume that $\mathscr{C}$ is spatially differentiable. Let α be a fixed real number > 1. Define a sequence $g(s)$, $s > 1$ of real numbers by

$$M_s = s^{s[\alpha+g(s)]} \quad \text{for all} \quad s > 1 \ .$$

Then $\mathscr{C}$ is an asymptotic Gevrey class α if and only if (5.4) and (5.5) hold:

$$g(s) \to 0 \quad \text{as} \quad s \to \infty \tag{5.4}$$

$$\overline{\lim_{s\to\infty}} \, |g(s)| \log s = \infty \ . \tag{5.5}$$

Notation. Suppose $\mathscr{C}(2,n,M_s,X)$ is spatially differentiable. In accordance with theorem 5.1 (and the obvious extension of remark 4.1), if $\mathscr{C}(p,n,M_s,X)$ is an asymptotic Gevrey class α, we may denote it by $\gamma_p^{n,(\alpha+g)}(X)$.

6. A MAJORIZATION TECHNIQUE.

We now return to the Cauchy problem (1.1). $v = v(t,x)$ is the non-homogeneous term and $b = b(t,x)$ is the coefficient.

Let $p \in \{2,\infty\}$, $n \geq 0$. Let M_s, $s \geq 0$ be a sequence of positive real numbers. Let $\mathscr{C} = \mathscr{C}(p,n,M_s,X)$ be spatially differentiable. Suppose that $\mathscr{C}$ is an asymptotic Gevrey class α for some $\alpha > 1$. Then we may assume that there is a sequence $g(s)$, $s > 1$ which determines $\mathscr{C}$ where g is described in the statement of theorem 5.1.

Now define two formal power series in a formal variable ρ, with coefficients depending on $t \in [0,T]$, by

$$V(t) = \sum_{s\geq 0} \frac{\rho^s}{s!} v_s(t) ,$$

$$B(t) = \sum_{s\geq 0} \frac{\rho^s}{s!} b_s(t)$$

where

$$v_s(t) = \sup_{\mu+\nu\leq n} \left(\int_{S_t} |D_t^\mu D_x^{s+\nu} v|^2 \, dx \right)^{\frac{1}{2}}$$

$$b_s(t) = \sup_{\mu+\nu\leq n} \operatorname{ess\,sup}_{S_t} |D_t^\mu D_x^{s+\nu} b| .$$

Let λ_s, $s \geq 0$ be a sequence of positive real numbers with $\lambda_0 = \lambda_1 = 1$. Define an operator λ which acts on formal power series in ρ with real coefficients

$$\Phi(\rho) = \sum_{s\geq 0} \frac{\rho^s}{s!} \Phi_s \quad \text{by} \quad \lambda\Phi = \sum_{s\geq 0} \frac{\rho^s}{s!} \lambda_s \Phi_s .$$

Consider the condition

(A) λV and λB are holomorphic in ρ at $\rho = 0$, uniformly in t for $t \in [0,T]$, i.e., there is an $r > 0$ for which λV, λB both have a radius of convergence $\geq r$ at $\rho = 0$ for each $t \in [0,T]$.

Then (B) and (C) are immediate:

(B) If $\lambda = \lambda(\delta)$, $\delta = \alpha - 1$ is given by $\lambda_s(\delta) = \frac{1}{s!^{\delta}}$, $s \geq 0$, then (A) $\Leftrightarrow$ $v \in \gamma_2^{n,(\alpha)}(X)$ and $b \in \gamma_\infty^{n,(\alpha)}(X)$.

(C) If $\lambda = \lambda(\delta,g)$, $\delta = \alpha - 1$ is given by $\lambda_s(\delta,g) = \frac{1}{s^{s[\delta+g(s)]}}$, $s > 1$, then (A) $\Leftrightarrow$ $v \in \gamma_2^{n,(\alpha+g)}(X)$ and $b \in \gamma_\infty^{n,(\alpha+g)}(X)$.

By means of the operator $\lambda(\delta)$ of (B), the majorization technique of [5] reduces the Cauchy problem (1.1) with data (1.2) and (1.3) to an analytic Cauchy problem. This majorization technique is based on the following two conditions which $\lambda(\delta)$ satisfies.

(I) $\lambda(\Phi\Psi) \ll (\lambda\Phi)(\lambda\Psi)$ for all formal power series $\Phi(\rho)$, $\Psi(\rho)$ which are $\gg 0$, i.e., with non-negative coefficients.

(II) Let q and r be fixed positive integers such that $q\delta < r$. Then to each $\eta > 0$, there exists $\eta' > 0$, sufficiently large, such that

$$\lambda\left(\frac{\partial}{\partial\rho}\right)^j \Phi \ll \text{const.} \left(\frac{\partial}{\partial\rho}\right)^j \left(\eta' + \eta\rho\frac{\partial}{\partial\rho}\right)^r (\lambda\Phi), \quad j \leq q$$

for all formal power series $\Phi(\rho)$ which are $\gg 0$. The constant is independent of η and Φ.

By comparing (B) and (C), it seems reasonable (and is in fact true, as may be seen by examining the majorization technique of [5]) that if $\lambda(\delta,g)$ of (C) satisfies (I) and (II), then the majorization technique of [5] can be effected for $v \in \gamma_2^{n,(\alpha+g)}(X)$ and $b \in \gamma_\infty^{n,(\alpha+g)}(X)$.

However $\lambda(\delta,g)$ depends on the choice of g. Now g satisfies the conditions of theorem 5.1 and these conditions do not in general guarantee that $\lambda(\delta,g)$ satisfies either (I) or (II). Therefore, we must *impose further conditions on* g so that (I) and (II) will be satisfied.

According to theorem 5.1, the sequence

$$\begin{aligned} M_s(\alpha,g) &= s^{s[\alpha+g(s)]}, \quad s > 1 \\ &= 1, \quad s = 0, 1 \end{aligned}$$

is logarithmically convex. To satisfy (I), we impose the stronger condition

(D) the sequence $1,1,s^{s[\alpha-1+g(s)]}$ for $s > 1$ is logarithmically convex.

To satisfy (II), we impose the condition

(E) $\overline{\lim_{s\to\infty}}\, s[g(s) - g(s-q)] \le 0$.

We note that (E) depends on the differential equation (1.1) as q is the order of the spatial derivative of u in the equation.

That (D) implies that $\lambda(\delta,g)$, $\delta = \alpha - 1$ satisfies (I) is immediate, since (D) implies that

$$\lambda_{r+s}(\delta,g) \leq \lambda_r(\delta,g)\lambda_s(\delta,g) \tag{6.1}$$

for all integers $r,s \geq 0$ by remark 3.1. Now (6.1) implies (and is in fact equivalent to) (I).

PROPOSITION 6.1. Let q be a fixed positive integer. Let $g(s)$, $s > 1$ be a sequence of real numbers satisfying (E) and

$$g(s) \to 0 \quad \text{as} \quad s \to \infty . \tag{6.2}$$

Let δ be a fixed positive real number. Then $\lambda(\delta,g)$ satisfies (II).

We first prove the

LEMMA 6.1. Assume the hypotheses of proposition 6.1. Then to each $\epsilon > 0$, there exists a positive constant $C(\epsilon)$ such that

$$\lambda_{s-q}(\delta,g) \leq C(\epsilon)s^{q\delta+\epsilon}\,\lambda_s(\delta,g) \tag{6.3}$$

for all $s \geq q$.

Proof of lemma 6.1. We have

$$\frac{\lambda_{s-q}(\delta,g)}{\lambda_s(\delta,g)} = \frac{s^{s[\delta+g(s)]}}{(s-q)^{(s-q)[\delta+g(s-q)]}}, \quad s > q + 1. \tag{6.4}$$

Applying the equality $s - q = s\left(\frac{s-q}{s}\right)$ to the base in the denominator in the right hand side of (6.4), we obtain

$$\frac{\lambda_{s-q}(\delta,g)}{\lambda_s(\delta,g)} = f_1(s)f_2(s), \quad s > q + 1 \tag{6.5}$$

where

$$f_1(s) = s^{q\delta+s[g(s)-g(s-q)]+qg(s-q)}, \quad s > q+1 \tag{6.6}$$

$$f_2(s) = \left(1 + \frac{q}{s-q}\right)^{(s-q)[\delta+g(s-q)]}, \quad s > q+1 \tag{6.7}$$

Since $\left(1 + \frac{q}{s-q}\right)^{(s-q)} \to e^q$ as $s \to \infty$, it follows from (6.2) and (6.7) that $f_2(s) \to e^{q\delta}$ as $s \to \infty$. Therefore there exists a constant $C' > 1$ such that

$$f_2(s) \le C' \quad \text{for all} \quad s > q+1 \tag{6.8}$$

From (6.6), (E) and (6.2), it follows that to each $\epsilon > 0$, we have

$$f_1(s) \le s^{q\delta+\epsilon} \quad \text{for all} \quad s \ge \text{some } s_0 = s_0(\epsilon) > q+1 \tag{6.9}$$

Let

$$C'(\epsilon) = 1 + \max_{q+1<s\le s_0} \frac{f_1(s)}{s^{q\delta+\epsilon}}. \tag{6.10}$$

From (6.9) and (6.10), we have that

$$f_1(s) \le C'(\epsilon)s^{q\delta+\epsilon} \quad \text{for all} \quad s > q+1. \tag{6.11}$$

From (6.5), (6.11) and (6.8), it follows that

$$\lambda_{s-q}(\delta,g) \le C(\epsilon)s^{q\delta+\epsilon}\lambda_s(\delta,g) \quad \text{for all} \quad s > q+1 \tag{6.12}$$

where $C(\epsilon) = C'(\epsilon)C'$. By increasing the constant $C(\epsilon)$ in (6.12), if necessary, (6.12) holds for all $s \ge q$.

Proof of proposition 6.1. Since $0 < q\delta < r$, it follows that $0 < q\delta < \sqrt{q\delta r} < r$. We use (6.3) with ϵ determined by the equation $q\delta + \epsilon = \sqrt{q\delta r}$. Let $C = C(\sqrt{q\delta r} - q\delta)$. By (6.3), we have that

$$\lambda_{s-q}(\delta,g) \le Cs^{\sqrt{q\delta r}}\lambda_s(\delta,g) \quad \text{for} \quad s \ge q \,. \qquad (6.13)$$

We now use the following fact, proved in [5].

(F) Let R' and R be fixed real numbers such that $0 < R' < R$. Let p' be the negative real number determined by

$$\frac{R}{R'} + \frac{1}{p'} = 1 \,.$$

Then to each $\eta > 0$, we have that

$$s^{R'} \le (\eta' + \eta s)^R \quad \text{for all} \quad s \ge 1$$

where $\eta' = \eta^{p'}$.

By letting $R' = \sqrt{q\delta r}$ and $R = r$ in (F), and using (6.13), it follows that to each $\eta > 0$, we have that

$$\lambda_{s-q}(\delta,g) \le C(\eta' + \eta s)^r \lambda_s(\delta,g) \quad \text{for all} \quad s \ge q \qquad (6.14)$$

(II) follows from (6.14).

REMARK 6.1. (6.3) implies for $\epsilon = 1$ that $M_s(\alpha,g) \le \text{const.}\ s^{q\alpha+1} M_{s-q}(\alpha,g)$ for all $s \ge q$. If $M_s(\alpha,g)$, $s \ge 0$ is non-decreasing, it follows that the classes $\mathscr{C}(p,n,M_s(\alpha,g),X)$, $p = 2,\infty$, $n \ge 0$ are spatially differentiable.

7. EXISTENCE THEOREM.

Given two positive integers p and q with $p > q$, let α be a fixed real number such that $1 < \alpha < \frac{p}{q}$. Let $g(s)$, $s > 1$ be a sequence of real numbers satisfying the following four conditions:

$$\lim_{s\to\infty} g(s) = 0 \tag{7.1}$$

$$\overline{\lim_{s\to\infty}} |g(s)| \log s = \infty \tag{7.2}$$

$$\overline{\lim_{s\to\infty}} s[g(s) - g(s-q)] \leq 0 \tag{7.3}$$

$s^{s[\alpha-1+g(s)]}$ is logarithmically convex for s large[4] (7.4)

Under these conditions, we have proved that the classes determined by

$$\overline{\lim_{s\to\infty}} \frac{1}{s^{\alpha+g(s)}} |f|_{p,n}^{\frac{1}{s}} < \infty$$

are spatially differentiable for $p = 2,\infty$; n any non-negative integer (remark 6.1). Furthermore, they are asymptotic Gevrey classes α (theorem 5.1).

Consider the Cauchy problem

$$(C) \qquad \begin{aligned} &(D_t^p - bD_x^q)u = v \\ &u|S_0 = \cdots = D_t^{p-1}u|S_0 = 0 \end{aligned}$$

of section 1. $b = b(t,x)$ and $v = v(t,x)$ are given on the strip X: $0 \leq t \leq T$ and $S_{t'}$, $t' \in [0,T]$ has equation $t = t'$.

Let n be an integer $\geq p$. We have the

[4]If (7.4) and (7.1) are satisfied, then by Newton's polygon method (see [8]), the sequence $g(s)$, $s > 1$ may be changed, if necessary, by a finite number of terms so that (D) of section 6 holds.

EXISTENCE THEOREM. Let $b \in \gamma_\infty^{n,(\alpha+g)}(X)$. Whenever $v \in \gamma_2^{n,(\alpha+g)}(X)$, there is a solution u of $(C) \in \gamma_2^{p+n,(\alpha+g)}(X)$.

We only indicate here how to start the proof of the existence theorem, as the double majorization technique needed is given in [5].

Consider the Cauchy problem

$$(C_0)\qquad \begin{aligned} &D_t^p u_0 = v \\ &u_0|S_0 = \cdots = D_t^{p-1} u_0|S_0 = 0 \end{aligned}$$

and the sequence of Cauchy problems

$$(C_k), k > 0\qquad \begin{aligned} &D_t^p u_k - bD_x^q u_{k-1} = 0 \\ &u_k|S_0 = \cdots = D_t^{p-1} u_k|S_0 = 0 \end{aligned}$$

To each problem (C_j), $j \geq 0$ is associated a formal problem (F_j) in t and a formal variable ρ, which in some sense majorizes the problem (C_j). The sequence (F_j), $j \geq 0$ leads to a formal problem (F) which majorizes (C). The majorization of $\sum_{j=0}^{\infty} u_j$ is effected by means of Gårding's inequality (for hyperbolic operators).

The problem (F) is in turn majorized by use of the operator $\lambda(\delta,g)$, $\delta = \alpha - 1$ (section 6). $\lambda(\delta,g)$ satisfies (I) and (II) of section 6, and these two conditions are sufficient to carry out the second majorization.

The following are two examples of sequences that satisfy (7.1) - (7.4), and consequently, to which the existence theorem applies. Define g_1 by $s^{g_1(s)} = \log s$, $s > 1$, i.e., $g_1(s) = \frac{\log \log s}{\log s}$, $s > 1$. g_1 determines the asymptotic Gevrey class α given by

$$\overline{\lim_{s\to\infty}} \frac{1}{s^{\alpha} \log s} |f|_{p,n}^{\frac{1}{s}} < \infty .$$

Define g_2 by $s^{g_2(s)} = \frac{1}{\log s}$, $s > 1$, i.e., $g_2(s) = - \frac{\log \log s}{\log s}$, $s > 1$. g_2 determines the asymptotic Gevrey class α given by

$$\overline{\lim_{s\to\infty}} \frac{\log s}{s^{\alpha}} |f|_{p,n}^{\frac{1}{s}} < \infty .$$

This class was discussed in section 1.

8. AN OPEN QUESTION.

The motivation for asymptotic Gevrey classes was to obtain finer estimates for solutions (see the Introduction — section 1). This point can be pushed much farther.

Let's first take a simpler case than asymptotic Gevrey classes. Suppose in the Cauchy problem (1.1), the non-homogeneous term v and coefficient b satisfy (1.2) and (1.3) respectively, but *in an effective way*. That is, suppose that

(A) $\quad v, b \notin \gamma_2^{n,(\alpha')}(X), \gamma_\infty^{n,(\alpha')}(X)$ respectively
for any α' such that $1 < \alpha' < \alpha$.

Then a natural question to consider is — does u satisfy (1.4) in an effective way, i.e., is it true that

(B) $\quad u \notin \gamma_2^{p+n,(\alpha')}(X)$ for any α' such that $1 < \alpha' < \alpha$.

A more delicate type of problem is: suppose that v and b satisfy

$$\overline{\lim_{s\to\infty}}\left(\frac{1}{s^{\alpha}\log s}\right)|D_x^s v|_{2,n}^{\frac{1}{s}} < \infty$$

$$\overline{\lim_{s\to\infty}}\left(\frac{1}{s^{\alpha}\log s}\right)|D_x^s b|_{\infty,n}^{\frac{1}{s}} < \infty,$$

but *in an effective way*. That is, suppose that

(A') $\quad v, b \notin \gamma_2^{n,(\alpha)}(X), \gamma_\infty^{n,(\alpha)}(X)$ respectively.

Then is it true that u satisfies

$$\overline{\lim_{s\to\infty}}\left(\frac{1}{s^{\alpha}\log s}\right)|D_x^s u|_{2,p+n}^{\frac{1}{s}} < \infty$$

in an effective way, i.e., is it true that

(B') $\quad u \notin \gamma_2^{p+n,(\alpha)}(X)$.

Such notions of effectiveness have appeared in the classical literature (e.g., Hadamard [2,p.29]).

Further references on the classical theory of classes are [1], [6] and [7].

REFERENCES

[1] T. Carleman, *Les fonctions quasi analytiques*, Gauthier-Villars, Paris, 1926.

[2] J. Hadamard, *Lectures on Cauchy's problem in linear*

partial differential equations, Dover, New York, 1952.

[3] L. Hörmander, *Linear partial differential operators*, Die Grundlehren der Math. Wissenschaften, Bd. 116, Springer-Verlag, Berlin and New York, 1969.

[4] A. Kolmogoroff, *On inequalities between the upper bounds of the successive derivatives of an arbitrary function on an infinite interval*, Amer. Math. Soc. Transl. no. 4 (1949).

[5] J. Leray et Y. Ohya, *Systèmes linéaires, hyperboliques non stricts*, Deuxième Colloq. l'Anal. Fonct. Liège, Centre Belge Recherches Math. (1964), 105-144.

[6] S. Mandelbrojt, *Séries de Fourier et classes quasi-analytiques de fonctions*, Gauthier-Villars, Paris, 1935.

[7] __________, *Analytic functions and classes of infinitely differentiable functions*, The Rice Institute Pamphlet, vol. 29, no. 1, 1942.

[8] __________, *Séries adhérentes, régularisation des suites, applications*, Gauthier-Villars, Paris, 1952.

[9] L. Nirenberg, *On elliptic partial differential equations*, Ann. Scuola Norm. Sup. Pisa, ser. 3, vol. 13, fasc. 2 (1959), 115-162.

BOUNDARY LAYER METHODS FOR ORDINARY DIFFERENTIAL EQUATIONS WITH SMALL COEFFICIENTS MULTIPLYING THE HIGHEST DERIVATIVES

by

R. E. O'Malley, Jr.*
University of Arizona
Department of Mathematics
Tucson, Arizona 85721

ABSTRACT

Many singular perturbation problems of applied mathematics involve differential equations with a small parameter multiplying the highest derivatives. Many of the asymptotic results obtained through the familiar boundary layer methods carry over to equations with small coefficients multiplying these derivatives. Moreover, these results can be readily obtained through numerical experimentation. Specific results are given for boundary value problems for certain higher order linear equations and for some second order quasilinear equations.

1. INTRODUCTION

Singular perturbation problems often arise in the solution of differential equations with a small parameter multiplying the highest derivatives. Many of the asymptotic

---

*This work supported in part by the Office of Naval Research under Grant No. N00014-67-A-0209-0022.

results of singular perturbation theory for ordinary differential equations carry over to the more general case that small coefficients multiply the highest derivatives. By asymptotic, we then mean results which hold in the limit as the size of the leading coefficients tends to zero. Our results will not rely on any explicit parameter dependence and will apply to many equations whose coefficients may vary considerably in magnitude.

Boundary value problems for singularly perturbed equations on finite intervals generally feature solutions which converge (under appropriate hypotheses) to the solution of a reduced boundary value problem within the interval. At the endpoints, nonuniform convergence (boundary layer behavior) can be expected. The reduced problem will consist of a lower order (limiting) differential equation and (the limiting forms of) an appropriate number of the original boundary conditions. A basic objective of singular perturbation theory is the determination of the reduced problem or, alternatively, of the cancellation law which states which of the original boundary conditions will be omitted in defining the reduced problem (cf. Wasow [18] or O'Malley [12]). We shall show that determining possible reduced problems can be accomplished through numerical experimentation.

Detailed boundary layer analysis will not be attempted, since solving the reduced problem is sufficient to obtain the limiting behavior within the open interval. We point out that blind numerical integration of the original boundary value problem is usually very awkward and easily leads to incorrect results, since the solution and/or its derivatives change rapidly in the boundary layers. Stability considerations must be introduced to ascertain whether any solution of a reduced problem can be a limiting solution in the interior of the interval.

Our most complete results will be for linear boundary value problems, paralleling the original work of Wasow [18] and its later generalizations. Further tentative results are given for a class of second order quasilinear equations. Linear boundary conditions which each link derivatives at an endpoint are used. More general boundary conditions coupling derivatives at both endpoints may also be examined (cf. Harris [7]). It is hoped that this work will help stimulate the considerable further study it deserves, both numerically and analytically. Numerical work on such problems is being done by Joseph Flaherty of Rensselaer Polytechnic Institute.

2. A GENERAL LINEAR BOUNDARY VALUE PROBLEM

Consider the linear differential equation

$$\sum_{j=0}^{m} a_j(x)\frac{d^j y}{dx^j} = 0 \tag{1}$$

on the bounded interval $0 \leq x \leq 1$ subject to the linear boundary conditions

$$(2)\quad \begin{cases} y^{(\lambda_i)}(0) + \sum_{j=0}^{\lambda_i - 1} b_{ij} y^{(j)}(0) = \ell_i, \quad i = 1, 2, \ldots, r, \\ \qquad \lambda_i\text{'s decreasing}, \quad \lambda_1 < m \\ y^{(\lambda_k)}(1) + \sum_{j=0}^{\lambda_k - 1} b_{kj} y^{(j)}(1) = \ell_k, \quad k = r+1, \\ r+2, \ldots, m, \quad \lambda_k\text{'s decreasing}, \quad \lambda_{r+1} < m. \end{cases}$$

We'll assume that all coefficients a_j are sufficiently smooth and that a_m <u>is small throughout</u> $0 \leq x \leq 1$, <u>but nonzero</u>. Such differential equations, with small coefficients but without explicit dependence on one or more small parameters, arise even more frequently than those usually discussed where the coefficient dependence on some parameter ϵ is assumed to be given asymptotically to all order ϵ^j (cf., e.g., O'Malley and Keller [13]). We'll also assume that the boundary conditions (2) have bounded coefficients b_{ij} and ℓ_i which may be small.

Let us numerically determine which coefficients $a_{m-1}(x)$, $a_{m-2}(x)$, ..., $a_{n+1}(x)$ <u>are</u> (like a_m) <u>small throughout</u> $0 \leq x \leq 1$ and <u>suppose</u> $a_n(x)$ <u>is not small</u> within the interval. Then, we naturally call

$$\sum_{j=0}^{n} a_j(x)y^{(j)}(x) = 0 \tag{3}$$

the reduced differential equation. Since a_n is nonzero, the reduced equation will be nonsingular and the full equation avoids having turning points.

The asymptotic solutions of the linear equation (1) are determined by the roots of the characteristic polynomial

$$\sum_{j=0}^{m} a_j(x)p^j = 0 \tag{4}$$

which has $m - n$ large roots determined asymptotically by the roots p^0 of the lower order polynomial

$$\sum_{j=0}^{m-n} a_{n+j}(x)(p^0)^j = 0. \tag{5}$$

(for more detailed discussion, see the appendix.)

We shall assume that the roots p^0 all have nonzero real parts and that they are distinct throughout $0 \leq x \leq 1$. This could be checked numerically. Further, suppose their real parts are ordered so that

$$\begin{cases} \operatorname{Re} p_i^0(x) < 0, \quad i = 1, 2, \ldots, \sigma \\ \\ \operatorname{Re} p_i^0(x) > 0, \quad i = \sigma + 1, \ \sigma + 2, \ldots, \sigma + \tau = m - n. \end{cases} \tag{6}$$

We shall let the corresponding large roots of (4) be denoted by $p_1(x), \ldots, p_{m-n}(x)$.

Generalizing Wasow's results (see the appendix), we readily obtain:

THEOREM. _If_

(i) _the reduced problem_

$$(7)\begin{cases} \sum_{j=0}^{n} a_j(x) y^{(j)}(x) = 0 \\ y^{(\lambda_i)}(0) + \sum_{j=0}^{\lambda_i - 1} b_{ij} y^{(j)}(0) = \ell_i, \quad i = \sigma + 1, \ldots, r \\ y^{(\lambda_k)}(1) + \sum_{j=0}^{\lambda_k - 1} b_{kj} y^{(j)}(1) = \ell_k, \quad k = r + \tau + 1, \ldots, m \end{cases}$$

has a unique solution $z(x)$,

(ii) _the two matrices_

$$(8) \quad \Sigma = \left(\left(\frac{p_i^0(0)}{\max_{s=1,\ldots,\sigma} |p_s^0(0)|} \right)^{\lambda_j - \lambda_\sigma} \right) \qquad i, \; j = 1, \ldots, \sigma$$

and

$$(9) \quad \mathcal{T} = \left(\left(\frac{p_{\sigma+i}^0(1)}{\max_{t=1,\ldots,\tau} |p_{\sigma+t}^0(1)|} \right)^{\lambda_{r+j} - \lambda_{r+\tau}} \right) \qquad i, \; j = 1, \ldots, \tau$$

are both nonsingular, then there is a unique solution of the problem (1)-(2) *of the form*

$$(10)\quad y(x) = \sum_{i=1}^{\sigma} \frac{A_i(x)\exp[\int_0^x p_i(s)ds]}{\max\limits_{s=1,\ldots,\sigma} \left|p_s^0(0)\right|^{\lambda_\sigma}}$$

$$+ \sum_{i=\sigma+1}^{m-n} \frac{A_i(x)\exp[\int_1^x p_i(s)ds]}{\max\limits_{t=1,\ldots,\tau} \left|p_{\sigma+t}^0(1)\right|^{\lambda_{r+\tau}}} + \tilde{z}(x)$$

where the $A_i(x)$ *are bounded and* $\tilde{z}(x)$ *tends asymptotically to* $z(x)$. *In particular,* $y(x) \to z(x)$ *within* $0 < x < 1$.

Remarks. *1*. The boundary conditions for the reduced problem (7) are obtained by cancelling the first σ boundary conditions (2) at $x = 0$ and the first τ boundary conditions at $x = 1$. Thus, stability considerations (signs of the large roots of the characteristic polynomial) are critical in defining the appropriate reduced problem. We note that the reduced problem will not be defined if the original problem (1)-(2) had fewer than σ boundary conditions at $x = 0$ or fewer than τ boundary conditions at $x = 1$. In those cases ($r < \sigma$ or $m - r < \tau$), there will generally be no limiting solution (cf. (Wasow [18]) for a list of counterexamples).

2. When Σ and T are both nonsingular, the limiting solution within $(0,1)$ is a particular solution of the reduced problem. A difficulty occurs when the reduced problem has a nonunique solution and the original problem has a unique solution. Aside from the condition that the matrix

$$(11)\quad \begin{bmatrix} y_{m-n+i}^{0(\lambda_{\sigma+j})}(0) + \sum_{k=0}^{\lambda_{\sigma+j}-1} b_{\sigma+j,k} y_{m-n+i}^{0(k)}(0) \\ \\ y_{m-n+i}^{0(\lambda_{r+\tau+\ell})}(1) + \sum_{k=0}^{\lambda_{r+\tau+\ell}-1} b_{r+\tau+\ell,k} y_{m-n+i}^{0(k)}(1) \end{bmatrix}$$

$$i = 1, \ldots, n \qquad j = 1, \ldots, r - \sigma$$

$$k = 1, \ldots, m - r - \tau = n - r - \sigma$$

be nonsingular where $y_{m-n+1}^{0}, \ldots, y_{m}^{0}$ form a fundamental set of solutions to the reduced equation (3), there seems to be no convenient numerical clue available when the reduced problem (7) has nonunique solutions. Numerical solution of (7) could easily yield the wrong solution unless one checked that (11) is nonsingular. An example is provided by

$$\begin{cases} \epsilon^2 y^{(iv)} - y'' - \pi^2 y = 0 \\ \text{with} \\ y'(0) \text{ and } y'(1) \text{ prescribed and } y(0) = y(1) = 0. \end{cases}$$

The reduced problem has the solution

$$z(x) = \frac{z'(0)}{\pi}\sin \pi x$$

with $z'(0)$ arbitrary, while the unique solution of the full problem has the asymptotic limit

$$\frac{1}{2\pi}(y'(0) - y'(1))\sin \pi x,$$

as $\epsilon \to 0$.

3. The matrices Σ and $\mathcal{T}$ are singular if two of their columns are equal. In particular, Σ would be singular if any of the roots $p_1^0(0), \ldots, p_\sigma^0(0)$ were allowed to be equal. When $\lambda_1, \lambda_2, \ldots, \lambda_\sigma$ are successive integers, Σ will be otherwise nonsingular since it is then a multiple of a Vandermonde matrix. Generally, however, the determinants of Σ and $\mathcal{T}$ will have other zeros (cf. Chapter 6 of Aitken [2] for a discussion of such determinants).

4. Analogous results hold for nonhomogeneous equations (cf. O'Malley [12] for equations with parameter).

5. Examples worth further study include two point problems for the equation

$$\epsilon y'' + \mu y' + y = c(x), \qquad 0 \leq x \leq 1$$

where μ is small and ϵ is small and positive. When either $\frac{\epsilon}{\mu^2} \to 0$ as $\mu \to 0$ or $\frac{\mu^2}{\epsilon}$ and $\frac{\epsilon}{\mu}$ both tend to zero, note that different cancellations would be appropriate for $\mu > 0$ then for $\mu < 0$. Note, however, that difficulties would occur if $\mu = \pm\epsilon$ or $\mu = 0$.

3. REMARKS ON A QUASILINEAR PROBLEM

To develop a general theory of boundary value problems for nonlinear singularly perturbed equations would seem difficult. It is well-known, for example, that

$$\epsilon^2 y'' + y' + (y')^3 = 0$$

$$y(0) = \alpha, \qquad y(1) = \beta$$

has no solution for ϵ sufficiently small when $\alpha \neq \beta$.

We shall restrict attention to boundary value problems for the quasilinear second order equation

$$(12) \qquad a_0(x,y)y'' + a_1(x,y)y' + a_2(x,y) = 0, \qquad 0 \leq x \leq 1$$

where the a_i's are bounded for y bounded and a_0 is nonzero but everywhere small compared to the larger of $|a_1|$ and $|a_2|$ (or $|a_{2y}|$ if $a_2 = 0$). The linear (separated) boundary conditions (2) can be rewritten in either of the six ways:

$$(13a) \qquad y'(0) = \ell_1, \qquad y(0) = \ell_2$$

$$(13b) \qquad y'(0) + b_{10}y(0) = \ell_1, \qquad y'(1) + b_{20}y(1) = \ell_2$$

$$(13c) \qquad y'(0) + b_{10}y(0) = \ell_1, \qquad y(1) = \ell_2$$

$$(13d) \qquad y(0) = \ell_1, \qquad y'(1) + b_{20}y(1) = \ell_2$$

$$(13e) \qquad y(0) = \ell_1, \qquad y(1) = \ell_2,$$

or

$$(13f) \qquad y'(1) = \ell_1, \qquad y(1) = \ell_2.$$

Our objective is to make preliminary observations only, though we anticipate that the ideas used have wider applicability.

We note that the asymptotic solution to certain boundary value probles for the equations (12) with $a_0(x,y) = \epsilon^2$

is well known (cf. O'Malley [12]). We mention only the linear problems

$$(14)\qquad \epsilon^2 y'' + b(x)y = c(x), \qquad y(0), \qquad y(1) \text{ prescribed}$$

and

$$(15)\qquad \epsilon^2 y'' + a(x)y' + b(x)y = c(x), \quad y(0), \quad y(1) \text{ prescribed.}$$

If $b(x) < 0$ throughout $0 \leq x \leq 1$, the solution of (14) satisfies

$$y(x) = \frac{c(x)}{b(x)} + \left(y(0) - \frac{c(0)}{b(0)}\right)\exp\left[\frac{1}{\epsilon}\int_0^x b(s)ds\right]$$

$$+ \left(y(1) - \frac{c(1)}{b(1)}\right)\exp\left[\frac{1}{\epsilon}\int_x^1 b(s)ds\right] + O(\epsilon)$$

as the small positive parameter $\epsilon \to 0$. Thus, the limiting solution within $(0,1)$ satisfies the reduced equation $b(x)y = c(x)$. If instead, $b(x) > 0$, the limiting solution is rapidly oscillatory and has no limit as $\epsilon \to 0$. Likewise, if $a(x) > 0$, the solution of (15) satisfies

$$y(x) = z(x) + (y(0) - z(0))\exp\left[-\frac{1}{\epsilon^2}\int_0^x a(s)ds\right] + O(\epsilon)$$

where z satisfies $a(x)z' + b(x)z = c(x)$, $z(1) = y(1)$. The solution for $a(x) < 0$ follows by replacing x by $1 - x$. Thus, the limiting solution within $(0,1)$ satisfies the reduced equation $a(x)y' + b(x)y = c(x)$ with the boundary value $y(1)$ if $a(x) > 0$ and $y(0)$ if $a(x) < 0$. When $b(x)$ in (14) or $a(x)$ in (15) has a zero within $[0,1]$, however, the complications of turning points occur (cf. Wasow [18] or O'Malley [12]).

Generally, when a limiting solution to a singular perturbation problem exists away from the endpoints, it satisfies the limiting (or reduced) equation obtained by setting the small parameter(s) equal to zeoro. (Theorems to this effect have been given by Eckhaus and others, while the oonclusion is obvious for most examples one encounters). Thus, for equation (12) without an explicit parameter, we must expect any bounded limiting solution within $(0,1)$ to satisfy either the limiting (algebraic) equation

$$a_2(x,y) = 0 \tag{16}$$

or the limiting (differential) equation

$$a_1(x,y)y' + a_2(x,y) = 0. \tag{17}$$

To determine all limiting solutions for the given boundary value problem (12)-(13), one should examine

(i) all solutions $Y_0(x)$ of (16) for which $|a_0(x,Y_0(x))|$ and $|a_1(x,Y_0(x))|$ are small relative to $|a_{2y}(x,Y_0(x))|$ throughout $0 \leq x \leq 1$ and

(ii) all solutions $Y_0(x)$ of (17) subject to one of the original boundary conditions such that $\left|\frac{a_1(x,Y_0(x))}{a_0(x,Y_0(x))}\right|$ is not small compared to $\left|\frac{a_2(x,Y_0(x))}{a_0(x,Y_0(x))}\right|$ throughout [0,1].

Many of these potential limiting solutions will be rejected as inappropriate. Our requirements will be more stringent than necessary, however, so may screen out some potentially valid limiting possibilities. For example, because of their inherent difficulties, we shall avoid turning points by requiring that $a_{2y}(x,Y_0(x)) \neq 0$ when (16) applies and $a_1(x,Y_0(x)) \neq 0$ when (17) applies.

Stability conditions must be satisfied to obtain boundary layer behavior. Thus, if the appropriate limiting equation is $a_2(x,Y_0(x)) = 0$, we shall require that $\left(\frac{a_2}{a_0}\right)_y$ be negative at both $x = 0$ and $x = 1$ for all values of y. (This condition can be motivated by the preceding discussion of the linear problem (14) and that of O'Malley [12] for quasilinear equations with parameter.) Unless $Y_0(x)$ happened to satisfy the full problem (12)-(13), positivity of $\left(\frac{a_2}{a_0}\right)_y$ at either $(x,y) = (0,Y_0(0))$ or $(1,Y_0(1))$ would make the requisite boundary layer behavior impossible. The endpoint sign restriction could be relaxed, however. For example, if $y(0)$ were prescribed, it would suffice that the partial derivative at $x = 0$ be negative for y values between $y(0)$ and $Y_0(0)$ only. Alternatively, a less restrictive integral condition could be used as in Fife [5]. We note that $\left(\frac{a_2(x,Y_0(x))}{a_0(x,Y_0(x))}\right)_y$ will be negative throughout $0 \leq x \leq 1$ since $a_0(x,Y_0)$ and $a_{2y}(x,Y_0)$ don't change sign.

Other stability conditions are appropriate if $Y_0(x)$ satisfies $a_1(x,y)y' + a_2(x,y) = 0$ and one of the original two boundary conditions (13). (Recall the preceding discussion of the linear problem (15) and that of O'Malley [12] for quasilinear problems with parameter). When (13a) or (13f) apply, it is appropriate to let Y_0 be the solution of (17) subject to the prescribed y value (i.e., we neglect the derivative boundary condition). Otherwise, we solve (17) subject to one of the boundary conditions, neglecting the boundary condition at the other endpoint. In all cases, if the boundary condition neglected is at $x = 0$, we shall require that $\frac{a_1(x,Y_0(x))}{a_0(x,Y_0(x))}$ be positive throughout $0 \le x \le 1$, while $\frac{a_1(x,Y_0(x))}{a_0(x,Y_0(x))}$ must be negative there if the neglected boundary condition is at $x = 1$. Unless $Y_0(x)$ happened to satisfy the original boundary value problem, $Y_0(x)$ could not be a limiting solution if $\frac{a_1}{a_0}$ had the opposite sign. Convergence of the solution of the original problem (12)-(13) to $Y_0(x)$ should be uniform at the endpoint with a retained boundary condition, while derivatives converge nonuniformly at the endpoint with cancellation.

When (13b)-(13e) apply, an additional stability condition for the solution Y_0 must hold at the endpoint where cancellation occurred. Specifically, if the neglected boundary condition specified y, we require that the sign condition on $\frac{a_1}{a_0}$ hold at the endpoint for all y values between the value of Y_0 there and the prescribed value (a less restrictive integral condition is also possible). Otherwise, if the cancelled boundary condition prescribes $y' + by$ at an endpoint, we ask that the sign condition hold there for all y.

The reduced problem has multiple solutions (cf. O'Malley [11]) if the boundary condition retained prescribes $y' + by$ at an endpoint. Corresponding multiple solutions of the full problem (12)-(13) then result. An example is provided by

$$\begin{cases} \epsilon^2 y'' - y' - y^2 = 0 \\ y'(0) - y(0) = 0, \quad y(1) = 1 \end{cases}$$

which has both $Y_0(x) = 0$ and $Y_0(x) = \frac{1}{x+1}$ as solutions of the reduced problem

$$Y_0' + Y_0^2 = 0, \quad Y_0'(0) - Y_0(0) = 0.$$

Each is a limiting solution of the full problem for $x < 1$.

The sign condition on $\frac{a_1}{a_0}$ eliminates the possibility of two limiting solutions satisfying (17), one with a cancelled boundary condition at $x = 0$ and another with cancellation at $x = 1$. It is easy, however, to find examples with one limiting solution satisfying (16) and another satisfying (17). One such problem is

$$\begin{cases} \epsilon^2 y'' + a_1(y)y' - y = 0 \\ y(0) = 1 = y(1) \end{cases}$$

where a_1 is a smooth function such that

$$a_1(y) = \begin{cases} 0, & y < \frac{1}{4} \\ 1, & y > \frac{3}{4} \end{cases}.$$

One limiting solution within $(0,1)$ is the trivial solution $y \equiv 0$; another is $y = e^{x-1}$. Our hypotheses on $\left(\frac{a_2}{a_0}\right)_y$ likewise prevents the possibility of two solutions of (16) being limiting solutions to (12) within $(0,1)$.

4. COMMENT ON THE LITERATURE

From the standpoint of numerical analysis, singular pertrubation problems pose considerable difficulty. Knowledge of the kind of asymptotic behavior common to such problems can be used to advantage in developing numerical methods (cf. Dorr [4] and Miranker [8], for example). Because the stepsize used in finite differencing is generally large compared to the thickness of the boundary layers, solving the reduced problem within the interval may often suffice (cf. Abrahamsson, Keller and Kreiss [1]). If the boundary layer behavior is important, numerical schemes for obtaining it must be developed (cf., e.g., Pearson [30], Murphy [9], or Yarmish [19] for current techniques). It is hoped that the preceding results might be utilized to develop further numerical methods. Once satisfactory numerical results are available, they should be useful in deducing the asymptotic behavior of solutions to more complicated problems which might then be subject to analysis.

5. APPENDIX: MORE DETAILS ON THE LINEAR PROBLEMS

A. THE ASYMPTOTIC SOLUTION OF THE CHARACTERISTIC POLYNOMIAL

For the differential equation (1), the large roots of the characteristic polynomial (4) will be shown to satisfy the lower order polynomial (5). If we artifically introduce a small positive parameter μ by defining

$$\mu^{m-n} \equiv \sup_{\substack{0<x\leq 1 \\ j=n+1,\ldots,m}} |a_j(x)| \tag{18}$$

and set

$$a_j(x) \equiv \mu^{m-n}\tilde{a}_j(x), \qquad j > n,$$

the $\tilde{a}_j$'s will be bounded. Multiplying (4) by μ^n and introducing $\tilde{p} = p\mu$, we have

$$\sum_{j=n+1}^{m} \mu^{m-j}\tilde{a}_j(x)(\tilde{p})^j + \sum_{j=0}^{n} a_j(x)\mu^{n-j}(\tilde{p})^j = 0.$$

Since a_n isn't small while $\mu^{n-j}a_j$ is small for $j < n$, this polynomial has $m - n$ (possibly large) roots $\tilde{p}$ which are asymptotically determined by the lower degree polynomial

$$\sum_{j=n+1}^{m} \mu^{m-j}\tilde{a}_j(x)\tilde{p}^{j-n} + a_n(x) = 0, \tag{19}$$

corresponding to (5).

Remark. In the special case that $\dfrac{1}{\tilde{a}_m(x)}$ is bounded, we can instead use the $m - n$ distinct asymptotic roots

$$p^0(x) = \frac{1}{\mu}\left(- \frac{a_n(x)}{\tilde{a}_m(x)}\right)^{1/m-n} = \left(- \frac{a_n(x)}{a_m(x)}\right)^{1/m-n}.$$

B. THE ASYMPTOTIC SOLUTIONS OF THE DIFFERENTIAL EQUATION

Under the conditions of our theorem, we will obtain m linearly independent asymptotic solutions of the differential equation (1) in the form

$$\begin{cases} y_i(x) = \mathcal{A}_i(x)\exp\left[\int_0^x p_i(s)ds\right], \quad \mathcal{A}_i(0) = 1, \quad i = 1, \ldots, \sigma \\ y_i(x) = \mathcal{A}_i(x)\exp\left[\int_1^x p_i(s)ds\right], \quad \mathcal{A}_i(1) = 1, \\ \qquad i = \sigma + 1, \ldots, \sigma + \tau = m - n \quad \text{and} \\ y_i(x) = \mathcal{A}_i(x), \quad i = m - n + 1, \ldots, m. \end{cases} \tag{20}$$

This generalizes the usual representation for constant coefficient linear equations (cf. Coddington and Levinson [3]) and for equations with parameter (cf. Turrittin [15]).

First note that $y_i(x) = \mathfrak{A}_i(x)\exp\left[\int^x p_i(s)ds\right]$ has the jth derivative

$$y_i^{(j)}(x) = [(p_i(x))^j\mathfrak{A}_i(x) + \{\frac{j(j-1)}{2}(p_i(x))^{j-2}p_i'(x)\mathfrak{A}_i(x)$$

$$+ j(p_i(x))^{j-1}\mathfrak{A}_i'(x) + \ldots + A_i^{(j)}(x)]\exp\left[\int^x p_i(s)ds\right].$$

Substituting into (1), then, $\mathfrak{A}_i$ must satisfy

$$\text{(21)} \qquad \left[\sum_{j=0}^{m} a_j(x)(p_i(x))^j\right]\mathfrak{A}_i(x)$$

$$+ \left[\sum_{j=0}^{m} ja_j(x)(p_i(x))^{j-1}\left(\mathfrak{A}_i'(x) + \left(\frac{j-1}{2}\right)\frac{p_i'(x)}{p_i(x)}\mathfrak{A}_i(x)\right)\right]$$

$$+ \ldots + \left[\sum_{j=0}^{m} a_j(x)\mathfrak{A}_i^{(j)}(x)\right] = 0.$$

For $i \leq m - n$, note that each bracketed term is successively smaller by $0\left(\frac{1}{p_i(x)}\right)$ with $p_i(x)$ large. The first bracket vanishes by (4). Since $p_i \to p_i^0$, then, the differential equation (21) tends to the linear first order equation

$$(22) \qquad \left(\sum_{j=0}^{m} j a_j(x)(p_i^0(x))^{j-1}\right)\chi_i'(x)$$

$$+ \left(\sum_{j=0}^{m} \frac{j(j-1)}{2} a_j(x)(p_i^0(x))^{j-1} \frac{p_i^{0\prime}(x)}{p_i^0(x)}\right)\chi_i(x) = 0$$

which is nonsingular since the p_i^0's are assumed to be distinct. Using the solution χ_i^0 of (22) with the appropriate boundary value as a first approximant, a solution χ_i of (21) can be readily obtained by successive approximations. (This requires some smoothness on the coefficients in (1).) If the large roots of (4) were not distinct, the representation (20) might still be correct, but the scheme would generally need modification (cf. Turrittin [15] and [16]). Since the first $m - n + 1$ coefficients of (1) are small, the last n solutions of (20) can be obtained as regular perturbations of any n linearly independent solutions y_i^0 of the reduced equation (3).

C. SOLUTION OF THE BOUNDARY VALUE PROBLEM

Using the asymptotic solutions (20), let us seek a solution to (1)-(2) of the form

$$(23) \qquad y(x) = \sum_{i=1}^{\sigma} c_i \frac{y_i(x)}{\max\limits_{s=1,\ldots,\sigma} \left|p_s^0(0)\right|^{\lambda_\sigma}}$$

$$+ \sum_{i=\sigma+1}^{m-n} c_i \frac{y_i(x)}{\max\limits_{t=1,\ldots,\tau} \left| p^0_{\sigma+t}(1) \right|^{\lambda_{r+\tau}}} + \sum_{j=m-n+1}^{m} c_i y_i(x)$$

with bounded coefficients c_i. Differentiating (23) we obtain

$$y^{(j)}(x) \sim \sum_{i=1}^{\sigma} \frac{c_i \mathcal{A}_i(x)(p^0_i(x))^j}{\max\limits_{s=1,\ldots,\sigma} \left| p^0_s(0) \right|^{\lambda_\sigma}} \exp\left[\int_0^x p_i(s)ds\right]$$

$$+ \sum_{i=\sigma+1}^{m-n} \frac{c_i \mathcal{A}_i(x)(p^0_i(x))^j}{\max\limits_{t=1,\ldots,\tau} \left| p^0_{\sigma+t}(1) \right|^{\lambda_{r+\tau}}} \exp\left[\int_1^x p_i(s)ds\right]$$

$$+ \sum_{i=m-n+1}^{m} c_i y^{(j)}_i(x).$$

Neglecting further asymptotically small terms, then,

$$y^{(j)}(0) \sim \sum_{i=1}^{\sigma} \frac{c_i (p^0_i(0))^j}{\max\limits_{s=1,\ldots,\sigma} \left| p^0_s(0) \right|^{\lambda_\sigma}} + \sum_{i=m-n+1}^{m} c_i y^{(j)}_i(0)$$

$$y^{(j)}(1) \sim \sum_{i=\sigma+1}^{m-n} \frac{c_i (p^0_i(1))^j}{\max\limits_{t=1,\ldots,\tau} \left| p^0_{r+t}(1) \right|^{\lambda_{r+\tau}}} + \sum_{i=m-n+1}^{m} c_i y^{(j)}_i(1).$$

Thus, (2) implies

$$(24)\quad \begin{cases} \ell_j \sim \sum_{i=1}^{\sigma} \dfrac{c_i(p_i^0(0))^{\lambda_j}}{\max_{s=1,\ldots,\sigma} \left|p_s^0(0)\right|^{\lambda_\sigma}} + \sum_{i=\sigma+1}^{m} c_i(\text{e.s.t.}) \\ \qquad + \sum_{i=m-n+1}^{m} c_i\left(y_i^{0^{(\lambda_i)}}(0) + \sum_{k=0}^{\lambda_j-1} b_{jk} y^{0(k)}(0)\right), \quad j = 1, \ldots \\ \text{and} \\ \ell_j \sim \sum_{i=1}^{\sigma} c_i(\text{e.s.t.}) + \sum_{i=\sigma+1}^{m-n} \dfrac{c_i(p_i^0(1))^{\lambda_j}}{\max_{t=1,\ldots,\tau} \left|p_t^0(1)\right|^{\lambda_{r+\tau}}} \\ \qquad + \sum_{i=m-n+1}^{m} c_i\left(y_i^{0^{(\lambda_j)}}(1) + \sum_{k=0}^{\lambda_j-1} b_{jk} y_i^{0(k)}(1)\right), \\ j = r+1, \ldots, m \end{cases}$$

were e.s.t. represents asymptotically exponentially small terms. The first σ equations imply

$$\sum_{i=1}^{\sigma} c_i\left(\frac{p_i^0(0)}{\max_{s=1,\ldots,\sigma}\left|p_s^0(0)\right|}\right)^{\lambda_j} \sim \max_{s=1,\ldots,\sigma} (|p_s^0(0)|)^{\lambda_\sigma-\lambda_j} L_j,$$

$$j = 1, \ldots, \sigma$$

where L_j is known in terms of $c_{\sigma+1}, \ldots, c_n$. The nonsingularity of $\sum$ (cf. (7)) allows $c_1, \ldots, c_\sigma$ to be uniquely determined as bounded functions of $c_{\sigma+1}, \ldots, c_m$. Likewise, we have

$$\sum_{i=\sigma+1}^{m-n} c_i \left(\frac{p_i^0(1)}{\max_{t=1,\ldots,\tau} \left|p_{r+t}^0(1)\right|} \right)^{\lambda_j} \sim \max_{t=1,\ldots,\tau} \left(|p_{r+t}^0(1)|^{\lambda_{r+\tau}-\lambda_j} \right) L_j$$

$$j = r + 1, \ldots, r + \tau$$

where L_j is known in terms of $c_{m-n+1}, \ldots, c_m$. Since $\mathcal{T}$ is nonsingular, $c_{\sigma+1}, \ldots, c_{m-n}$ become uniquely specified. The remaining n conditions can be rewritten as

$$\begin{cases} \ell_j \sim \displaystyle\sum_{i=m-n+1}^{m} c_i \left[\left(y_i^{0(\lambda_j)}(0) + \sum_{k=0}^{\lambda_j-1} b_{jk} y_i^{0(k)}(0) \right) \right. \\ \qquad \left. + \, 0\left(\max_{s=1,\ldots,\sigma} |p_s^0(0)|^{\lambda_j-\lambda_\sigma} \right) \right], \quad j = \sigma + 1, \ldots, r \\ \ell_j \sim \displaystyle\sum_{i=m-n+1}^{m} c_i \left[\left(y_i^{0(\lambda_j)}(1) + \sum_{k=0}^{\lambda_j-1} b_{jk} y_i^{0(k)}(1) \right) \right. \\ \qquad \left. + \, 0\left(\max_{t=1,\ldots,\tau} |p_{r+t}^0(1)|^{\lambda_j-\lambda_{r+\tau}} \right) \right], \quad j = r + \tau + 1, \ldots, m. \end{cases}$$

Since the reduced problem has a unique solution, say

$$z(x) = \sum_{i=m-n+1}^{m} c_i^0 y_i^0(x),$$

the matrix (11) is nonsingular. This implies that the last n equations for $c_{m-n+1}, \ldots, c_m$ have a unique solution which tends to $c_{m-n+1}^0, \ldots, c_m^0$ asymptotically. Thus, our theorem is proved.

We note that it also follows that if the problem (1)-(2) has a solution of the form (23) with bounded c_i's, the limiting solution within $(0,1)$ will satisfy the reduced problem (7).

BIBLIOGRAPHY

1. L. R. Abrahamson, H. B. Keller, and H. O. Kreiss, "Difference approximations for singular perturbations of systems of ordinary differential equations". (To appear).

2. A. C. Aitken, Determinants and Matrices, Oliver and Boyd, Edinburgh, 1956.

3. E. A. Coddington and N. Levinson, Theory of Ordinary Differential Equations, McGraw-Hill, New York, 1965.

4. F. W. Dorr, "The numerical solution of singular perturbations of boundary value problems," SIAM J. Numer. Anal. 7 (1970), pp. 281-313.

5. P. C. Fife, "Semilinear elliptic boundary value problems with small parameters," Arch. Rational Mech. Anal. 52 (1973), pp. 205-232.

6. N. Fröman and P. O. Fröman, JWKB Approximation, North Holland, Amsterdam, 1965.

7. W. A. Harris, Jr., "Singularly perturbed boundary value problems revisited," Lecture Notes in Mathematics 312, Springer-Verlag, Berlin, 1973, pp. 54-64.

8. W. L. Miranker, "Numerical methods of boundary layer type for stiff systems of differential equations," Computing 11 (1973), pp. 221-234.

9. W. D. Murphy, "Numerical analysis of boundary layer problems in ordinary differential equations," Math. Comp. 21 (1967), pp. 583-596.

10. R. E. O'Malley, Jr., "On the asymptotic solution of boundary value problems for nonhomogeneous ordinary differential equations containing a parameter," J. Math. Anal. Appl. 28 (1968), pp. 450-460.

11. R. E. O'Malley, Jr., "On multiple solutions of a singular perturbation problem," Arch. Rational Mech. Anal. 49 (1972), pp. 89-98.

12. R. E. O'Malley, Jr., Introduction to Singular Perturbations, Academic Press, New York, 1974.

13. R. E. O'Malley, Jr., and J. B. Keller, "Loss of boundary conditions in the asymptotic solution of linear differential equations: II. Boundary value problems," Comm. Pure Appl. Math. 21 (1968), pp. 263-270.

14. C. E. Pearson, "On a differential equation of boundary layer type," J. Math. and Physics 67 (1968), pp. 134-154.

15. H. L. Turrittin, "Asymptotic solutions of certain ordinary differential equations associated with multiple roots of the characteristic equation," Amer. J. Math. 58 (1936), pp. 364-378.

16. H. L. Turrittin, "Asymptotic expansions of solutions of systems of ordinary linear differential equations containing a parameter," Contrib. Theory Nonlinear Oscillations 2 (1952), pp. 81-116.

17. W. Wasow, "On the asymptotic solution of boundary value problems for ordinary differential equations containing a parameter," J. Math. and Physics 23 (1944), pp. 173-183.

18. W. Wasow, Asymptotic Expansions for Ordinary Differential Equations, Interscience, New York, 1965.

19. J. Yarmish, Aspects of the Numerical and Theoretical Treatment of Singular Perturbations, Doctoral Dissertation, New York University, 1972.

FIXED POINT ITERATIONS USING INFINITE MATRICES, II

By

B. E. Rhoades

We shall be concerned with the iteration scheme

(1) $x_0 = \bar{x}_0 \in I = [0,1]$

(2) $\bar{x}_{n+1} = f(x_n)$, $n \geq 0$

(3) $x_{n+1} = (1 - c_n)x_n + c_n f(x_n)$,

where $f \in C[I]$, $f: I \to I$, and the $\{c_n\}$ satisfies: (i) $c_0 = 1$, (ii) $0 \leq c_n \leq 1$ for all n, and (iii) Σc_n diverges.

It was shown in [2] that if $\{c_n\}$ also satisfies: (iv) $\lim_n c_n = 0$, then the scheme (i) - (3) converges to a fixed point of f.

Under the additional assumption that f is monotone increasing, we shall show that condition (iv) is not needed for the convergence of (1) - (3). We shall also examine the behavior of $\{x_n\}$ for different choices of $\{c_n\}$.

THEOREM I. Let $f: I \to I$, f continuous and nondecreasing. If $\{c_n\}$ satisfies (i) - (iii), then (1) - (3) converges to a fixed point of f.

Conditions (i) - (iii) on $\{c_n\}$ guarantee that the matrix $A = (a_{nk})$ which $a_{nk} = 0$, $k > n$, $a_{nn} = c_n$, $a_{nk} = c_k \Pi_{j=k+1}^{n} (1 - c_j)$, $k < n$, is regular. (See e.g. [1].) Consequently, it will be sufficient to establish the convergence of $\{x_n\}$.

Let $M = \sup \{x \in I | f(x) = x\}$ and $m = \inf \{x \in I | f(x) = x\}$. We shall assume that $0 < m \leq M < 1$, since either $m = 0$ or $M = 1$ gives us immediate convergence to a fixed point by choosing $x_0 = 0$ or 1. From [1], $0 \leq x < m$ implies $f(x) > x$ and $M < x \leq 1$ implies $f(x) < x$.

If $x_0 > M$, then $x_n \geq M$ for all n, and $\{x_n\}$ is non-increasing. The proof is by induction. From (3), $x_1 - x_0 = c_0 (f(x_0) - x_0) \leq 0$ since $x_0 > M$. Also, $x_1 - M = (1 - c_0) + c_0(f(x_0) - M) \geq 0$ since $x_0 > M$ and f is nondecreasing. Assume the induction hypothesis. Then $x_{n+1} - x_n = c_n(f(x_n) - x_n) \leq 0$ since $x_n \geq M$, and $x_{n+1} - M = (1 - c_n)(x_n - M) + c_n(f(x_n) - M) \geq 0$ since f is nondecreasing.

Similarly, $x_0 < m$ implies $x_n \leq m$ for all n and $\{x_n\}$ nondecreasing.

If f has a unique fixed point, or if $M - m > 0$ and every point of $[m,M]$ is a fixed point, then the proof is complete. Suppose f has at least two distinct adjacent fixed points $p < q$ in $[m,M]$. In the interval (p,q), $f(x) - x$ will be of constant sign. Let $x_0 \in (p,q)$. If $f(x) > x$ for $p < x < q$, then it can be shown by induction that $p \leq x_n \leq q$ and $x_{n+1} \geq x_n$ for each n. On the other hand, if $f(x) < x$ for each x in (p,q), then $p \leq x_n \leq q$ and $x_{n+1} \leq x_n$ for each n.

Since $\{x_n\}$ is bounded and monotone, it converges.

THEOREM 2. Let $f: I \to I$, f continuous and nondecreasing, $c_0 = d_0 = 1$, $0 \leq c_n \leq d_n \leq 1$ for all n, $\Sigma\, c_n$ diverges, $x_0 = y_0$, and, for $n \geq 0$,

$$(4)\quad x_{n+1} = (1 - c_n)x_n + c_n\, f(x_n)$$

$$(5)\quad y_{n+1} = (1 - d_n)y_n + d_n\, f(y_n)$$

If $x_0 > M$, then $x_n \geq y_n$ for all n. If $x_0 < m$, then $x_n \leq$ for all n. If there exist a pair of distinct adjacent fixed points p,q, satisfying $m \leq p < q \leq M$, and $x_0 \in (p,q)$, then $f(x) > x$ for $x \in (p,q)$ implies $x_n \leq y_n$ and $f(x) < x$ for $x \in (p,q)$ implies $x_n \geq y_n$.

Suppose $x_0 > M$. Then, from (4) and (5), $x_1 = \bar{x}_1 = y_1 = \bar{y}_1$,

$x_2 - y_2 = (d_1 - c_1)(x_1 - f(x_1))$. From Theorem 1, $x_1 \geq M$, so that $x_2 \geq y_2$. Assume the induction hypothesis. We may write

$x_{n+1} - y_{n+1} = x_n - y_n + d_n(y_n - f(y_n)) - c_n(x_n - f(x_n))$.

From Theorem 1, $x_n \geq M$, so that $y_n \geq f(y_n)$. Since $d_n \geq c_n$,

$x_{n+1} - y_{n+1} \geq x_n - y_n + c_n(y_n - f(y_n)) - c_n(x_n - f(x_n)) =$

$(1 - c_n)(x_n - y_n) + c_n(f(x_n) - f(y_n))$. By the induction hypothesis $x_n \geq y_n$. Since f is nonincreasing, we have $x_{n+1} \geq y_{n+1}$.

A similar argument takes care of the other cases.

Let z be any fixed point of f. From Theorem 2, in conjunction with Theorem 1, we can conclude that $|y_n - z| \leq |x_n - z|$. For, if $z = M$, then we have $0 \leq y_n - M \leq x_n - M$. If $z = m$, then $0 \leq m - y_n \leq m - x_n$. If $x_0 \in (p,q)$, then $f(x) > x$, $p < x < q$ implies $q - x_n \geq q - y_n \geq 0$ and $f(x) < x$ for $p < x < q$ implies $0 \leq y_n - p \leq x_n - p$.

The following Corollary is an immediate consequence of Theorems 1 and 2.

COROLLARY. Let f, $\{c_n\}$ satisfy the hypotheses of Theorem 1. Then, for each n, $|f^n(x_0) - z| \leq |x_n - z|$, where z is the fixed point of f to which (1) - (3) converges.

If f is monotone decreasing, then nothing can be said about the relative sizes of x_n and y_n. For example, let $f(x) = 1 - x$, $x \in I$. Pick $x_0 = 0$ and select $c_n = \lambda d_n$, $0 < \lambda < 1$. A simple calculation verifies that, if $0 < \lambda < 1/6$, then $x_3 > y_3$, whereas $1/6 < \lambda < 1$ gives $x_3 < y_3$.

REFERENCES

[1] J. Reinermann, Über Toeplitzsche Iterationsverfahren undeinige ihre Anwendungen in der konstruktiven Fixpunktheorie. Studia Math. 32 (1969), pp. 209 - 227.

[2] B. E. Rhoades, Fixed point theorems using infinite matrices, to appear in Trans. Amer. Math. Soc.

Indiana University

Bloomington

Indiana

The Line Method for Parabolic Differential Equations
Problems in Boundary Layer Theory and Existence of Periodic Solutions

Wolfgang Walter

ABSTRACT

The (longitudinal) line method for parabolic differential equations is considered, where the space variable only is discretized. An initial-boundary value problem for a parabolic differential equation is transformed into an initial value problem for a system of ordinary differential equations of first order. Using the theory of ordinary and parabolic differential inequalities, constructive existence proofs for nonlinear problems by means of the line method can be given. Recent work on several problems in boundary layer theory along these lines is described. The main part of this exposition deals with periodic solutions of the equation $u_t = u_{xx} + f(t,x,u)$ (periodic means periodic in t, f is assumed to be periodic). Under suitable conditions on f, there exists exactly one periodic solution u. It is shown that for the corresponding line method approximation, there exists also exactly one periodic solution. The approximations converge uniformly to u, their t-derivatives converge uniformly to u_t, their first and second order differences in the x-direction converge uniformly to u_x and u_{xx}, respectively.

Introduction

We consider the nonlinear parabolic DE for $u = u(t,x)$ $(t,x \in R$, one space variable)

$$(1) \qquad u_t = f(t,x,u,u_x,u_{xx}) \quad \text{in} \quad J \times D.$$

In the first boundary value problem, J is the t-interval [0,T], D is the x-interval [0,a] (a,T > 0), and the boundary values are given by

$$(2) \qquad u(t,0) = \phi_0(t), \quad u(t,a) = \phi_1(t) \quad \text{for} \quad t \in J,\ u(0,x) = \phi(x) \quad \text{for } x \in D.$$

Here, ϕ_0, ϕ_1 and ϕ are given functions. It is always assumed that $f = f(t,x,z,p,r)$ is increasing in r.

In the (longitudinal) line method considered here, the space variable only is discretized. Let n be a fixed positive integer, let $h = a/n$, $x_i = ih$ $(i = 0,1,\dots,n)$. Then the discretized version of problem (1)(2) is a system of ODEs for the functions $u_i(t) \approx u(t,x_i)$

$$(3) \qquad u_i' = f(t,x_i,u_i,\delta u_i,\delta^2 u_i) \quad \text{for } t \in J \quad (i = 1,\dots,n-1),$$

$$(4) \qquad u_i(0) = \phi(x_i) \quad (i = 1,\dots,n-1),$$

where

$$(5) \qquad \delta u_i = \frac{1}{h}(u_{i+1}-u_i) \ [\approx u_x], \quad \delta^2 u_i = \frac{1}{h^2}(u_{i+1}-2u_i+u_{i-1}) \ [\approx u_{xx}]$$

and

$$(6) \qquad u_0(t) = \phi_0(t), \quad u_n(t) = \phi_1(t).$$

Let us note that in the discretized problem the boundary conditions at $x = 0$ and $x = a$ are contained in the first and last Eq.(3) by means of definition (6), which is in fact a definition for $\delta^2 u_1$, $\delta^2 u_{n-1}, \ldots$, e.g., $\delta^2 u_1 = (u_2 - 2u_1 + \phi_0)/h^2$, $\delta u_{n-1} = (\phi_1 - u_{n-1})/h$.

If the Cauchy problem (or any other problem with unbounded D) is considered, the system of ODEs (3) obtained by means of the line method is infinite. In this case, (3) (4) is considered as an initial value problem for an ODE in an appropriate Banach space.

The line method has been used successfully in proving existence theorems for boundary value problems and the Cauchy problem for nonlinear parabolic DEs. An outline of the method and a bibliography are given in Chapters 35 and 36 of the author's book "Differential and Integral Inequalities" [9].

1. The Boundary Layer Equations

The boundary layer equations

$$(7)\qquad \begin{aligned} uu_x + vu_y &= \nu u_{yy} + U(x)U'(x) \\ u_x + v_y &= 0 \end{aligned}$$

in the domain $0 \le x \le a$, $0 < y < \infty$ describe the steady two-dimensional flow of a viscous fluid near a wall (which coincides with the x-axis). Here, $\nu > 0$ is the viscosity coefficient, the given function $U(x)$ is the free stream velocity (velocity of the outer flow), u and v are the velocity components in the direction of the wall (x-axis) respectively perpendicular to the wall (y-axis). The boundary conditions are given by

$$u(x,0) = v(x,0) = 0 \quad \text{for} \quad 0 \le x \le a, \quad u(0,y) = \tilde{u}(y) \quad \text{for } 0 \le y < \infty$$

(8)

$$\lim_{y\to\infty} u(x,y) = U(x) \quad \text{uniformly in} \quad 0 \le x \le a,$$

where $\tilde{u}(y)$ is the given initial velocity.

Using the von Mises transformation [9; Chap.30]

$$\xi = x, \quad \eta = \eta(x,y) = \int_0^y u(x,s)ds, \quad w(\xi,\eta) = u^2(x,y), \quad W(\xi) = U^2(x),$$

problem (7)(8) is transformed into

$$(9) \qquad w_\xi = W'(\xi) + \nu\sqrt{w}\, w_{\eta\eta} \quad \text{in} \quad 0 \le \xi \le a, \ 0 < \eta < \infty$$

$$(10) \qquad w(\xi,0) = 0 \quad \text{for} \quad 0 \le \xi \le a, \quad w(0,\eta) = \tilde{w}(\eta) \quad \text{for} \quad 0 \le \eta < \infty$$

$$\lim_{\eta\to\infty} w(\xi,\eta) = W(\xi) \quad \text{uniformly in} \quad 0 \le \xi \le a,$$

where the initial values $\tilde{w}$ are defined by $\tilde{w}(\eta) = \tilde{u}^2(y)$.

Equation (9) is a quasi-linear parabolic equation with ξ being the "time" variable and η the "space" variable. The line method approximation corresponding to problem (9)(10) is obtained by discretizing η. Let $h > 0$ be a fixed real number, let $\eta_i = hi$ $(i = 0,1,\ldots)$ and let δ^2 be the central second order difference operator defined in (5). Then problem (9)(10) yields the following countably infinite system of ODEs

$$(11) \qquad \begin{aligned} &w_i'(\xi) = W'(\xi) + \nu\sqrt{w_i}\,\delta^2 w_i \quad \text{for} \quad 0 \le \xi \le a \\ &w_i(0) = \tilde{w}(\eta_i), \quad w_0(\xi) = 0. \end{aligned} \qquad (i = 1,2,\ldots)$$

There is no way of transforming the "boundary condition at infinity" ($w \to W$ as $\eta \to \infty$) into the discrete problem, but it turns out that the corresponding discrete condition at infinity is automatically satisfied by the solution (w_i) of (11).

System (11) is considered as an ODE in the Banach space ℓ_∞ of bounded sequences $y = (y_1, y_2, \ldots)$ with the maximum norm

$$||y|| = \sup \{|y_i| : i = 1,2,\ldots\}.$$

As a consequence of an elementary theorem about initial value problems for ODEs with a Lipschitz continuous right hand side, problem (11) has exactly one solution.

The main part of the existence proof deals with a priori bounds independent of h for the functions (w_i), (w_i'), (δw_i), where δw_i is defined by (5). These bounds are obtained by using theorems on ordinary differential inequalities. The rest of the proof follows a well known pattern. If, for fixed $h > 0$, the function $w^h(\xi,\eta)$ is obtained from (w_i) by linear interpolation in the η direction, then it follows from the a priori bounds that the family of functions $\{w^h\}$ is uniformly bounded and uniformly Lipschitz continuous in $(\xi,\eta) \in [0,a] \times [0,\infty)$. By the Ascoli-Arzelà theorem, there exists a null sequence (h_k) such that $w^{h_k} \to w$ as $k \to \infty$. Then it can be shown that this limit function w is a (strong) solution of problem (9)(10), that it is unique and that $w^h \to w$ as $h \to 0$ uniformly in $[0,a] \times [0,\infty)$.

In this way the author [8] has proved the following:

<u>Existence Theorem</u>. Let $a > 0$, $U(x) \in C^2[0,a]$, $U > 0$ and $U' \geq 0$ in $[0,a]$, $\tilde{u}(y) \in C^2[0,\infty)$, $\tilde{u}(0) = 0$, $\tilde{u}'(0) > 0$, $\tilde{u}''(y) \leq 0$ and locally Hölder continuous for $y > 0$. Furthermore, let $\tilde{u}(y) \to U(0)$ as $y \to \infty$ and

$$\tilde{u}''(y) + U(0)U'(0) = O(y^2) \quad \text{as } y \to 0.$$

Then the boundary value problem (9)(10) has exactly one solution w such that w, w_ξ and w_η are bounded and continuous in G, $w > 0$ in G_0, $|w_\xi| \le C\eta$ in G, $w_\eta \ge 0$ and $w_\xi \le W(\xi)$ in G, $w_\eta(\xi,0) \ge \alpha > 0$ in $[0,a]$, where $G = [0,a] \times [0,\infty)$ and $G_0 = [0,a] \times (0,\infty)$. The differential equations is satisfied in G_0.

For given $h > 0$, problem (11) has exactly one solution $(w_i) = (w_i^h)$. It converges uniformly to the solution of problem (9)(10) in the following sense

$$\sup\{|w_i^h(\xi) - w(\xi,ih)| : 0 \le \xi \le a,\ i = 1,2,\ldots\} \to 0 \text{ as } h \to 0.$$

As far as existence is concerned, this result is not new. The first existence proof has been given by Oleinik [3][4].

We report briefly on some new results in boundary layer theory which were obtained by means of the line method. Lemmert [2] has proved an existence and convergence theorem similar to the one cited above, but with the two conditions $U' \ge 0$ and $\tilde{u}'' \le 0$ removed. It is clear that under those more general hypotheses only existence in the small, i.e. in $G' = [0,a_o] \times [0,\infty)$, $0 < a_o \le a$, can be proved. He also proved existence and convergence for the stagnation point problem [2], where $U(0) = 0$ and $\tilde{u}(y) = 0$ and $cx \le U(x) \le Cx$ for some positive constants c,C. An existence proof for the instationary boundary layer equations

$$\begin{aligned} &u_t + uu_x + vu_y = \nu u_{yy} + U_t + UU_x \\ (12)\qquad &(ru)_x + (rv)_y = 0 \end{aligned}$$

by means of the line method was given by Silaev [6].

2. Periodic solutions

Our aim is to give a constructive proof for the existence of periodic solutions of the parabolic differential equation

(13) $$u_t = u_{xx} + f(t,x,u) \quad \text{in } R \times D, \quad \text{where } D = [0,1],$$

subject to boundary conditions

(14) $$u(t,0) = u(t,1) = 0 \quad \text{for } t \in R.$$

We assume that the following condition (A) is satisfied:

Condition (A). $f(t,x,z) \in C^2\ (R \times D \times R)$ and $f_z \leq L < \pi^2$ in $R \times D \times R$. Furthermore, f is periodic in t with period $T > 0$,

$$f(t + T,x,z) = f(t,x,z) \quad \text{in} \quad R \times D \times R.$$

It is well known that under this condition the differential equation (13) has exactly one solution satisfying the boundary condition (14) and the initial condition

(15) $$u(0,x) = u_o(x) \quad \text{for} \quad x \in J, \text{ where } u_o(0) = u_o(1) = 0.$$

Reasoning in the usual way, we consider the map

(16) $$S{:}u_o(x) \to u(T,x),$$

where $u(t,x)$ is the solution of problem (13)-(15). This map is well defined since the solution u exists for all positive t. A fixed point of S, i.e. an initial value $u_o(x)$ such that $u_o(x) = u(0,x) = u(T,x)$, leads to a periodic solution $u(t,x)$. In order to obtain a fixed point of S and also to find an a priori bound for this fixed point, let us introduce the Banach space $E \subset C^o(D)$ of all functions $y(x) \in C^o(D)$ with finite norm

(17) $$||y|| := \sup\{|y(x)|/\phi(x);\ x \in D\}.$$

Here, $D = [0,1]$ and $\phi(x)$ is a fixed function in $C^2(D)$ satisfying

(18) $$\phi'' + (L+\alpha)\phi + 1 \leq 0 \text{ in } J,\ \phi(0) = \phi(1) = 0,\quad \phi > 0 \text{ in } (0,1),$$

where $\alpha > 0$ is fixed in such a way that $L < L+\alpha < \pi^2$. For example,

(19) $$\phi(x) = \frac{\cos[\sqrt{L+\alpha}(x-\frac{1}{2})]}{(L+\alpha)\cos[\sqrt{L+\alpha}/2]} - \frac{1}{L+\alpha}.$$

The following lemma is easily proved.

<u>Lemma</u>

(a) S is a contraction in E.

(b) If $u_o \in D$ is the fixed point of S, then the corresponding solution $u(t,x)$ of (13)-(15) is periodic in t with period T.

(c) For any periodic solution $v(t,x)$ of (13)(14) (continuous in $R \times D$ and with continuous derivatives v_t, v_x, v_{xx} in $R \times (0,1)$), the a prior bound

(20) $$|v(t,x)| \leq M\phi(x)e^{\alpha T} \text{ in } R \times D,\quad M := \max\,[|f(t,x,0)|:(t,x) \in R \times D]$$

holds.

From (c) it follows that any periodic solution has initial values belonging to E, hence there is exactly one periodic solution. We give a proof of (a) and (c).

Proof of (a).

Let $u_o, v_o \in E$, let $u(t,x)$, $v(t,x)$ be the corresponding solutions of (13)-(15) and let $w_o = u_o - v_o$, $w = u-v$. Then

(21) $$w_t = w_{xx} + f_z w,$$

where we consider f_z as a given function of t and x satisfying $f_z \leq L$. We want to show that the function

$$\sigma(t,x) = ||w_o|| e^{-\alpha t} \phi(x)$$

is an upper bound for w in $[0,T] \times D$. Using Nagumo's Lemma [9; 24.VI] we have to show that

$$\sigma_t \geq \sigma_{xx} + f_z \sigma \quad \text{in } [0,T] \times D$$

and $w \leq \sigma$ on the parabolic boundary of $[0,T] \times D$. The latter is clear since $w = \sigma = 0$ for $x = 0$ and $x = 1$ and $w(0,x) = w_o(x) \leq ||w_o|| \, \phi(x) = \sigma(0,x)$ for $x \in D$. The differential equation for σ follows from the inequalities

$$\sigma_{xx} + f_z \sigma - \sigma_t \leq e^{-\alpha t} ||w_o|| (\phi'' + L\phi + \alpha\phi) < 0.$$

Since the same proof applies to the difference v-u, we have proved that

$$|v(t,x) - u(t,x)| \leq ||v_o - u_o|| e^{-\alpha t} \phi(x) \quad \text{in } [0,T] \times D.$$

This ineq. is, for $t = T$, equivalent to

(22) $$||Su_o - Sv_o|| \leq q ||u_o - v_o||, \text{ where } q = e^{-\alpha T} < 1.$$

Thus, S is a contraction.

Proof of (c).

Let $v(t,x)$ be any periodic solution and let $\sigma(t,x) = A(\phi + \varepsilon)e^{-\alpha t}$, where A and ε are positive constants to be determined. Again, we want to show by Nagumo's Lemma that σ is an upper bound for v in $[0,T] \times D$. Obviously, $v \le \sigma$ for $x = 0$ and $x = 1$, and there exists (for given $\varepsilon > 0$) a minimal $A \ge 0$ such that $v(0,x) \le \sigma(0,x)$. Hence, σ is an upper bound if the differential inequality [note that $f(t,x,\sigma) \le M + L\sigma$]

$$\sigma_{xx} + f(t,x,\sigma) - \sigma_t \le Ae^{-\alpha t}[\phi'' + Me^{\alpha t}/A + L(\phi+\varepsilon) + \alpha(\phi+\varepsilon)] \le 0$$

is satisfied. Due to (18), this is certainly true if

$$(*) \qquad (L+\alpha)\varepsilon + Me^{\alpha T}/A \le 1, \quad \text{e.g.,} \quad \varepsilon \le 1/2(L+\alpha), \quad A \ge 2Me^{\alpha T}.$$

Thus far, the following fact has been established. If the minimal $A \ge 0$ such that $v(0,x) \le \sigma(0,x)$ satisfies $A \ge 2Me^{\alpha T}$, then σ is an upper bound for v, and especially $v(T,x) \le \sigma(T,x) = q\sigma(0,x)$, where $q < 1$ is given by (22). But this is a contradiction since $v(T,x) = v(0,x)$ and A is minimal. Therefore, for any small positive ε, we have the bound $v(t,x) \le 2Me^{\alpha T}(\phi(x)+\varepsilon)$ in $R \times D$. In this inequality, we can replace ε by 0, and thereafter $2M$ by M (see ineq. (*)). Since the same reasoning applies to $-v$, Proposition (c) is proved.

Remarks. The above proof for the a priori bound (c) goes through if f satisfies a one-sided Lipschitz condition

$$(23) \qquad f(t,x,z) - f(t,x,z') \le L(z - z') \quad \text{for } z \ge z', \text{ where } L < \pi^2.$$

No other regularity assumption is needed. If, in addition, f has smoothnass properties such that an existence theorem for the initial-boundary value problem holds, then (a) is true, and there is exactly one periodic solution.

We also note that inequality (20) holds for any small $\alpha > 0$. Hence, the corresponding inequality for $\alpha = 0$ is true,

$$|v(t,x)| \le M\phi_o(x) \quad \text{in} \quad R \times D, \tag{20'}$$

where ϕ_o is the function given by (19) for $\alpha = 0$.

As far as the above exposition is concerned, we do not claim any originality. It has been given mainly because it carries over to the line method approach which we shall outline now. In this approach, we do not use any facts from existence theory for parabolic differential equations. The use of successive approximation in connection with periodic solutions goes back to Karimov [1].

3. The Line Method Approximation

In what follows, we assume that f(t,x,z) satisfies Condition (A), and we use the notation explained in the introduction. We are looking for periodic solutions of (3), which in the present case reads

$$u_i' = \delta^2 u_i + f(t,x_i,u_i) \quad \text{for} \quad t \in R \ (i = 1,\ldots,n-1;\ u_o = u_n = 0). \tag{24}$$

Here, $x_i = i/n$, δ^2 is defined by (5), and in (6) we have $\phi_o = \phi_1 = 0$. Let ϕ be the function defined by (19), let $\phi_i := \phi(x_i)$ and let

$$||y||_n := \max_i |y_i|/\phi_i, \quad \text{where } y = (y_1,\ldots,y_{n-1}) \in R^{n-1}. \tag{25}$$

Let E_n be the Banach space R^{n-1} with the above norm. The operator S_n from E_n into itself is defined by

$$(26) \qquad S_n: u_o \to u(T),$$

where $u_o \in E_n$ and $u(t) = (u_1(t),\ldots,u_{n-1}(t))$ is the solution of (24) satisfying $u(0) = u_o$. This solution exists for all positive t. Again, we have the following

Lemma

(a) For any large n, S_n is a contraction in E_n.

(b) The solution u(t) of (24) with initial value $u(0) = u_o$, where u_o is the fixed point of S_n, is periodic with period T.

(c) To a given $\delta > 0$ there corresponds a $N = N(\delta) > 0$ such that any periodic solution $v(t) = (v_i(t))_1^{n-1}$ of (24) with $n > N$ satisfies

$$(27) \qquad ||v(t)||_n \le M + \delta \quad \text{for } t \in R, \text{ where } M = \max|f(t,x,0)|.$$

Proof of (a).

As before, let $u_o, v_o \in E_n$, let u(t), v(t) be the corresponding solutions of (24) and let $w_o = u_o - v_o$, $w = u - v$. The function w(t) satisfies

$$(28) \qquad w_i' = \delta^2 w_i + f_z w_i \quad (i = 1,\ldots,n-1;\ w_o = w_n = 0).$$

Since the right hand side of this equation is quasimonotone increasing [9; 6.II], we can apply M. Müller's theorem on differential inequalities.

[9; 12.V and 12.X] in order to obtain an upper bound $\sigma(t) = (\sigma_i)$ for $w(t)$. More explicitly, if σ satisfies

$$\sigma_i' \geq \delta^2\sigma_i + f_z\sigma_i \quad \text{for} \quad t \geq 0 \quad \text{and} \quad w_i(0) \leq \sigma_i(0) \; (i = 1,\ldots,n-1),$$

then $w_i(t) \leq \sigma_i(t)$ for $t > 0$, $i = 1,\ldots,n-1$. Using the discretized version of the function $\sigma(t,x)$ considered in Section 2,

$$\sigma_i(t) = ||w_o||_n e^{-\alpha t}\phi_i,$$

we arrive at

$$\delta^2\sigma_i + f_z\sigma_i - \sigma_i' \leq ||w_o||_n e^{-\alpha t}(\delta^2\phi_i + L\phi_i + \alpha\phi_i) \leq 0,$$

if n is so large that

$$\delta^2\phi_i \leq \phi''(x_i) + 1.$$

Since

$$\delta^2\phi_i = \phi''(x_i) + \frac{h^2}{12}\phi^{(4)}(\xi_i), \quad \text{where} \quad |x_i - \xi_i| < h = 1/n, \tag{29}$$

This is certainly true for large n, more explicitly, if $n > N_o = [12(L+\alpha)/\cos\sqrt{L+\alpha}/2]$. Since the same proof applies to $-w = v-u$, we obtain, as in Section 2, the estimate

$$||S_n u_o - S_n v_o||_n \leq q||u_o - v_o||_n \quad \text{for } n > N_o, \text{ where } q = e^{-\alpha T} < 1.$$

<u>Proof of (c)</u>

We assume that $v(t)$ is a periodic solution of (24) and proceed as in Section 2 (actually it is somewhat simpler since it is not necessary to introduce a constant $\varepsilon > 0$ in the upper bound). The function $\sigma(t)$, $\sigma_i = Ae^{-\alpha t}\phi_i$, is an upper bound for $v(t)$ in $[0,T]$, if $A \geq ||v(0)||_n$ and if

$$\delta^2\sigma_i + f(t,x_i,\sigma_i) - \sigma_i' \le Ae^{-\alpha t}(\delta^2\phi_i + Me^{\alpha t}/A + (L+\alpha)\phi_i) \le 0.$$

The latter inequality holds if, for given $\varepsilon \in (0,1)$, $A \ge Me^{\alpha T}/(1-\varepsilon) =: A_o$ and if n is so large that $\delta^2\phi_i \le \phi''(x_i) + \varepsilon$. Now, if $||v(0)||_n \ge A_o$, we may choose $A = ||v(0)||_n$ and obtain the inequalities

$$||v(0)||_n = ||v(T)||_n \le ||\sigma(T)||_n = q||v(0)||_n,$$

which is a contradiction. Hence, $||v(0)||_n < A_o$, $|v_i(t)| \le A_o e^{-\alpha t}\phi_i$, and (c) is proved.

4. A priori bounds

In this section we shall derive a priori bounds for derivatives and finite differences of u(t), where u(t) is the periodic solution of (24). It turns out that these bounds are easily obtained from the following lemma.

Lemma

Let $y(t) = (y_1(t),\ldots,y_{n-1}(t) \in C^1(R)$ be a periodic solution with period $T > 0$ of the system of equations

$$y_i' = \delta^2 y_i + c_i(t)y_i + d_i(t) \quad \text{for } t \in R, \ (i = 1,\ldots,n-1;\ y_o = y_n = 0 \tag{30}$$

where $c_i(t)$, $d_i(t)$ are periodic with period T and

$$c_i(t) \le L < \pi^2, \quad |d_i(t)| \le D \quad \text{for } t \in R \quad (i = 1,\ldots,n-1).$$

Then, for n sufficiently large,

$$||y(t)||_n \le 2De^{\alpha T} \quad \text{for } t \in R$$

More precisely, this inequality is true for all n such that $\delta^2\phi_i \leq \frac{1}{2} + \phi''(x_i)$. The proof follows the same pattern as the proof of Proposition (c) in the foregoing section. The function $\sigma(t)$, $\sigma_i = Ae^{-\alpha t}\phi_i$, is an upper bound for y if $A \geq ||y(0)||_n$ and

$$\delta^2\sigma_i + c_i(t)\sigma_i + d_i(t) - \sigma_i' \leq Ae^{-\alpha t}(\delta^2\phi_i + (L+\alpha)\phi_i + De^{\alpha t}/A) \leq 0.$$

Because of (18), this inequality is true for $0 \leq t \leq T$ if n is so large tha $\delta^2\phi_i \leq \phi''(x_i) + \frac{1}{2}$ and $De^{\alpha T}/A \leq \frac{1}{2}$ or $A \geq A_o := 2De^{\alpha T}$. The assumption that $||y(0)||_n \geq A_o$ leads to a contradiction exactly as in the proof of (c) in Sec.3. Hence, $||y(0)||_n \leq A_o$ and $y_i(t) \leq A_o e^{-\alpha t}\phi_i$ for $0 \leq t \leq T$. Since the same reasoning applies to $-y$, the lemma is proved.

Let u(t) be, for fixed n, the unique periodic solution of (24). Since f is of class C^2, we obtain the following differential equations for u' and u''

(31) $\quad v = u': \quad v_i' = \delta^2 v_i + f_t + f_z v_i$

(32) $\quad w = u'': \quad w_i' = \delta^2 w_i + f_{tt} + 2f_{tz}v_i + f_{zz}v_i^2 + f_z w_i.$

Here, i runs from 1 to n-1, $v_o = v_n = w_o = w_n = 0$, and the derivatives of f are taken at (t,x_i,u_i). In Sec.3, an a priori bound for u independent of n has been established. Hence, the functions $f_t(t,x_i,u_i),\ldots,f_{zz}(t,x_i,u_i)$ can be considered as given functions of t which are also bounded by a bound independent of n. It follows from the lemma above that $||v(t)||_n$, $||w(t)||_n \leq B$, where B is a bound independent of n. Let B be chosen in such a way that $||u(t)||_n \leq B$ holds, too.

In the next step we derive bounds for $\delta u_i = (u_{i+1}-u_i)/h$, $h = 1/n$, and δv_i. Since

$$\delta u_i = \delta u_o + h(\delta^2 u_1+\ldots+\delta^2 u_i)$$

and

$$|\delta u_o| = |u_1|/h \leq B\phi(x_1)/h \leq \phi'(0)$$

and since from (24) and the bound for v_i a bound for $\delta^2 u_i$ follows, we have also a bound for δu_i (a bound is always understood to be a bound valid for $i = 1,\ldots,n-1$ and independent of n). In exactly the same way, we can obtain a bound for $\delta^2 v_i$ and δv_i, using (31) and the bound for $w_i = v_i'$. Summing up, we have the following result:

There exists a constant $C > 0$ independent of n such that

$$|u_i|,\ |\delta u_i|,\ |\delta^2 u_i|,\ |u_i'|,\ |\delta u_i'|,\ |\delta^2 u_i'|,\ |u_i''| \leq C.$$

Furthermore, we can conclude from (24) that a bound independent of n exists for $\delta(\delta^2 u_i)$, since we have such a bound for $\delta u_i'$ and for $\delta f(t,x_i,u_i) = f_x + f_z$

5. Construction of a periodic solution

The rest of the existence proof is simple, and it runs as in [9; 36.XI]. Let n be fixed, let $u(t) = (u_1,\ldots,u_{n-1})$ be the periodic solutio of (24) and let $u^n(t,x)$ be the function obtained from $(u_i(t))$ be linear interpolation in the x-direction,

$$u^n(t,\alpha x_i + (1-\alpha)x_{i+1}) = \alpha u_i(t) + (1-\alpha)u_{i+1}(t) \quad \text{for} \quad 0 \leq \alpha \leq 1,$$

where $u_o = u_n = 0$. The function $u^n(t,x)$ is continuous and piecewise continuo differentiable in $R \times [0,1]$, and the partial derivatives u_t^n, u_x^n are bounded independent of n, since this is true for u_i' and δu_i. In the same way we define functions $p^n(t,x)$ from (δu_i), $r^n(t,x)$ from $(\delta^2 u_i)$ and $v^n(t,x)$ from (u_i'). All these functions are continuous and piecewise continuousely differentiable, and their partial derivatives of first order are bounded independent of n. Hence, the functions u^n, v^n, p^n, r^n are bounded and satisfy a Lipschitz condition in $R \times [0,1]$, where the bound and the Lipschitz constant are independent of n. It follows from the Ascoli-Arzelà theorem that there exists an increasing sequence $(n(k))$ of integers such that

$$u^{n(k)}(t,x) \to u(t,x),\ p^{n(k)}(t,x) \to p(t,x),\ r^{n(k)}(t,x) \to r(t,x),$$

$$v^{n(k)}(t,x) \to v(t,x) \quad \text{as } k \to \infty.$$

The functions u, p, r, v are periodic in t with period T, uniformly Lipschitz constituous in $R \times [0,1]$, and $u_t = v$, $u_x = p$, $u_{xx} = r$. Finally, u satisfies the differential equation (13) and the boundary condition (14). Our result can be summed up in the following theorem.

Existence and Convergence Theorem

Let $f = f(t,x,z)$ satisfy Condition (A) of Sec.3. Then there exists one and only one solution $u(t,x)$ of (13)(14) periodic in t with period T. The solution u and its derivatives u_x, u_{xx}, u_t are uniformly Lipschitz continuous in $R \times [0,1]$, and they satisfy the inequalities

$$(33) \qquad |u(t,x)| \le M\phi(x),\ |u_t(t,x)| \le B\phi(x),\ |u_x(t,x)| + |u_{xx}(t,x)| \le C.$$

Here, ϕ is the function given by (19), M is the maximum of $|f(t,x,0)|$, B and C depend only on ϕ (i.e., on L) and on upper bounds for f and its first and second order partial derivatives in $[0,T] \times [0,1] \times [-A,A]$, where A is a bound for $|u|$.

For sufficiently large n, equation (24) has one and only one periodic solution with period T, say, $u^n(t) = (u_1^n,\dots,u_{n-1}^n)$. The following four relations hold

$$u_i^n \to u, \ \delta u_i^n \to u_x, \ \delta^2 u_i^n \to u_{xx}, \quad (u_i^n)' \to u_t \quad \text{as } n \to \infty ,$$

where $u_i^n \to u$ means that

$$\sup\{|u_i^n(t) - u(t,i/n)| \ : t \in R, \ 0 < i < n\} \to 0 \quad \text{as } n \to \infty,$$

and similarly for the other three relations.

The assertion about convergence of $(u_i^n)\dots$ is proved in a well known way. It follows from the fact that any subsequence has a subsequence which is convergent and whose limit function is a solution of (13)(14), this limit being unique.

The existence part of the above theorem is contained in a result of Prodi [5]. For related results, see Vaghi [7]. Periodic solutions for quasilinear parabolic differential equations have been treated by several authors. In a forthcoming paper by R. Gaines and the present author, periodic solution of the equation $u_t = a(t,x,u,u_x)u_{xx} + b(t,x,u,u_x)$ will be considered, using the line method.

[1] Dz. H. Karimov, On periodic solutions of nonlinear differential equations of parabolic type. Doklady Akad. Nauk SSSR (N.S.) 58, 969-972 (1947). [Russian].

[2] R. Lemmert, unpublished (the results are part of Mr. Lemmert's doctoral dissertation which will soon be published).

[3] Oleinik, O.A., On the system of equations of boundary layer theory. Z. Vycisl. Mat. i Mat. Fiz. 3, 489-507 (1963).

[4] Oleinik, O.A., The Prandtl system of equations in boundary layer theory. Dokl. Akad. Nauk SSSR 150 - Soviet Math. 4, 583-586 (1963).

[5] G. Prodi, Soluzioni periodiche di equazioni alle derivate parziali di tipo parabolico e non lineari. Rivista Mat. Univ. Parma 3, 265-290 (1952).

[6] D.A. Silaev, Construction of a solution for a nonstationary Prandtl system using a method of straight lines in time. Uspehi Mat. Nauk, 28:2(170), 243-244 (1973). [Russian].

[7] C. Vaghi, Soluzioni limitate, o quasi-periodiche, di un'equazione di tipo parabolico non lineare. Boll. Un. Mat. Ital. (4) 1 (1968), 559-580.

[8] W. Walter, Existence and convergence theorems for the boundary layer equations based on the line method. Archive Rat. Mech. Anal. 39 (1970), 169-188.

[9] W. Walter, Differential and integral inequalities. Ergebnisse der Math. und ihrer Grenzgebiete, Vol.55, Springer Verlag, Berlin-Heidelberg-New York 1970.

Professor Dr. Wolfgang Walter
Mathematisches Institut I
Universität Karlsruhe
75 Karlsruhe / Germany

and

Department of Mathematics
Colorado State University
Fort Collins, Colorado / USA

An Integral Equation Method for Generalized Analytic Functions [+]

by

Wolfgang L. Wendland [*]

Technische Hochschule Darmstadt

Germany

Abstract. An integral equation method is presented for solving the standard boundary value problem for generalized analytic functions. The method combines a Fredholm equation of the second kind in the domain with Fredholm equations of the first and the second kind on the boundary for single boundary and surface layers. An approximation method for solving these equations is prescribed. The method provides an application for solving semilinear problems by an imbedding method combined with Newton's approximation. For the standard problem a few numerical results are given in the appendix.

[+] An invited address at the Conference on Constructive and Computational Methods for Differential and Integral Equations, Research Center for Applied Science, Indiana University, Bloomington, February 1974.

---

[*] This investigation was carried out while the author was a Visiting Unidel Chair Professor at the University of Delaware.

1. Introduction

Many problems of the classical Mathematical Physics can be modeled by elliptic boundary value problems in the plane (see I. N. Vekua's book [12]). If only two unknown functions are involved, then one is led to consider Riemann Hilbert problems for generalized analytic functions. These problems can be reduced to the standard form:

$$(1.1) \qquad w_{\bar{z}} = aw + b\bar{w} + c \qquad \text{in} \quad G ,$$

$$(1.2) \qquad \operatorname{Re} w = \psi \quad \text{on} \quad \partial G \qquad \text{and} \qquad \int_{\partial G} \operatorname{Im} w \, \sigma \, ds = \kappa \quad .$$

Here G is a bounded plane domain with a smooth boundary ∂G; the coefficients a, b, c are complex valued functions; ψ, σ are specified real valued functions; κ is a given real constant; w is the unknown complex valued solution and

$$\partial/\partial\bar{z} = (1/2)(\partial/\partial x + i\partial/\partial y).$$

The reduction to the standard form involves only elementary algebraic methods and one smooth extension of a given vector field on ∂G to $\bar{G} := G \cup \partial G$ (see [6], chap. 10.2, 11, 12 or [14] § 6). The solvability theory for these boundary value problems was developed by I. N. Vekua,

W. Haack and G. Hellwig and many others in the 50's, see [12], [6]. In [12] and [6] several constructive methods are given which involve either singular integral equations on ∂G, or the conformal mapping from G onto the unit disk, if G is simply connected.

In this paper an integral equation method is presented which is a combination of a Fredholm equation of second kind in G, the old Neumann integral equation on ∂G, and a Fredholm equation of the first kind on the boundary.

In §2, the solution is represented by the sum of a single layer potential and a plane Newtonian potential:

$$(1.3) \quad w(z) = \iint_G p(\tilde{z}) \log|z-\tilde{z}|\, d\tilde{x}\, d\tilde{y} + \int_{\partial G} q(s) \log|z-\tilde{z}(s)|\, ds,$$

$$z = x + iy \in \overline{G} = G + \partial G$$

where $p = p_1 + ip_2$ and $q = q_1 + iq_2$ denote the layers. The use of Newtonian potentials for equation (1.1) was originally due to I. N. Vekua [12] and was recently extended to quasilinear equations in more than two independent variables by R. P. Gilbert and G. C. Hsiao [5]. The boundary integral equation of the first kind was used y N. I. Muskhelishvili [11] for treating Dirichlet's roblem for the Laplace equation in the plane. G. ichera [4], and G. C. Hsiao and R. C. MacCamy [7] xtended this technique to a quite large class of oundary value problems. The presented method is a imple combination of these methods.

In §3 a successive approximation technique is ised which converges if a and b are sufficiently mall. Using an analytic continuation of the resolvent, s in the book of L. V. Kantorovich and V. I. Krylov 8], and in [3], one can modify the successive approxi- ation technique so that it converges for arbitrary a nd b. An a priori estimate for (1.1) (1.2) leads to a onstruction of this continuation.

In §4 some remarks are made for the numerical reatment of this problem.

In §5 we consider a more general problem by eplacing (1.1) with

$$w_{\bar{z}} = H(z,\bar{z},w,\bar{w}). \tag{1.4}$$

For continuous H with continuous first and second order derivatives with respect to $w,\bar{w}$, an imbedding method along with Newton's method according to H. Wacker [13] leads to a construction of w. Under boundedness conditions for H an existence and uniqueness Theorem for (1.4)(1.2) is established. By applying Newton's method, at each step one obtains linear boundary value problems of the type (1.1)(1.2) which can be solved by the layer method.

For ease of reading we collect three known Lemmas of the potential theory in the Appendix A.

In the Appendix B first numerical results are presented. The numerical computations were made by Mrs. L. Jones with the assistance of F. Pehrson. An error analysis and additional computations will be published in a different paper.

The author wishes to acknowledge the helpful discussions with G. C. Hsiao.

2. The Integral Equations

Let G be a simply connected bounded domain with a α-Hölder continuously differentiable boundary $\partial G (0 < \alpha < 1)$. $C^{r+\alpha}(\overline{G})$ denotes the space of r times Hölder continuously differentiable functions furnished with the usual Hölder norm $||\cdot||_{r+\alpha}$ and $C^{r+\alpha}(\partial G)$ denotes the corresponding boundary function space. $C^{o}(\overline{G})$ denotes the space of continuous functions with the

aximum norm $||\cdot||_o$. We assume for the coefficients
, b, c $\varepsilon\ C^{\alpha}(\overline{G})$ with $a_z, b_z, c_z \ \varepsilon\ C^{o}(\overline{G})$, and further
, $\sigma\ \varepsilon\ C^{1+\alpha}(\partial G)$ with $\sigma > 0$, $\kappa\ \varepsilon\ R$. In order to
void the case of eigensolutions belonging to the
ntegral equation of the first kind ([7] Remark 2.5) we
ssume without loss of generality:

$$\text{(2.1)} \quad \text{diameter } (G) < 1.$$

irst of all we show that every solution of (1.1)(1.2)
s of the form (1.3).

emma 1: The problem (1.1)(1.2) has a unique solution
$w\ \varepsilon\ C^{1+\alpha}(\overline{G})$ which can be represented by (1.3) with
niquely determined layers $p\ \varepsilon\ C^{o}(\overline{G})$ and $q\ \varepsilon\ C^{\alpha}(\partial G)$.

roof: The existence and uniqueness of $w\ \varepsilon\ C^{1+\alpha}(\overline{G})$ is
ell known [6], [12]. Thus, the differential equation
1.1) implies $w_{\bar{z}z}\ \varepsilon\ C^{o}(\overline{G})$ where
$\partial/\partial z = (1/2)(\partial/\partial x - i\partial/\partial y)$, and Green's theorem provides
he representation:

$$\text{(2.2)} \quad w(z) = (2/\pi)\iint_G w_{\bar{\tilde{z}}\tilde{z}} \log|z-\tilde{z}|\,d\tilde{x}d\tilde{y}$$

$$- (2\pi)^{-1}\int_{\partial G} \log\,|z-\tilde{z}|\,(dw/dn_{\tilde{z}})\,ds_{\tilde{z}}$$

$$+ (2\pi)^{-1}\int_{\partial G} w\{(d/dn_{\tilde{z}})\log\,|z-\tilde{z}|\}\,ds_{\tilde{z}}\ .$$

The last term defines a harmonic function $\phi_1 + i\,\phi_2 \in C^{1+\alpha}(\overline{G})$. Therefore, it suffices to solve the first kind equations:

$$(2.3) \qquad \phi_j(z) = \int_{\partial G} \tilde{q}_j \log|z-\tilde{z}|\,ds_{\tilde{z}} \quad , \ j = 1,2 \quad \text{on} \quad \partial G.$$

In Lemma 3 in the Appendix, it will be shown that the assumption (2.1) excludes eigensolutions of (2.3). Then the equations (2.3) have each exactly one solution in $C^\alpha(\partial G)$ ([11] p. 180-186). The uniqueness of p and q follows immediately from (1.3) and (2.2) by differentiation on one hand, and from the unique solvability of (2.3) on the other hand.

For convenience, let us use the following abbreviations for the potentials:

$$(2.4) \qquad Pp: = \iint_G p(\tilde{z}) \log|z-\tilde{z}|\,d\tilde{x}\,d\tilde{y} \ ,$$

$$(2.5) \qquad Qq: = \int_{\partial G} q(\tilde{z}) \log|z-\tilde{z}|\,ds_{\tilde{z}} \ ,$$

$$(2.6) \qquad Kp: = (\partial/\partial\overline{z})Pp = (1/2)\iint_G p(\tilde{z})(\overline{z}-\overline{\tilde{z}})^{-1}d\tilde{x}\,d\tilde{y} \ ,$$

$$(2.7) \qquad Sq: = (\partial/\partial\overline{z})Qq = (1/2)\int_{\partial G} q(\tilde{z})(\overline{z}-\overline{\tilde{z}})^{-1}ds_{\tilde{z}} \ .$$

If the potential (1.3) solves the boundary value problem (1.1)(1.2), then it will satisfy the following system:

$$(2.8) \quad Kp + Sq - a(Pp + Qq) - b(P\overline{p} + Q\overline{q}) = c \quad \text{in } G ,$$

$$(2.9) \quad Pp_1 + Qq_1 = \psi \quad \text{on } \partial G ,$$

$$(2.10) \quad - \int_{\partial G} p_2 \tilde{\sigma}\, ds - \iint_G q_2 \tilde{\sigma}\, dx\, dy = \kappa$$

where $\tilde{\sigma}$ is defined by the potential

$$(2.11) \quad \tilde{\sigma}(z) = -P\sigma .$$

$\tilde{\sigma}$ is positive in $\overline{G}$ because $\log|z-\tilde{z}| < 0$ for all $(z,\tilde{z}) \in \overline{G} \times \overline{G}$ and $\sigma > 0$. If we differentiate (2.8) with respect to z then it becomes

$$(2.12) \quad (\pi/2)p = a_z(Pp + Qq) + b_z(P\overline{p} + Q\overline{q}) + a\left(\overline{K\overline{p}} + \overline{S\overline{q}}\right) + b\left(\overline{Kp} + \overline{Sq}\right) + c_z .$$

This is a second kind equation, and in the special case $a = b = 0$, it determines p uniquely. Then q_1 is determined by (2.9), but q_2 is not uniquely determined from (2.10). Therefore, we need a new equation for q_2. It can be found from (2.8), if z approaches boundary points. Using (2.9) on the boundary and the special combination of (2.8):

$$w_{\bar{z}}\dot{\bar{z}} + \bar{w}_z \dot{z} = (12)\dot{\bar{z}} + (\overline{12})\dot{z} = 2 \operatorname{Re} w_{\bar{z}}\dot{\bar{z}}$$

where $z \to \partial G$ and $\dot{z}$ approaches the tangent in the boundary point, we obtain with the jump conditions for S, the equation:

$$(2.13) \qquad \pi q_2 - \int_{\partial G} \left\{ q_2 (d/dn_z) \log|z-\tilde{z}| \right\} ds_{\tilde{z}}$$

$$= \iint_G p_2 \left\{ (d/dn_z) \log|z-\tilde{z}| \right\} d\tilde{x}\, d\tilde{y} + \dot{\psi}$$

$$- \left(a\dot{\bar{z}}+\bar{b}\dot{\bar{z}}\right)\left(Pp+Qq\right) - \left(\bar{a}\dot{z}+b\dot{\bar{z}}\right)\left(P\bar{p}+Q\bar{q}\right) - \left(c\dot{\bar{z}}+\bar{c}\dot{z}\right).$$

The left hand side defines on the boundary the well known Neumann integral operator, which has one eigen-solution and is therefore not invertable. This can be overcome by the separation of this eigenvalue with a degenerated kernel. We use the condition (2.10) and replace the equation (2.13) by

$$Nq_2 = \pi q_2(z) - \int_{\partial G} q_2(\tilde{z}) \left\{ \left((d/dn_z) \log|z-\tilde{z}| \right) - \tilde{\sigma}(\tilde{z}) \right\} ds_{\tilde{z}}$$

$$(2.14) \qquad = -\kappa - \iint_G p_2 \tilde{\sigma}\, d\tilde{x}\, d\tilde{y} + \iint_G p_2 (d/dn_z) \log|z-\tilde{z}|\, d\tilde{x}\, d\tilde{y} + \dot{\psi}$$

$$- \left(a\dot{\bar{z}}+\bar{b}\dot{\bar{z}}\right)\left(Pp+Qq\right) - \left(\bar{a}\dot{z}+b\dot{\bar{z}}\right)\left(P\bar{p}+Q\bar{q}\right) - \left(c\dot{\bar{z}}+\bar{c}\dot{z}\right).$$

rom Lemma 4 in the Appendix it follows that the
perator N is invertalbe. Furthermore, it is well
nown that the kernel

$$(d/dn_z) \log|z-\tilde{z}| - \tilde{\sigma}(\tilde{z})$$

n (2.14) is Hölder continuous.

The equations (2.12), (2.9) and (2.14) form a
ystem of integral equations for the layers p, q.
or simplicity we denote this system by

$$(2.15) \quad T(p,q) = \left\{c_z, \psi, -\kappa + \dot{\psi} - c\bar{\dot{z}} - \bar{c}\dot{z}\right\},$$

here T maps the pair of layers (p,q) into the triple
n the right hand side. To solve the boundary value
roblem (1.1)(1.2) by making use of (2.15) one requires
o establish the equivalence between the integral
quations (2.15) and the boundary value problem (1.1)(1.2).
his can be seen from the following theorem.

<u>heorem 1</u>: Every solution $(p,q) \in C^0(\overline{G}) \times C^\alpha(\partial G)$ of
he system of integral equations (2.15) generates by (1.3)
he solution w of the boundary value problem (1.1)(1.2).

<u>roof</u>: The potential w satisfies from (2.9), the
econd equation in (2.15), the boundary condition

$$\text{(2.16)} \qquad \operatorname{Re} w = \psi \qquad \text{on} \quad \partial G .$$

Equation (2.12), which is the first equation of (2.15), shows that the function

$$\text{(2.17)} \qquad \phi := w_{\bar{z}} - aw - b\bar{w} - c$$

is antiholomorphic:

$$\text{(2.18)} \qquad \phi_z = 0 \qquad \text{in} \quad G .$$

If we replace $\dot{\psi}$ in (2.14) by the derivative of (2.16) then (2.14) with (2.17) becomes

$$\text{(2.19)} \quad 2 \operatorname{Re} \phi \, \bar{\dot{z}} = \kappa + \iint_G p_2 \tilde{\sigma} \, d\tilde{x} \, d\tilde{y} + \int_{\partial G} q_2 \tilde{\sigma} ds$$

$$= \kappa - \int_{\partial G} \operatorname{Im} w \sigma ds \quad \text{on} \quad \partial G$$

Integrating of (2.19), one can show by Cauchy's theorem, that the constant on the right hand side will vanish because ϕ is antiholomorphic:

$$\text{(2.20)} \qquad 2 \operatorname{Re} \int_{\partial G} \phi(\bar{z}) d\bar{z} = 0 = \int_{\partial G} ds \cdot \left(\kappa - \int_{\partial G} \operatorname{Im} w \sigma \, ds \right) .$$

Thus, w satisfies both boundary conditions in (1.2). Furthermore ϕ satisfies the homogeneous boundary condition:

(2.21) $$2 \operatorname{Re} \phi \, \bar{\dot{z}} = 0 \quad \text{on} \quad \partial G$$

and is, therefore, a continuous solution of an homogeneous elliptic boundary value problem of characteristic + 1. (Regard $\partial/\partial z$ in (2.18) instead of $\partial/\partial \bar{z}$.) Thus, ϕ vanishes identically ([6] p. 306), and (2.17) shows that w solves (1.1).

In the following theorem we shall show that the integral equations (2.15) can be solved not only for the special right hand sides in (2.15) but also for general triples.

<u>Theorem 2</u>: $T : (p,q) \to (d,\psi,\phi)$ maps $C^{o}(\bar{G}) \times C^{\alpha}(\partial G)$ bijective onto $C^{o}(\bar{G}) \times C^{1+\alpha}(\partial G) \times C^{\alpha}(\partial G)$.

<u>Proof</u>: We shall show that T is Fredholm of index 0 and that its nullity is zero, where T is a linear mapping if all spaces are considered as <u>real</u> spaces over the field of real numbers. P and K map bounded subsets of $C^{o}(\bar{G})$ into uniformly Hölder-continuous bounded subsets of $C^{o}(\bar{G})$. By Arzela-Ascoli's Theorem, they define compact mappings in (2.12). Q and S map bounded subsets of $C^{\alpha}(\partial G)$, by Privaloff's Theorem, into bounded subsets of $C^{\alpha}(\bar{G})$ which are compact in $C^{o}(\bar{G})$.

Therefore, all mappings on the right hand side of (2.12) are compact.

In (2.9) P maps bounded subsets of $C^o(\overline{G})$ into bounded subsets of $C^{1+\gamma}(\partial G)$ $(0 < \gamma < 1$ arbitrary$)$, which are compact subsets of $C^{1+\alpha}(\partial G)$ if we choose $\gamma > \alpha$.

In (2.14) P maps bounded subsets of $C^o(\overline{G})$, and Q maps bounded subsets of $C^{1+\alpha}(\partial G)$ into bounded subsets of $C^{1+\alpha}(\partial G)$ which are compact in $C^{\alpha}(\partial G)$. The weak singular integral operator

$$\iint_G p_2(d/dn_z)\log|z-\tilde{z}|d\tilde{x}\ d\tilde{y}$$

maps $C^o(\overline{G})$ continuously into any $C^{\gamma}(\partial G)$ which is compact imbedded into $C^{\alpha}(\partial G)$ for $\gamma > \alpha$.

Therefore, $T - T_o$ is compact where T_o is defined by

(2.22) $$T_o(p,q) := \left(\frac{\pi}{2}\ p,\ Qq_1,\ Nq_2\right) .$$

Each of the three operators defining T_o has a continuous inverse: The first is essentially the identity, Q^{-1} is equivalent to solving equation (2.3) for $\phi_1 \in C^{1+\alpha}$, $q_1 \in C^{\alpha}$; and N is a classical

Fredholm operator of second kind, which has an inverse in $C^{\alpha}(\partial G)$. Thus, T_o is Fredholm with index 0. Because $T-T_o$ is compact, Atkinson's Theorem ([9] Theorem 5.26) implies that T is Fredholm with index 0 also. Consequently, for the bijectiveness of T we have only to show T has no eigensolutions. If $T(p_o,q_o) = (0,0,0)$ for any densities p_o,q_o then the corresponding potential w_o becomes a solution of the homogeneous boundary value problem (1.1)(1.2) (Theorem 1) and; consequently, w_o vanishes identically ([6] p. 322). From Lemma 1 it then follows that the densities p_o,q_o vanish too. Hence, T has no eigensolutions.

3. A Successive Approximation

Corresponding to the boundary value problem (1.1)(1.2), we consider the sequence of problems

$$(3.1) \qquad w_{n+1\bar{z}} = aw_n + b\bar{w}_n + c \ , \ \mathrm{Re}\, w_{n+1} = \psi, \ \int_{\partial G} \mathrm{Im}\, w_{n+1} \sigma ds = \kappa \ ,$$

and define accordingly a sequence of densities by

$$(\pi/2) p_{n+1} := a_z\left(Pp_n + Qq_n\right) + b_z\left(P\bar{p}_n + Q\bar{q}_n\right)$$

$$+ a\left(\overline{K\bar{p}_n} + \overline{S\bar{q}_n}\right) + b\left(\overline{Kp_n} + \overline{Sq_n}\right) + c_z \ ,$$

$$(3.2)\qquad Qq_{1,n+1} := \psi - Pp_{n+1} \quad \text{on} \quad \partial G ,$$

$$Nq_{2,n+1} := -\kappa - \iint_G p_{2,n+1}\, \tilde{\sigma}\, d\tilde{x}\, d\tilde{y}$$

$$+ \iint_G p_{2,n+1} \Big((d/dn_z)\log|z-\tilde{z}|\Big)\, d\tilde{x}\, d\tilde{y}$$

$$- \Big(a\dot{\bar{z}}+\bar{b}\dot{\bar{z}}\Big)\Big(Pp_n+Qq_n\Big) - \Big(\bar{a}\dot{z}+b\dot{\bar{z}}\Big)\Big(P\bar{p}_n+Q\bar{q}_n\Big)$$

$$- \Big(c\dot{\bar{z}}+\bar{c}\dot{z}\Big) + \dot{\psi} \qquad \text{on} \quad \partial G$$

with $p_o = 0,\ q_o = 0$.

Substituting p_{n+1} in the second and third equation, and multiplying by T_o^{-1}, we get an equation

$$(3.3)\qquad \Big(p_{n+1},\, q_{n+1}\Big) = U\Big(p_n,\, q_n\Big) + F ,$$

where $||U||$ becomes small if $||a||_\alpha$, $||b||_\alpha$ and $||a_z||_\alpha$, $||b_z||_\alpha$ are small. Therefore, the following Lemma holds:

<u>Lemma</u> <u>2</u>: If $||a||_\alpha$, $||b||_\alpha$, $||a_z||_\alpha$, $||b_z||_\alpha$ are sufficiently small, then the successive approximation (3.2) converges in $C^o(\bar{G}) \times C^\alpha(\partial G)$.

Remarks: 1) Instead of small coefficients we can assume, that the domain G is small like in [6] §11.5.

2) In (3.2) one needs the inverse of N, which can be found by a successive approximation itself (Appendix Lemma 4).

3) If we use an a priori estimate in (3.1) then the convergence of (3.1) will be guaranteed even for sufficiently small $||a||_{\alpha}$, $||b||_{\alpha}$, which implies the convergence of (3.2).

On the other hand, if a and b are large, one can obtain a modified iterative scheme by using the analytic continuation of the Neumann series defined by (3.2), provided, that a suitable estimate for the eigenvalues of U is known. Such estimates can be found with a priori estimates for the solutions of the boundary value problem (1.1)(1.2). To see this, we write the eigenvalue problem for the densities,

$$(3.4) \qquad (p,q) = \lambda \; U \; (p,q) \; ,$$

first with a real λ as a system of four equations for the real valued functions $\mathrm{Re}\, p$, $\mathrm{Im}\, p$, $\mathrm{Re}\, q$ and $\mathrm{Im}\, q$

because U is linear only over real vector spaces. These four equations are equivalent to the following boundary value problem for the potential $w = u + iv$:

$$\frac{1}{2}\left(u_x - v_y\right) = \lambda\left\{a_r u - a_i v + b_r v + b_i v\right\} ,$$

$$\frac{1}{2}\left(v_x + u_y\right) = \lambda\left\{a_i u + a_r v + b_i v - b_i v\right\} , \tag{3.5}$$

$$u = 0 \quad \text{on} \quad \partial G , \qquad \int_{\partial G} v\sigma \, ds = 0 ,$$

where $a_r = \text{Re } a$, $a_i = \text{Im } a$, $b_r = \text{Re } b$, $b_i = \text{Im } b$.

For the eigenvalue problem, λ can become a complex eigenvalue and hence u and v have to be extended from real valued to complex valued functions:

$$u = u_r + iu_i \ , \ v = v_r + iv_i , \ \lambda = \lambda_1 + i\lambda_2 \ . \tag{3.6}$$

In terms of the new unknown complex valued functions

$$w_1 = u_r + iv_r \ , \quad w_2 = u_i + iv_i \tag{3.7}$$

the eigenvalue problem (3.5) takes the form of a coupled system

$$w_{1\bar{z}} = \lambda_1\left(aw_1 + b\bar{w}_1\right) - \lambda_2\left(aw_2 + b\bar{w}_2\right) ,$$

(3.8) $$w_{2\bar{z}} = \lambda_1\left(aw_2 + b\bar{w}_2\right) + \lambda_2\left(aw_1 + b\bar{w}_1\right) ,$$

$$\operatorname{Re} w_1 = \operatorname{Re} w_2 = 0 \quad \text{on} \quad \partial G,$$

$$\int_{\partial G} \operatorname{Im} w_1\sigma \, ds = \int_{\partial G} \operatorname{Im} w_2\sigma \, ds = 0 ,$$

so that one can estimate λ_2 by making use of an a priori estimate for solutions of boundary value problems of the original kind:

(3.9) $$w_{\bar{z}} = \lambda_1 aw + \lambda_1 b\bar{w} + c , \operatorname{Re} w = 0 \quad \text{on} \quad \partial G ,$$

$$\int_{\partial G} \operatorname{Im} w\sigma \, ds = 0.$$

If we assume that the function c in (3.4) is given then (3.9) has for each real λ_1 a unique solution. Consequently, an a priori estimate

(3.10) $$||w||_o \leq \gamma(\lambda_1) \; ||c||_o$$

holds (see Lemma 5 in the Appendix), where γ can be chosen as a continuous function on the real axis. This is a very crude estimate because (1.1)(1.2) is elliptic and there hold the sharper estimates from Agmon, Douglis, Nirenberg ([1] Theorem 9.3 and Theorem 10.5), which can also be used for our eigenvalue estimate; nevertheless, (3.10) is good enough for providing a simple estimate for λ_2. It can be stated as follows:

<u>Theorem 3</u>: The strip

$$\underset{\sim}{R} := \left\{ \lambda = \lambda_1 + i\lambda_2 \,\middle|\, |\lambda_2| < \gamma(\lambda_1)(||a||_o + ||b||_o) \right\}$$

belongs to the resolvent set of (3.5) or U, respectively.

<u>Proof</u>: If λ is an eigenvalue then it follows from (3.8) with (3.9) and (3.10) that

$$(3.11) \quad ||w_1||_o \leq \gamma(\lambda_1)|\lambda_2|(||a||_o + ||b||_o)||w_2||_o$$

$$\leq \gamma^2(\lambda_1)|\lambda_2|^2(||a||_o + ||b||_o)^2||w_1||_o$$

holds where $||w_1||_o \neq 0$.

For the construction of a suitable analytic continuation of the Neumann series along the real axis, a conformal mapping $\lambda(\eta)$ is used which maps the unit disk $|\eta| < 1$ onto a domain in $\underset{\sim}{R}$ containing 0 and 1, and which is a polynomial with real coefficients:

$$(3.12) \qquad \lambda(\eta) = \sum_{j=1}^{M} \alpha_j \eta^j \ , \quad \lambda(\eta_0) = 1 \ , \quad \eta_0 > 0 \ .$$

Then Taylor's series of the analytic operator valued function $\left(I - \lambda(\eta)U\right)^{-1}$ about 0 converges at η_0 in the operator norm. The series defines an approximation [3], which can be formulated as:

The terms p_n *and* q_n $\left(\textit{not } p_{n+1} \textit{ and } q_{n+1}\right)$ *in* (3.2) *should always be replaced by*

$$\tilde{p}_n = \sum_{j=1}^{M} \alpha_j \eta^j p_{n+1-j} \ , \ \tilde{q}_n = \sum_{j=1}^{M} \alpha_j \eta_o^j q_{n+1-j} \ .$$

4. Some Remarks on the Numerical Procedure

Because the Cauchy integral S is discontinuous across the boundary curve, to compute p in (3.2) one should only use the interior points of G in the quadrature formula for the double integrals. Although this computation involves a large number of grid points in G (and, consequently, a large number of discretized equations), because of the explicit form of the first equation in (3.2), fortunately, one is not required to solve a system of algebraic equations in G. On the other hand, from the numerical quadrature of the second and third equation in (3.2), there are only a few linear equations on ∂G. Thus, one can apply the procedure easily in practice.

With respect to the numerical convergence, we observe, that all operators in the first and the last equation of (3.2), except the identities, are compact, while the second equation is of the first kind. Therefore, the replacement of the first and third equation by quadrature formulae will lead to collectively compact approximations, from which the convergence properties can be obtained by using P. Anselone's approach [2]. By following the idea of R. Kussmaul and P. Werner [10], G. Hsiao (in an unpublished paper) made an error analysis for the first kind equation (2.3).

Some numerical experiments of the proposed method are presently under investigation. The details will be presented in a different paper. In Appendix B we show some first results.

5. A Simple Semilinear Equation

Let us now consider the semilinear boundary value problem

(5.1) $$w_{\bar{z}} = H(z, \bar{z}, w, \bar{w}) \quad \text{in} \quad G,$$

$$\operatorname{Re} w = \psi \quad \text{on} \quad \partial G \quad \text{and} \quad \int_{\partial G} \operatorname{Im} w\sigma \, ds = \kappa \ .$$

Here we assume that

(5.2) $\quad H, H_w, H_{\overline{w}}, H_{ww}, H_{w\overline{w}}, H_{\overline{w}\overline{w}} \in C^0(\overline{G} \times C)$

and all these functions are bounded by a constant K. In addition, for each fixed value of $w \in C$, $H(z,\overline{z}, w,\overline{w}) \in C^\alpha(\overline{G})$.

Our main concern is the existence of the solution to the equation (5.1) as well as the approximation to the solution. The results can be stated as follows:

Theorem 4: The semilinear boundary value problem (5.1) has exactly one solution $w \in C^{1+\alpha}(\overline{G})$, which can be constructed by the method of imbedding according to H. Wacker [13].

Remarks. If the existence of a solution to (5.1) is assumed, then the boundedness condition on H is not necessary because H in (5.1) can be multiplied by a function $\chi(w)$ which has a sufficiently large compact support. The essences of the proof of Theorem 4 are the imbedding of the given problem (5.1) into a family of problems and an a priori estimate for the solutions of the whole family.

(i) Imbedding:

(5.3) $$w_{\bar{z}}(z,t) = tH\left(z,\bar{z},w,\bar{w}\right) \quad \text{in } G \text{ with } t \in [0,1] \text{ and}$$

$$\operatorname{Re} w = \psi \quad \text{on} \quad \partial G \;, \quad \int_{\partial G} \operatorname{Im} w\sigma \, ds = \kappa \;.$$

(ii) A priori estimate (see Lemma 5 in the Appendix):

For all $w \in C^1(\bar{G})$ satisfying the homogeneous boundary conditions

$$\operatorname{Re} w = 0 \quad \text{on} \quad \partial G \;, \quad \int_{\partial G} \operatorname{Im} w\sigma \, ds = 0$$

the a priori estimate

(5.4) $$||w||_o \leq \gamma \, ||w_{\bar{z}} - aw - b\bar{w}||_o$$

holds for any a and b with $||a||_o, \; ||b||_o \leq K$, where the constant γ depends only on K, G and σ .

We return now to the

Proof of Theorem 4:

For $t = 0$ the solution $w(z,0)$ can easily be constructed by making use of the layer method in the preceding chapters. Using this solution as the initial approximation, we solve (5.3) for $t = t_1 > 0$ by ewton's method. Then $w(z,t_1)$ is known and defines he initial approximation for (5.3) with $t = t_2 > t_1$. e shall show that after finite steps the solution for $= 1$, the original problem, can be found. To this nd let us consider Newton's method for a fixed value f $t = t_j$:

$$
(5.5)\qquad w_{n+1\bar{z}} = t_j H_w(z,\bar{z},w_n,\bar{w}_n)(w_{n+1}-w_n) + t_j H_{\bar{w}}(z,\bar{z},w_n,\bar{w}_n)(\bar{w}_{n+1}-\bar{w}_n) + t_j H(z,\bar{z},w_n,\bar{w}_n) ,
$$

$$
\operatorname{Re} w_{n+1} = \psi \text{ on } \partial G , \int_{\partial G} \operatorname{Im} w_{n+1}\, \sigma ds = \kappa , \; n=0,1,\ldots .
$$

or each iteration we have a linear problem (1.1)(1.2) or w_{n+1}, which can be solved by the method of the revious sections. For the convergence of (5.5) we investigate $w_{n+1} - w_n$, which satisfies the equation

$$
(5.6)\qquad w_{n+1\bar{z}} - w_{n\bar{z}} = t_j H_w(z,w_n,\bar{w}_n)(w_{n+1}-w_n) + t_j H_{\bar{w}}(\bar{w}_{n+1}-\bar{w}_n)
$$

$$
+ t_j \tfrac{1}{2} H_{(w,w)}(w_n-w_{n-1})^2 + t_j H_{(w,\bar{w})}|w_n-w_{n-1}|^2
$$

$$
+ t_j \tfrac{1}{2} H_{(\bar{w},\bar{w})}(\bar{w}_n-\bar{w}_{n-1})^2 ,
$$

$$
\operatorname{Re}(w_{n+1}-w_n) = 0 \text{ on } \partial G, \int_{\partial G} \operatorname{Im}(w_{n+1}-w_n)\sigma ds = 0 ,
$$

vhere $H_{(w,w)}$ is defined by

$$(5.7) \qquad H_{(w,w)} = \int_0^1 H_{ww}\left(z, tw_n + (1-t)w_{n-1}, t\overline{w}_n + (1-t)\overline{w}_{n-1}\right) dt .$$

and $H_{(w,\overline{w})}$, $H_{(\overline{w},\overline{w})}$ are defined similarly.

Using (5.2) and (5.4), we get the estimate

$$(5.8) \qquad ||w_{n+1} - w_n||_o \leq 2Kt_j\gamma||w_n - w_{n-1}||_o^2 .$$

Thus (5.5) converges if the initial approximation w_o is chosen such that

$$(5.9) \qquad 2Kt_j\gamma||w_1 - w_o||_o < 1 .$$

For w_o we use the solution of (5.4) corresponding to $t_{j-1} < t_j$. Then $w_1(t_j,z) - w_o$ satisfies

$$(5.10) \qquad w_{1\overline{z}} - w_{o\overline{z}} = t_j H_w\left(z, w_o, \overline{w}_o\right)\left(w_1 - w_o\right) + t_j H_w\left(\overline{w}_1 - \overline{w}_o\right) + \left(t_j - t_{j-1}\right) H\left(z, w_o, \overline{w}_o\right),$$

and the homogeneous boundary conditions. Therefore, under the assumptions of H, the a priori estimate (5.4) yields the inequality

$$(5.11) \qquad ||w_1 - w_o||_o \leq \left(t_j - t_{j-1}\right) K\gamma ||H\left(\cdot,\cdot,w(\cdot,t_{j-1}),\overline{w}(\cdot,t_{j-1})\right)||_o .$$

Thus, according to (5.9), t_j has to be chosen such that

$$(5.12) \qquad t_j\left(t_j - t_{j-1}\right) < \left(2\gamma^2 K^2 ||H\left(\cdot,\cdot,w(\cdot,t_{j-1}),\overline{w}(\cdot t_{j-1})\right)||_o\right)$$

holds. The assumed boundedness of H provides a uniform bound for the right hand side.

Letting $t_o = 0$, we solve (5.4) for a finite set of t_j-values, by using Newton's method described by (5.5) and end up with $t_E = 1$ and $w(z,1)$, the solution of the desired problem (5.1).

The uniqueness of w can be obtained easily as follows: Let w_1 and w_2 be any two solutions of (5.1). Then the difference $w^* = w_1 - w_2$ satisfies a homogeneous linear boundary value problem:

$$(5.13) \qquad w^*_{\overline{z}} = H_{(w)} w^* + H_{(\overline{w})} \overline{w}^* , \quad \text{Re } w^* = 0 \quad \text{on} \quad \partial G,$$

$$\int_{\partial G} \text{Im } w^* \sigma \, ds = 0 ,$$

where $H_{(w)}$ and $H_{(\overline{w})}$ are defined similar to (5.7). By the uniqueness of this problem it follows that $w^* \equiv 0$ ([6] p. 322).

It remains to consider the regularity of the solution. This can be obtained from (5.1). Since $w_{\bar{z}} = H$ is continuous it follows that w is Hölder continuous, and hence H becomes $C^{\alpha}(\overline{G})$, which implies $w \in C^{1+\alpha}(\overline{G})$.

Appendix A:

Lemma 3: If diam $(G) < 1$ then the first kind equation (2.3) has no eigensolutions.

Proof: According to [11] p. 181 ff., any eigensolution ν_o of (2.3) must be the "natural layer" which is the eigensolution of Neumann's integral equation on ∂G; ν_o generates a potential u, which is constant in G and grows for $z \to \infty$:

$$u = -\int_{\partial G} \nu_o \log|z-\tilde{z}|\, ds_{\tilde{z}} = -\log|z| \int_{\partial G} \nu_o ds_{\tilde{z}} + O\left(\frac{1}{|z|}\right)$$

where

$$\int_{\partial G} \nu_o ds \neq 0 .$$

Thus the outer potential has a constant maximum on ∂G and from E. Hopf's second Lemma, it follows that

$$2\pi\nu_o = (d/dn)u^- < 0 ,$$

where $(d/dn)u^-$ denotes the exterior normal derivative. The condition diam $(G) < 1$, together with $\nu_o < 0$

implies that $u \neq 0$ on ∂G and hence ν_o is not an eigensolution.

Lemma 4: The operator N in (2.14) has no eigensolutions. Moreover, the equation

$$Nu = f$$

can be solved by the successive approximation

$$\text{(a.1)} \qquad u_{j+1} = \frac{1}{\pi}\int_{\partial G} u_j (d/dn_z)\log|z-\tilde{z}|ds_{\tilde{z}} - \nu_o \int_{\partial G} u_j\tilde{\sigma}ds$$

$$+ f + \kappa(\nu_o - 1)$$

$$= Ku_j + f + \kappa(\nu_o - 1)$$

if $\kappa = \int_{\partial M} u\tilde{\sigma}\, ds$ is known.

Proof: If q_o is an eigenvolustion then $\int Nq_o\, ds = 0$ implies

$$\text{(a.2)} \qquad \int_{\partial G} q_o\tilde{\sigma}\, ds = 0 \ .$$

Hence

$$\pi q_o = \int q_o (d/dn_z)\log|z-\tilde{z}|ds_{\tilde{z}}$$

holds which implies

$$\text{(a.3)} \qquad q_o = \alpha\nu_o$$

for some constant α. Substituting (a.3) into (a.2) one obtains $\alpha = 0$ hence $q_o = 0$.
In the same way it follows that in $I - \lambda K$ the value $\lambda = 1$ is no eigenvalue. A simple decomposition of the other eigensolutions into $\alpha\nu_o$ and its orthogonal complement shows that all the other eigenvalues λ_j are the eigenvalues of Neumann's integral operator. Because they fulfill $|\lambda_j| > 1$ ([6] p. 134 ff.), the spectral radius of K is smaller than 1. Hence (a.1) converges.

<u>Lemma 5</u>: Let w be the solution of

$$w_{\bar{z}} = aw + b\bar{w} + c \ ,$$

(a.4) $$\operatorname{Re} w = 0 \quad \text{on} \quad \partial G \ , \quad \int_{\partial G} \operatorname{Im} w\sigma \, ds = 0$$

where $|a|, |b| \leq K$. Then there exists a constant γ depending only on G, K, σ, such that

(a.5) $$||w||_o \leq \gamma ||c||_o \ .$$

<u>Proof</u>: Let $\tilde{w}$ be any function satisfying the boundary condition (a.4). One can show by using the representation formula (10.4.3) in [6], that

(a.6) $$||w||_o \leq \hat{\gamma} \ ||\tilde{w}_{\bar{z}}||_o$$

holds where $\hat{\gamma}$ depends only on G and σ. Now let v be the solution of

$$v_{\bar{z}} = \begin{cases} a + b\,\frac{\bar{w}}{w} & \text{if } w \neq 0 \\ a & \text{if } w = 0 \end{cases}$$

and let v satisfy (9.4). Then clearly v fulfills inequality

$$(a.7) \qquad ||v||_o \leq 2\hat{\gamma}\, K .$$

Setting

$$(a.8) \qquad f_o = w \exp(-v) - i\alpha$$

one can easily show that f_o satisfies the boundary value problem

$$(a.9) \qquad f_{o\bar{z}} = c \exp(-v), \ \operatorname{Re} f_o = o \ \text{ on } \ \partial G, \int_{\partial G} \operatorname{Im} f_o \sigma \, ds = 0$$

with α defined by

$$\alpha = -\left(\int_{\partial G} \exp(-v)\sigma ds\right)^{-1}\left(\int_{\partial G} \operatorname{Im} f_o \exp(-v)\sigma ds\right).$$

Then it follows from (a.6) with the help of (a.7) that

$$(a.10) \qquad ||f_o||_o \leq \hat{\gamma} \exp(2\hat{\gamma}K)\,||c||_o \,.$$

A simple computation shows that

$$(a.11) \quad |\alpha| \le \hat{\gamma}\, \exp(\sigma\hat{\gamma}K)\,||\sigma||_o \left(\int \sigma ds\right)^{-1} ||c||_o .$$

The result (a.5) then follows immediately from (a.7), (a.8), (a.10) and (a.11) with γ defined by

$$\gamma = \hat{\gamma}\, \exp(4\hat{\gamma}K)\left\{1 + \exp(4\hat{\gamma}K)\,||\sigma||_o \left(\int \sigma ds\right)^{-1}\right\} .$$

Appendix B:

Dr. Louise Jones (Department of Statistics and Computer Science, University of Delaware) made the numerical experiments with the assistance of F. Pehrson, who wrote a program in FORTRAN IV.

For the domain G we chose ellipses

$$x = r\cos\Theta \ , \ y = \varepsilon r \sin\Theta \ , \quad \Theta\varepsilon[0,2\pi] \ , \quad r\varepsilon[0,0.4]$$

(Here r is restricted to be less than 0.4 because of condition (2.1)) with $\varepsilon = 1, 0.5, 0.2$. The boundary value problem were

$$w_{\overline{z}} = \delta(1-i)w + \delta(1+i)\overline{w} + c \text{ in } G, \ \text{Re } w = \psi \text{ on } \partial G, \ \kappa = 0$$

with $\delta = 0$, o.1,1, where the functions c and ψ were chosen by $c = z - 2\delta z\overline{z}$, $1 - 2\delta(z+\overline{z})$, $2\overline{z} - 2\delta(z^2+\overline{z}^2)$ and $\psi = z\overline{z}$, $z+\overline{z}$, $z^2+\overline{z}^2$ on ∂G such that the exact solutions are $w = z\bar{z}$, $z+\bar{z}$, $z^2+\bar{z}^2$.

The integrals with singularities were replaced by the following regularized expressions:

$$\iint p(\tilde{z})(z-\tilde{z})^{-1}d\tilde{x}d\tilde{y} = \iint_G (p(\tilde{z})-p(z))(z-\tilde{z})^{-1}d\tilde{x}d\tilde{y}$$

$$+ p(z) \int_{\tilde{z}\varepsilon\partial G} \exp(-i \arg (\tilde{z}-z))|\tilde{z}-z| d \arg (\tilde{z}-z)$$

where $(p(\tilde{z})-p(z))(z-\tilde{z})^{-1}$ for $z = \tilde{z}$ was replaced by a suitable finite difference quotient; and

$$\iint_G p(\tilde{z})\log|z-\tilde{z}|d\tilde{x}d\tilde{y} = \iint (p(\tilde{z})-p(z))\log|z-\tilde{z}|d\tilde{x}d\tilde{y}$$

$$+ p(z) \int_{\tilde{z}\varepsilon\partial G} (1/4)|z-\tilde{z}|^2(2\log|z-\tilde{z}|-1)d \arg (\tilde{z}-z)$$

where the integrands are zero for $z = \tilde{z}$.

In the neighborhood of z, $z\varepsilon\partial G$, the singular boundary integral of Q was replaced by

$$\int_{s_{j-1}}^{s_{j+1}} q \log|\tilde{z}-z(s_j)|ds \approx q(s_j)2\Delta s(\log\Delta s-1) \quad \text{with}$$

$$s_{j\pm 1} = s_j \pm \Delta s .$$

For the remaining boundary integral Simpson's rule was used. The double integrals were evaluated as iterated integrals, integrated first with respect to Θ using the trapezoidal rule with 20 node points corresponding to equidistant Θ intervalls and then with respect to r using an open quadrature formula with 9 equidistant node points and the weights $0,\frac{3}{2},\frac{3}{2},\frac{1}{3},\frac{4}{3},\frac{1}{3},\frac{3}{2},\frac{3}{2},0.$

The numerical results are collected in the Tables 1, 2, 3, where w_c denotes the computed value and w_E denotes the exact solution. w_c is computed at points on the positive x-axis.

TABLE 1: $w_E = z\bar{z}$

	x	w_C	w_E
domain ε = 1	0.1	0.01017	0.01000
δ = 0	0.2	0.04017	0.04000
no iterations	0.3	0.09024	0.09000
δ = 0.1	0.1	0.01017	0.01000
3 iterations	0.2	0.04018	0.04000
	0.3	0.09026	0.09000
δ = 1	0.1	0.01017	0.01000
3 iterations	0.2	0.04018	0.04000
	0.3	0.09026	0.09000
domain ε = 0.5	0.1	0.0128	0.0100
δ = 0	0.2	0.0417	0.0400
no iterations	0.3	0.0899	0.0900
δ = 0.1	0.1	0.0119	0.0100
3 iterations	0.2	0.0412	0.0400
	0.3	0.0895	0.0900
δ = 1	0.1	0.0106	0.0100
3 iterations	0.2	0.0430	0.0400
	0.3	0.0924	0.0900
domain ε = 0.2	0.1	0.0128	0.0100
δ = 0	0.2	0.0417	0.0400
no iterations	0.3	0.0899	0.0900
δ = 0.1	0.1	0.0130	0.0100
3 iterations	0.2	0.0420	0.0400
	0.3	0.0901	0.0900
δ = 1	0.1	0.0136	0.0100
3 iterations	0.2	0.0440	0.0400
	0.3	0.0979	0.0900

TABLE 2: $w_E = z+\overline{z}$:

	x	w_C	w_E
domain $\varepsilon = 1$	0.1	0.1982	0.2
$\delta = 0$	0.2	0.3965	0.4
no iterations	0.3	0.5951	0.6
$\delta = 0.1$	0.1	0.2284	0.2
iterations	0.2	0.4209	0.4
	0.3	0.6094	0.6
domain $\varepsilon = 0.5$	0.1	0.1988	0.2
$\delta = 0$	0.2	0.3975	0.4
no iterations	0.3	0.5957	0.6
$\delta = 0.1$	0.1	0.2109	0.2
3 iterations	0.2	0.4072	0.4
	0.3	0.6014	0.6
$\delta = 1$	0.1	0.31	0.2
3 iterations	0.2	0.50	0.4
	0.3	0.66	0.6
domain $\varepsilon = 0.2$	0.1	0.1998	0.2
$\delta = 0$	0.2	0.3985	0.4
no iterations	0.3	0.5975	0.6
$\delta = 0.1$	0.1	0.2021	0.2
3 iterations	0.2	0.4004	0.4
	0.3	0.5986	0.6
$\delta = 1$	0.1	0.223	0.2
3 iterations	0.2	0.419	0.4
	0.3	0.610	0.6

TABLE 3: $w_E = z^2 + \bar{z}^2$

	x	w_c	w_E
domain $\varepsilon = 1$	0.1	0.0188	0.02
$\delta = 0$	0.2	0.0752	0.08
no iterations	0.3	0.1695	0.18
$\delta = 0.1$	0.1	0.0289	0.02
3 iterations	0.2	0.0795	0.08
	0.3	0.1736	0.18
domain $\varepsilon = 0.5$	0.1	0.0256	0.02
$\delta = 0$	0.2	0.0823	0.08
no iterations	0.3	0.1763	0.18
$\delta = 0.1$	0.1	0.0271	0.02
3 iterations	0.2	0.0850	0.08
	0.3	0.1787	0.18
$\delta = 1$	0.1	0.0213	0.02
3 iterations	0.2	0.0974	0.08
	0.3	0.1972	0.18
domain $\varepsilon = 0.2$	0.1	0.0261	0.02
$\delta = 0$	0.2	0.0843	0.08
no iterations	0.3	0.1798	0.18
$\delta = 0.1$	0.1	0.0265	0.02
3 iterations	0.2	0.0850	0.08
	0.3	0.1809	0.18
$\delta = 1$	0.1	0.0283	0.02
3 iterations	0.2	0.0909	0.08
	0.3	0.1865	0.18

Because the boundary values can always be approximated by piecewise analytic functions, the use of Gaussian's quadrature formulas for parts of the Θ intervall and the r - integration promises better results. (In the method in [6] p. 128 the Gaussian quadratures provided a good improvement.) The numerical experiments are not finished at the present time.

REFERENCES

[1] S. Agmon, A. Douglis and L. Nirenberg, Estimates Near the Boundary for Solutions of Elliptic Partial Differential Equations Satisfying General Boundary Conditions II. Comm. Pure App. Math, XVII (1964) 35-92.

[2] P. M. Anselone, Collectively Compact Operator Approximation Theory, Prentice-Hall, 1971.

[3] G. Bruhn und W. Wendland, Über die näherungsweise Lösung von linearen Funktionalgleichungen, ISNM Vol. 7 (1967) Funktionalanalysis, Approximationstheorie, Numerische Marhematik, Birkhäuser Basel, S. 136-164.

[4] G. Fichera, Linear elliptic equations of higher order in two independent variables and singular integral equations. Proc. Conference on Partial Differential Equations and Continuum Mechanics (Madison, Wis.) Univ. of Wisconsin Press, Madison, 1961.

[5] R. P. Gilbert and G. C. Hsiao, On Dirichlet's Problem for Quasi-Linear Elliptic Equations. Proceedings to the Conference on Constructive and Computational Methods for Differential and Integral Equations, Bloomington, 1974.

[6] W. Haack and W. Wendland: Lectures on Partial and Pfaffian Differential Equations, Pergamon Press, Oxford, 1972.

[7] G. C. Hsiao and R. C. MacCamy, Solution of Boundary Value Problems by Integral Equations of the First Kind. SIAM Review Vol. 15, No. 4 (1973) 687-705.

[8] L. V. Kantorovich and V. I. Krylov, Approximate Methods of Higher Analysis, Interscience, 1958.

[9] T. Kato, Perturbation theory for linear operators. Springer, 1966.

[10] R. Kussmaul and P. Werner, Fehlerabschätzungen für ein numerisches Verfahren zur Auflösung linearer Integralgleichungen mit schwach singulären Kernen. Computing 3 (1968) 22-46.

[11] N. I. Muskhelishvili, Singular Integral Equations Noordhoff Groningen 1953 (Moscow 1946).

[12] I. N. Vekua, Generalized Analytic Functions. Pergamon Press, 1962.

[13] H. J. Wacker, Eine Lösungsmethode zur Behandlung nichtlinearer Randwertprobleme. Iterationsverfahren, Numerische Mathematik, Approximationstheorie ISNM Vol. 15, Birkhäuser 1970, S. 245-257.

[14] W. Wendland, Elliptic Boundary Value Problems in the Plane. Lecture Notes, University of Delaware 1974, to appear.

Solving Partial Differential Equations using ILLIAC IV

by

Robert Wilhelmson

Laboratory for Atmospheric Research
Center for Advanced Computations
University of Illinois

ABSTRACT

The solution of the Benard-Rayleigh convection problem using ILLIAC IV is reviewed and comments about solving partial differential equations using this computer are made. This includes a brief look at possible uses of parallelism in designing arithmetic logic units, at the Institute for Advanced Computations, at the ARPA communications network and at the ILLIAC IV. An explicit solution of the prognostic convection equations is considered and the use of iterative and direct methods for solving the Poisson streamfunction equation discussed. Storage considerations are emphasized. Closing remarks on algorithmic parallelism, others using ILLIAC and I-O and programming language considerations for ILLIAC are included.

1. The Benard-Rayleigh convection problem

Despite the increase in computer speed during the past 20 years the demands for faster computers continue. Vector processing machines such as the ILLIAC IV, the Control Data Corporations' STAR and the Texas Instruments' ASC belong to a class of parallel machines that provide increased speed for problems in which vector operations are important. The recent availability of ILLIAC IV has permitted the testing of codes written for ILLIAC IV and previously simulated on a Burrough's B6700.

The ILLIAC IV is not a general purpose machine and therefore a natural question for a potential user is 'Can my problems be solved using ILLIAC IV and can they be solved efficiently?'. This question motivated an efficiency study of ILLIAC IV for doing hydrodynamic calculations [7]. The numerical solution of the Benard-Rayleigh convection equations was considered to be a representative hydrodynamic problem. The non-dimensional Boussinesq equations in two-dimensions describing this flow in a rectangular region are for vorticity (η), temperature (T) and streamfunction (ψ)

$$\frac{\partial \eta}{\partial t} + u \frac{\partial \eta}{\partial x} + w \frac{\partial \eta}{\partial z} = \frac{\partial \eta}{\partial t} + J(\psi, \eta) = \frac{R_a}{P_r} \frac{\partial T}{\partial x} + \nabla^2 \eta, \qquad u = -\frac{\partial \psi}{\partial z}, \quad w = \frac{\partial \psi}{\partial x} \tag{1}$$

$$\frac{\partial T}{\partial t} + J(\psi, T) = \frac{1}{P_r} \nabla^2 T \tag{2}$$

$$\eta = \frac{\partial w}{\partial x} - \frac{\partial u}{\partial z} = \nabla^2 \psi \tag{3}$$

The temperature was held fixed at the top and bottom boundaries, with the lower value being hotter. When this temperature difference is sufficiently large convective motion occurs. Along the side boundaries $\partial T/\partial x = 0$. Both η and ψ were zero on all boundaries. The steady-state solution for air and for the constant Rayleigh (R_a) and Prandtl (P_r) numbers given in [7] appears in Fig. 1. The motion is cellular with advective cooling in the center of the domain. The solution of this problem using the vector capabilities of ILLIAC IV will be discussed later.

2. ILLIAC IV

a. Hardware parallelism

The functional relationships in a conventional computer as taken from either [1] or [2] are given in Fig. 2. Parallelism in the ALU can take different forms provided that the other functional units are appropriately designed. These are briefly described in order to distinguish the parallel approach used in ILLIAC IV from other approaches. They are:

1. concurrent use of two components such as the adder and multiplier,
2. the replication of a component coupled with concurrent usage. For example, an ALU might have two multiplier components such as the CDC 7600.

3. the use of a pipeline approach in which a particular process such as addition is broken into parts. For example, an adder can be divided according to the basic operations: adjust exponent, add mantissa, normalize result and round result. Provided that temporary storage is used to hold intermediate results of a basic operation until all other basic operations are complete a pipeline is set up. Two operands can be entering the pipe for exponent adjustment as two others whose exponents are already adjusted are having their mantissas added. An important consideration in the usefulness of this approach is the ability to keep the pipe full. With sufficiently long one-dimensional vectors providing the operands this can be accomplished apart from the first and last few operands added. Both the CDC STAR and TI ASC use this approach.
4. replication of the ALU [1,2]. This will be discussed further in Section c with regard to ILLIAC IV,
5. combinations of these forms.

b. The Institute for Advanced Computations (IAC)

The Institute for Advanced Computation, located at Ames Research Center in Moffett Field, California, was formed in 1971 to develop and operate a computer system which includes two one-of-a-kind devices. These are the ILLIAC IV Processor and the UNICON Laser Memory. The latter is a write-once-only carrousel device with removable slides or strips. It has an on-line storage capacity of 700 billion bits or 85 billion characters. A diagram showing the basic system components and access to the system is given in Fig. 3. The central system manages and provides services for the ILLIAC and the UNICON and consists of several computers, a central memory and disk and drum storage. A more detailed description of all IAC components and their functions can be found in an IAC press release [12].

Communication to and from the system's communications processor is handled through the ARPA network via remote terminals or remote computer systems. This network was initiated by the Advanced Research Projects Agency of the Department of Defense for the purpose of connecting via 50,000 baud high speed data transmission lines a number of research centers and projects around the United States. The reliability of this store-and-forward message transmission network is estimated to be a one bit error going undetected per year. A developmental report has been written by Roberts and Wessler [8]. Experience using the Network at the University of Illinois has been reported by Sher [9] and at the University of California at Los Angeles by Kehl [5].

c. ILLIAC IV

A functional block diagram of ILLIAC IV is given in Fig. 4. Some observations about it will be useful in later discussions. For further information about ILLIAC IV see Bouknight *et al* [1].

First, one notices that there are 64 processing elements (PE_i's) or ALU's. Associated with each processing element is a random access memory of 2048 64-bit words, the PEM_i. Access by one PE_i to another PEM_j is not permitted. Data is transferred to PE_i by first fetching it to PE_j from PEM_j and then routing it from PE_j to the desired PE_i. The CU has access, however, to all of memory, where both programs and data are stored. The CU also has a small ALU within it, which has a different instruction set than the PE's do. CU instructions and PE instructions can be executed simultaneously.

Execution of a PE instruction such as an add is carried out by all PE's unless they are deactivated by mode control. When the mode bit for a PE is off certain registers in that PE are protected and storage of data from that PE to its PEM is prohibited. The mode status can be tested and changed during program execution.

When a PE seeks an element in its PEM the base address it uses is provided by the CU. This base address can be modified in each PE by use of the PE index register. This permits simultaneous addressing of different storage locations in different PEM's through their corresponding PE's.

3. Numerical integration of the Benard-Rayleigh convection equations

In this section the relation between the way data is stored in the PEM's and its subsequent retrieval for parallel computations will be briefly discussed from the viewpoint of solving the convective equations (1), (2) and (3). The references for this section are Ogura, Sher and Ericksen [7] and Ericksen [3]. Since these reports became available several in-core convective codes have been successfully run using ILLIAC IV.

For simplicity in the following discussion the dimensions of the η, ψ and T arrays including boundary values are taken to be 64 x 64. Consideration of input-output requirement for large data arrays that require auxiliary storage is not included here.

<u>a. The prognostic equations</u>

Equations (1) and (2) were integrated using effectively explicit finite difference approximations. The temperature T, for example, at time step $\tau+1$ is computed from T at time steps $\tau-1$ and τ and from ψ at τ. Without specifying the details of the Jacobian approximation the scheme is

$$T_{i,k}^{(\tau+1)} = \frac{1}{c}\left[c' T_{i,k}^{(\tau-1)} + 2\Delta t \left\{ -J_{i,k}(\psi^{(\tau)}, T^{(\tau)}) + \frac{1}{P_r}\left(\frac{T_{i+1,k}^{(\tau)} + T_{i-1,k}^{(\tau)}}{(\Delta x)^2} + \frac{T_{i,k+1}^{(\tau)} + T_{i,k-1}^{(\tau)}}{(\Delta z)^2}\right)\right\}\right]. \tag{4}$$

The use of the DuFort-Frankel scheme for the diffusion term gives rise to the constants c and c' depending on Δt, Δx, Δz and P_r.

One appropriate storage scheme for this integration is given in Fig. 5. Here U stands for η, ψ or T and represents a straight mapping of the two-dimensional arrays into ILLIAC's two-dimensional memory. With this scheme simultaneous calculation of a row of 62 values of T at time step $\tau+1$ is possible. For example, temperature on rows i+1 and i-1 at time step τ can be added to form a partial sum in the diffusion calculation. PE's 0 and 63 could be deactivated so that this sum is not done by them. Then routing can be used so that each PE can simultaneously access temperatures in columns k+1 or in columns k-1 at time step τ to continue the diffusion calculation.

The Neumann lateral boundary values of T (k=1 and k=64), however, must be calculated sequentially. Because of their simplicity in this case little extra time and work is needed. If they became more complicated a storage scheme that would allow simultaneous access to both rows and columns might be considered. For the present problem, transposing all arrays before storage would suffice, as the top and bottom temperature values are not Neumann but held constant and boundary values for η and ψ are everywhere zero.

b. The diagnostic equation

Both iterative and direct methods for solving the streamfunction equation (3) have been considered by Ericksen [3]. Some of the codes at the end of this report have been modified and errors found when debugging them on ILLIAC IV.

A modified version of SOR called MSOR (modified successive over-relaxation) given by Young [11] permits simultaneous calculation of 62 iterate values. The equations are

$$\text{"red"} = \{\psi_{i,j} \mid i + j \text{ is even}\}$$

$$\text{"black"} = \{\psi_{i,j} \mid i + j \text{ is odd}\}$$

$$\text{RED} \quad \psi_{i,j}^{(\ell+1)} = \alpha(\psi_{i,j+1}^{(\ell)} + \psi_{i,j-1}^{(\ell)}) + \beta(\psi_{i-1,j}^{(\ell)} + \psi_{i+1,j}^{(\ell)}) - \gamma\eta_{i,j} + (1-\omega)\,\psi_{i,j}^{(\ell)}.$$

$$\text{BLACK} \quad \psi_{i,j}^{(\ell+1)} = \alpha(\psi_{i,j+1}^{(\ell+1)} + \psi_{i,j-1}^{(\ell+1)}) + \beta(\psi_{i-1,j}^{(\ell+1)} + \psi_{i+1,j}^{(\ell+1)} - \gamma\eta_{i,j} + (1-\omega)\,\psi_{i,j}^{(\ell)} .$$

where α, β and γ are constants dependent on Δx, Δz and the over relaxation parameter ω. A row of "red" elements is indicated in Fig. 5 by dark boxes. These elements can be accessed simultaneously through PE indexing. This and other iterative methods require at least two arrays, one for ψ and for η. Two other iterative methods, namely SLOR and ADI, investigated by Ericksen appear to have limited value on ILLIAC IV in terms of their adaptability to parallel computation.

The direct method of Hockney [4] without odd-even reduction can also be used to solve equation (3). Fourier analysis of the 62 columns can be performed simultaneously using the storage in Fig. 5. This results in 62 independent and tridiagonal systems of ordinary differential equations, one per row. If these rows are skewed as in Fig. 6, columns can be accessed simultaneously through PE indexing and the 62 ordinary differential systems solved in parallel. The values of ψ can then be found in parallel through Fourier synthesis after the rows are unskewed.

If skewed storage was originally used for all arrys then access to both rows and columns could be accoomplished without data shuffling. However, with this storage for all variables additional routing may be required. For example, in the temperature diffusion term of equation (2), i+1 and i-1 values are no longer stored in the same PEM as the i value being calculated.

In the convection problem Hockney's approach provided the fastest solution. It also requires only 1 array in memory as the source term can be overwritten as Fourier analysis is carried out. This is not true for iterative methods. For problems in which auxiliary storage is used and in which the storage of an array requires most of the memory this may be an important consideration.

4. Closing remarks--the use of vector processor like ILLIAC IV

The usefulness of a vector machine such as ILLIAC IV depends on the parallelism inherent in a particular algorithm the user wants to employ. In the past few years there has been an increasing interest in parallel algorithms. A paper by Miranker [6] in 1971 reviews parallelism in optimization, root finding, differential equations and solutions of linear systems. Stone [10] has pointed out four areas of concern in developing parallel algorithms. These are:

"1. Data in parallel computers must be arranged in memory for efficient parallel computation. In some cases, the data must be rearranged during the execution of an algorithm. The corresponding problem is nonexistent for serial computers.

2. Efficient serial algorithms are not necessarily efficient for parallel computers. Moreover, inefficient serial algorithms may lead to efficient parallel algorithms.

3. Serial algorithms can have severe serial constraints that appear to be inherent, but actually are removable. In some cases, transformations applied to these algorithms can remove these serial constraints to create efficient parallel algorithms.

4. In questions that pertain to the numerical behavior of an algorithm, a serial algorithm and its counterpart parallel algorithm may behave quite differently. The most important examples of these are the rate of convergence of iterative algorithms and the round-off error of numerical algorithms."

He goes on to discuss these areas in some detail.

Many algorithms for solving partial differential systems can be expressed in vector or parallel form. For solving partial differential equations algorithm investigation for and use of ILLIAC IV are continuing at the University of Illinois. The Rand Corporation is also using the ILLIAC IV to integrate the partial differential equations of the Mintz-Arakawa general weather circulation mode [12]. Other users are studying wave propagation using ILLIAC.

For these simulations storage considerations and efficient parallel processing are important as indicated in the last section and in area 1. above. Further, algorithms for the use of auxiliary storage are becoming more important as the increasing reliability of the ILLIAC system permits significant research to be completed using it. This is because many of the research problems requiring faster computers also require more storage than the PEM's offer. Careful consideration must be given to the input-output requirements of a problem and the efficient use of ILLIAC. At present it appears that such considerations must be made on the basis of a particular problem and on the size and number of data arrays involved.

In preparing a code for ILLIAC IV the user presently has a choice of using assembly language (ASK) or two somewhat higher level languages known as GLYPNIR and CFD. A version of FORTRAN known as IVTRAN is being implemented [13]. Further, there is some effort in automatic detection of parallelism in a serial code by the same group working on IVTRAN. In any case, the writing of code for ILLIAC is still the job of a specialist.

Finally, a question that is often asked is 'What kind of speedup over the machine X might I expect if I use ILLIAC IV?' This question often implies that some part of an algorithm can be done in parallel. An incomplete but somewhat informative answer is that if the serial code using 1 PE takes time S and if only half of the code (parts that take time S/2 serially) can be parallelized the speedup will be less than 2 for the overall code when compared with running time of the serial code on ILLIAC. This ignores the possible simultaneous execution of CU and PE instructions on ILLIAC but may give a good reference point for deciding on the use of ILLIAC IV for a problem. However, one should remember area 3. above. What appears to be only a serial process may in fact be expressed for efficient parallel execution.

From the preceeding discussion it should be clear that research on parallel algorithms and the use of parallel computers, particularly with regard to the vector computers, is underway. We may confidently look in the future for increased understanding of when these computers are useful.

References

1. W. J. Bouknight, S. A. Denenberg, D. E. McIntyre, J. M. Randall, A. H. Sameh, and D. L. Slotnick, "The ILLIAC IV System" Proc. of the IEEE, 60, 1972, pp. 369-388.
2. S. A. Denenberg, "An Introduction to the ILLIAC IV System", Document No. 10, Center for Advanced Computation, University of Illinois at Urbana, 1972.
3. J. H. Ericksen, "Iterative and Direct Methods for Solving Poisson's Equation and Their Applicability to ILLIAC IV," Document No. 60, Center for Advanced Computation, University of Illinois at Urbana, 1972.
4. R. W. Hockney, "The Potential Calculation and Some Applications," Methods in Computational Physics, 9, 1970, pp. 136-211.
5. W. Kehl, "The UCLA Campus Computing Network: An ARPANET Resource," EDUCOM, Bulletin of the Inter-university Communications Council, 8, No. 4, 1973, pp. 10-17.
6. W. L. Miranker, "A Survey of Parallelism in Numerical Analysis", SIAM Review, 13, 1971, pp. 524-547.
7. M. Ogura, M. S. Sher and J. H. Ericksen, "A Study of the Efficiency of ILLIAC IV in Hydrodynamic Calculations", Document No. 59, Center for Advanced Computation, University of Illinois at Urbana, 1972.
8. L. Roberts and B. Wessler, "Computer Network Development to Achieve Resource Sharing", Proc. Spring Joint Computer Conference, 36, 1970, pp. 543-549.

9. M. S. Sher, "Experience in Networking - A Case Study", EDUCOM, Bulletin of the Interuniversity Communications Council, 8, No. 3, 1973, pp. 8-13.

10. H. S. Stone, "Problems of Parallel Computation", Complexity of Sequential and Parallel Numerical Algorithms, Academic Press, New York, 1973, pp. 1-16.

11. D. M. Young, Iterative Solution of Large Linear Systems, Academic Press, New York, 1971, p. 271.

12. Press Seminar Handout as Prepared by the Institute for Advanced Computation, Ames Research Center, Moffett Field, California, August 22, 1973.

13. The IVTRAN Manual, Massachusetts Computer Associates, Inc., Lakeside Office Park, Wakefield, Massachusetts, 1973.

Figure Captions

Fig. 1. Temperature Field (solid lines at 0.1 intervals) and the Stream Function Field (dashed lines at 1 intervals) at Time Step 500.

Fig. 2. Functional relations within a conventional computer. The CU has the function of fetching instructions which are stored in memory, decoding or interpreting these instructions, and finally generating the microsequences of electronic pulses which cause the instruction to be performed. The performance of the instruction may entail the use of "driving" of the three other components. The CU may also contain a small amount of memory called registers that can be accessed faster than the main memory. The ALU contains the electronic circuitry necessary to perform arithmetic and logical operations. The ALU may also contain register storage. Memory is the medium by which information (instructions or data) is stored. The I/O accepts information which is input or putput from Memory. The I/O hardware may also take care of converting the information from one coding scheme to another. The CU and ALU taken together are sometimes called a CPU.

Fig. 3. The IAC System and access to it via the ARPA Network.

Fig. 4. A functional block diagram of ILLIAC IV.

Fig. 5. Straight storage allocation of U array for a 64 x 64 set of mesh points.

Fig. 6. Skewed storage allocation of U array for a 64 x 64 set of mesh points.

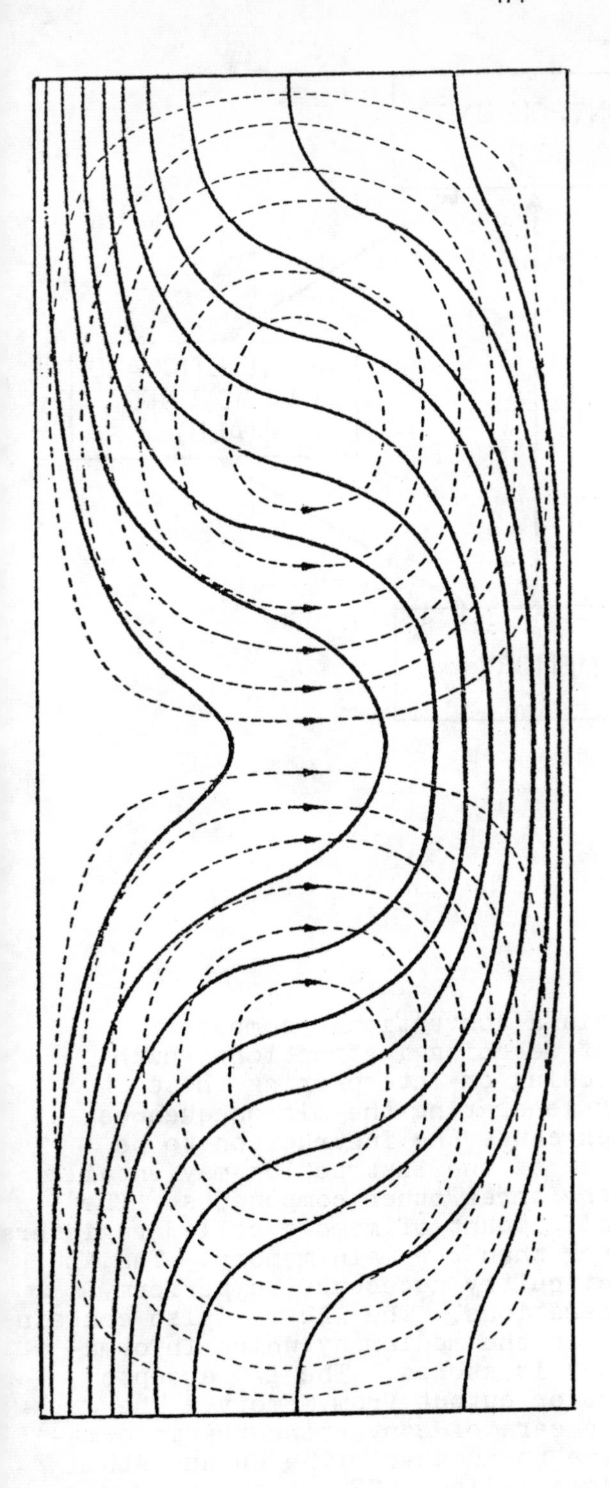

Fig. 1. Temperature Field (solid lines at 0.1 intervals) and the Stream Function Field (dashed lines at 1 intervals) at Time Step 500.

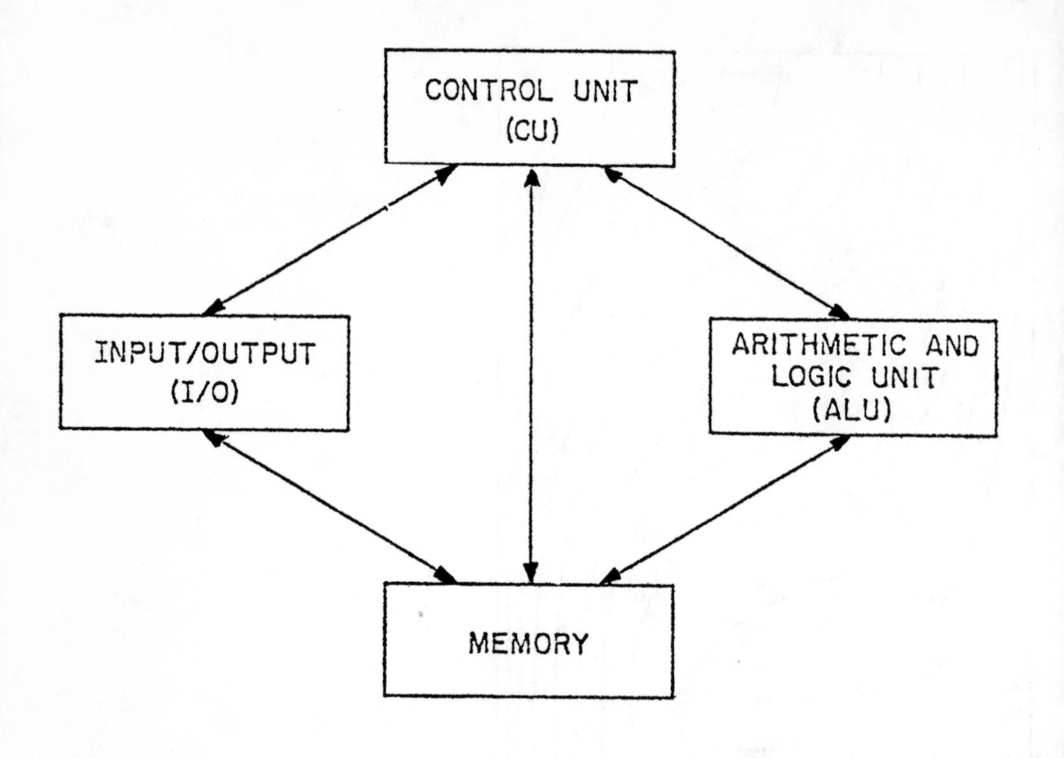

Fig. 2.

Functional relations within a conventional computer. The CU has the function of fetching instructions which are stored in memory, decoding or interpreting these instructions, and finally generating the microsequences of electronic pulses which cause the instruction to be performed. The performance of the instruction may entail the use of "driving" of the three other components. The CU may also contain a small amount of memory called registers that can be accessed faster than the main memory. The ALU contains the electronic circuitry necessary to perform arithmetic and logical operations. The ALU may also contain register storage. Memory is the medium by which information (instructions or data) is stored. The I/O accepts information which is input or output from Memory. The I/O hardware may also take care of converting the information from one coding scheme to another. The CU and ALU taken together are sometimes called a CPU.

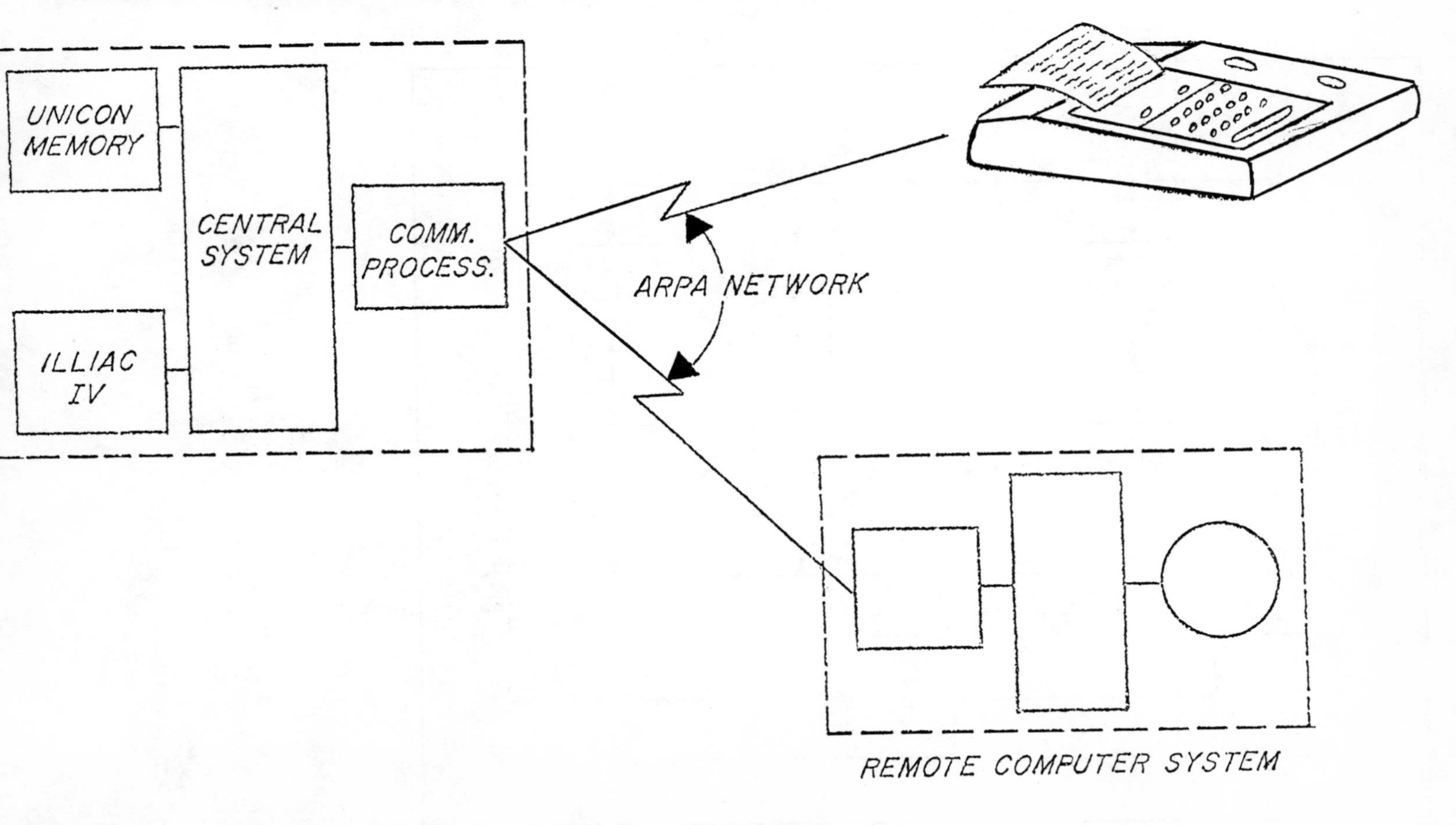

Fig. 3. The IAC System and access to it via the ARPA Network

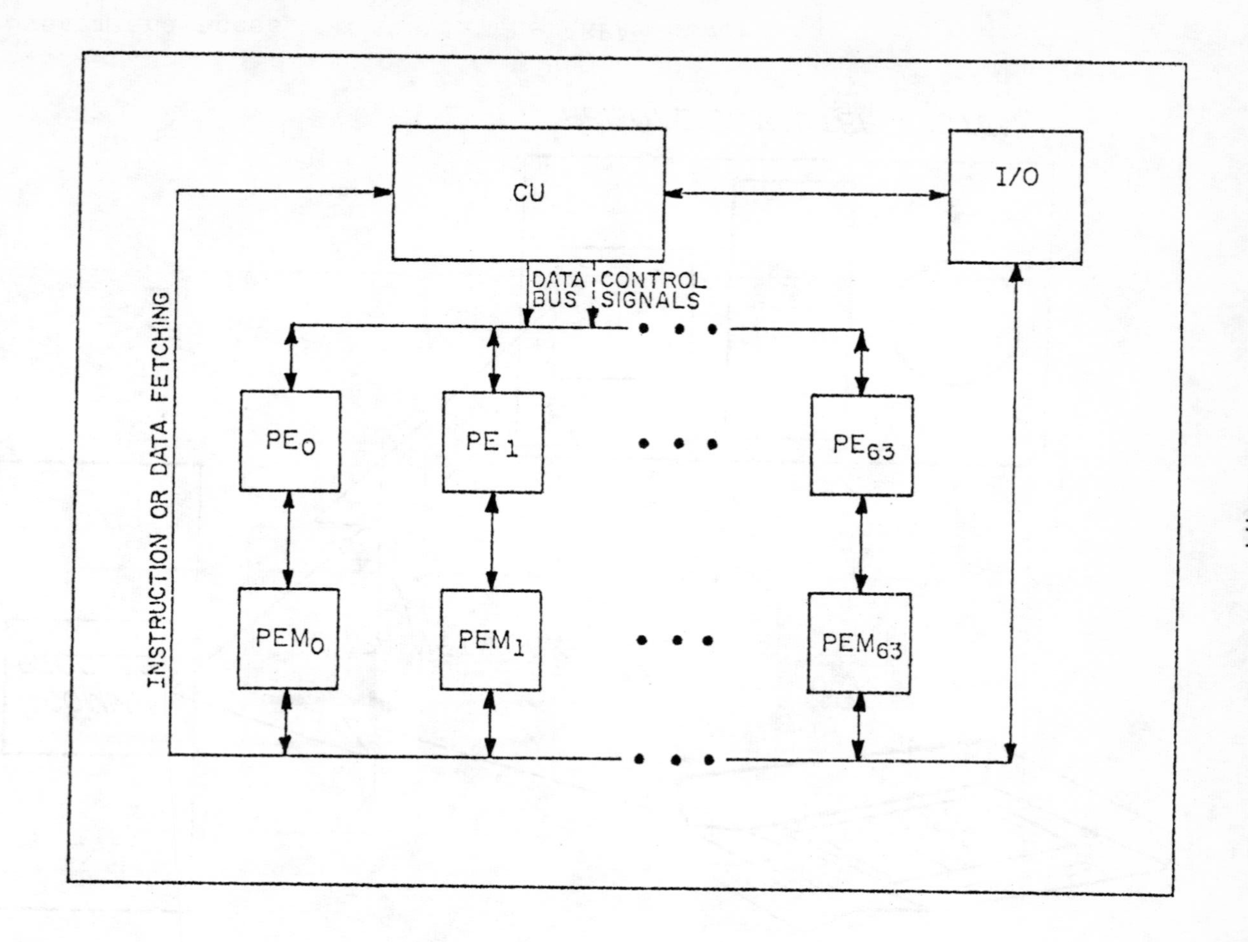

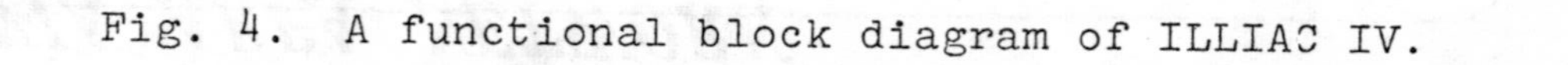

Fig. 4. A functional block diagram of ILLIAC IV.

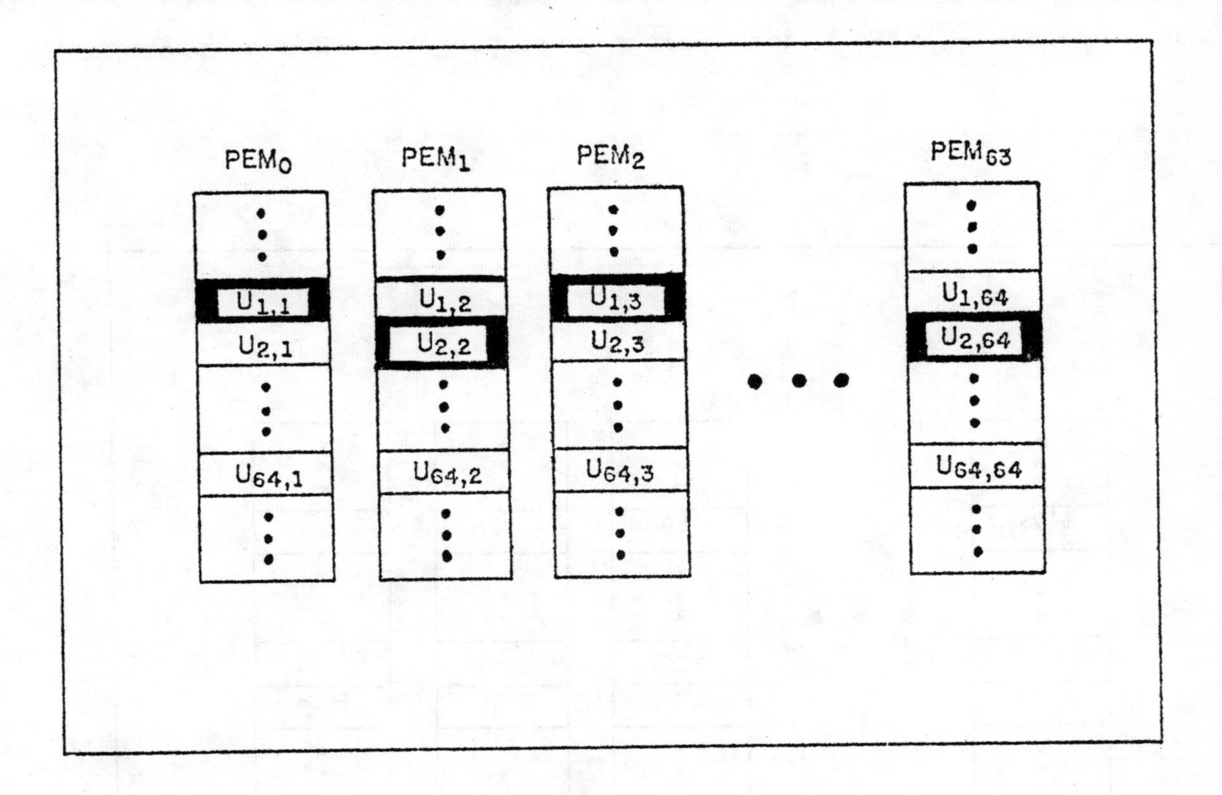

Fig. 5. Straight storage allocation of U array for a 64 x 64 set of mesh points.

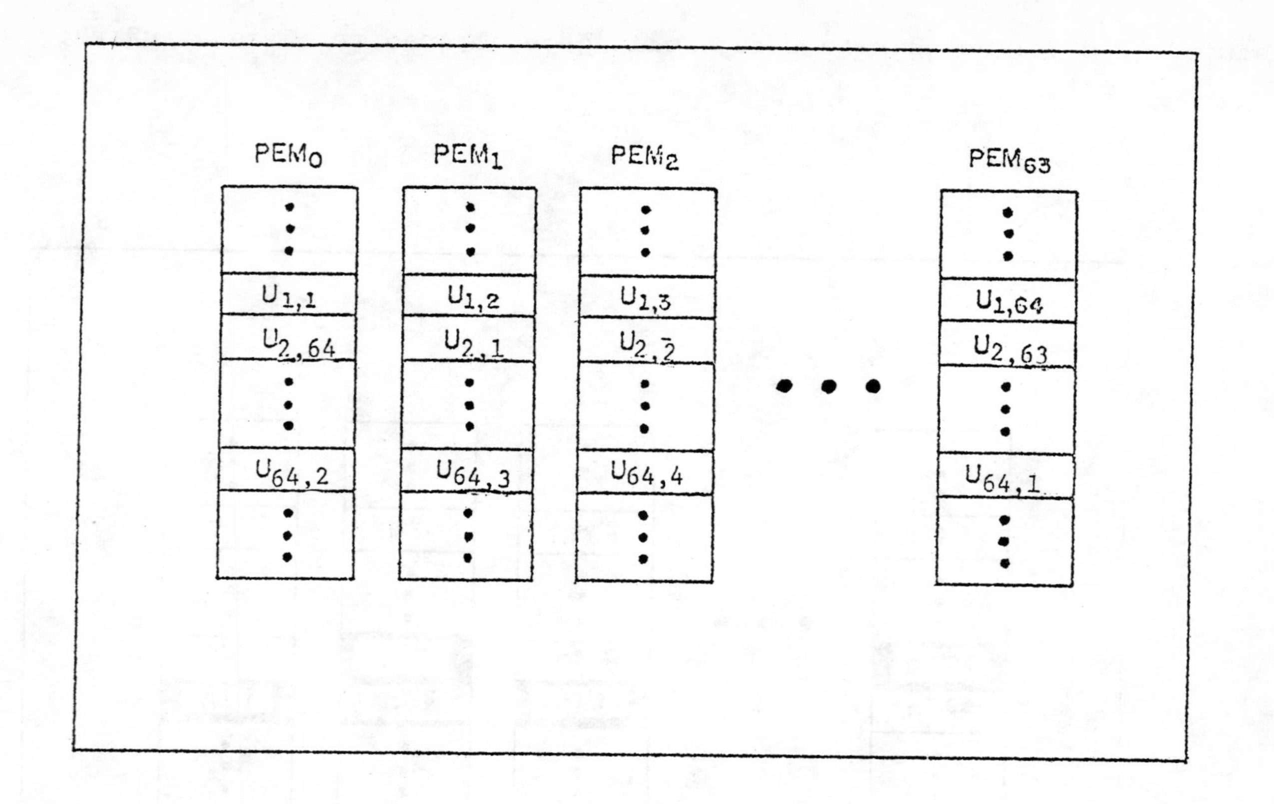

Fig. 6. Skewed storage allocation of U array for a 64 x 64 set of mesh points.

Vol. 277: Séminaire Banach. Edité par C. Houzel. VII, 229 pages. 1972. DM 20,–

Vol. 278: H. Jacquet, Automorphic Forms on GL(2). Part II. XIII, 142 pages. 1972. DM 16,–

Vol. 279: R. Bott, S. Gitler and I. M. James, Lectures on Algebraic and Differential Topology. V, 174 pages. 1972. DM 18,–

Vol. 280: Conference on the Theory of Ordinary and Partial Differential Equations. Edited by W. N. Everitt and B. D. Sleeman. XV, 367 pages. 1972. DM 26,–

Vol. 281: Coherence in Categories. Edited by S. Mac Lane. VII, 235 pages. 1972. DM 20,–

Vol. 282: W. Klingenberg und P. Flaschel, Riemannsche Hilbertmannigfaltigkeiten. Periodische Geodätische. VII, 211 Seiten. 1972. DM 20,–

Vol. 283: L. Illusie, Complexe Cotangent et Déformations II. VII, 304 pages. 1972. DM 24,–

Vol. 284: P. A. Meyer, Martingales and Stochastic Integrals I. VI, 89 pages. 1972. DM 16,–

Vol. 285: P. de la Harpe, Classical Banach-Lie Algebras and Banach-Lie Groups of Operators in Hilbert Space. III, 160 pages. 1972. DM 16,–

Vol. 286: S. Murakami, On Automorphisms of Siegel Domains. V, 95 pages. 1972. DM 16,–

Vol. 287: Hyperfunctions and Pseudo-Differential Equations. Edited by H. Komatsu. VII, 529 pages. 1973. DM 36,–

Vol. 288: Groupes de Monodromie en Géométrie Algébrique. (SGA 7 I). Dirigé par A. Grothendieck. IX, 523 pages. 1972. DM 50,–

Vol. 289: B. Fuglede, Finely Harmonic Functions. III, 188. 1972. DM 18,–

Vol. 290: D. B. Zagier, Equivariant Pontrjagin Classes and Applications to Orbit Spaces. IX, 130 pages. 1972. DM 16,–

Vol. 291: P. Orlik, Seifert Manifolds. VIII, 155 pages. 1972. DM 16,–

Vol. 292: W. D. Wallis, A. P. Street and J. S. Wallis, Combinatorics: Room Squares, Sum-Free Sets, Hadamard Matrices. V, 508 pages. 1972. DM 50,–

Vol. 293: R. A. DeVore, The Approximation of Continuous Functions by Positive Linear Operators. VIII, 289 pages. 1972. DM 24,–

Vol. 294: Stability of Stochastic Dynamical Systems. Edited by R. F. Curtain. IX, 332 pages. 1972. DM 26,–

Vol. 295: C. Dellacherie, Ensembles Analytiques, Capacités, Mesures de Hausdorff. XII, 123 pages. 1972. DM 16,–

Vol. 296: Probability and Information Theory II. Edited by M. Behara, K. Krickeberg and J. Wolfowitz. V, 223 pages. 1973. DM 20,–

Vol. 297: J. Garnett, Analytic Capacity and Measure. IV, 138 pages. 1972. DM 16,–

Vol. 298: Proceedings of the Second Conference on Compact Transformation Groups. Part 1. XIII, 453 pages. 1972. DM 32,–

Vol. 299: Proceedings of the Second Conference on Compact Transformation Groups. Part 2. XIV, 327 pages. 1972. DM 26,–

Vol. 300: P. Eymard, Moyennes Invariantes et Représentations Unitaires. II. 113 pages. 1972. DM 16,–

Vol. 301: F. Pittnauer, Vorlesungen über asymptotische Reihen. VI, 186 Seiten. 1972. DM 18,–

Vol. 302: M. Demazure, Lectures on p-Divisible Groups. V, 98 pages. 1972. DM 16,–

Vol. 303: Graph Theory and Applications. Edited by Y. Alavi, D. R. Lick and A. T. White. IX, 329 pages. 1972. DM 26,–

Vol. 304: A. K. Bousfield and D. M. Kan, Homotopy Limits, Completions and Localizations. V, 348 pages. 1972. DM 26,–

Vol. 305: Théorie des Topos et Cohomologie Etale des Schémas. Tome 3. (SGA 4). Dirigé par M. Artin, A. Grothendieck et J. L. Verdier. VI, 640 pages. 1973. DM 50,–

Vol. 306: H. Luckhardt, Extensional Gödel Functional Interpretation. VI, 161 pages. 1973. DM 18,–

Vol. 307: J. L. Bretagnolle, S. D. Chatterji et P.-A. Meyer, Ecole d'été de Probabilités: Processus Stochastiques. VI, 198 pages. 1973. DM 20,–

Vol. 308: D. Knutson, λ-Rings and the Representation Theory of the Symmetric Group. IV, 203 pages. 1973. DM 20,–

Vol. 309: D. H. Sattinger, Topics in Stability and Bifurcation Theory. VI, 190 pages. 1973. DM 18,–

Vol. 310: B. Iversen, Generic Local Structure of the Morphisms in Commutative Algebra. IV, 108 pages. 1973. DM 16,–

Vol. 311: Conference on Commutative Algebra. Edited by J. W. Brewer and E. A. Rutter. VII, 251 pages. 1973. DM 22,–

Vol. 312: Symposium on Ordinary Differential Equations. Edited by W. A. Harris, Jr. and Y. Sibuya. VIII, 204 pages. 1973. DM 22,–

Vol. 313: K. Jörgens and J. Weidmann, Spectral Properties of Hamiltonian Operators. III, 140 pages. 1973. DM 16,–

Vol. 314: M. Deuring, Lectures on the Theory of Algebraic Functions of One Variable. VI, 151 pages. 1973. DM 16,–

Vol. 315: K. Bichteler, Integration Theory (with Special Attention to Vector Measures). VI, 357 pages. 1973. DM 26,–

Vol. 316: Symposium on Non-Well-Posed Problems and Logarithmic Convexity. Edited by R. J. Knops. V, 176 pages. 1973. DM 18,–

Vol. 317: Séminaire Bourbaki – vol. 1971/72. Exposés 400–417. IV, 361 pages. 1973. DM 26,–

Vol. 318: Recent Advances in Topological Dynamics. Edited by A. Beck, VIII, 285 pages. 1973. DM 24,–

Vol. 319: Conference on Group Theory. Edited by R. W. Gatterdam and K. W. Weston. V, 188 pages. 1973. DM 18,–

Vol. 320: Modular Functions of One Variable I. Edited by W. Kuyk. V, 195 pages. 1973. DM 18,–

Vol. 321: Séminaire de Probabilités VII. Edité par P. A. Meyer. VI, 322 pages. 1973. DM 26,–

Vol. 322: Nonlinear Problems in the Physical Sciences and Biology. Edited by I. Stakgold, D. D. Joseph and D. H. Sattinger. VIII, 357 pages. 1973. DM 26,–

Vol. 323: J. L. Lions, Perturbations Singulières dans les Problèmes aux Limites et en Contrôle Optimal. XII, 645 pages. 1973. DM 42,–

Vol. 324: K. Kreith, Oscillation Theory. VI, 109 pages. 1973. DM 16,–

Vol. 325: Ch.-Ch. Chou, La Transformation de Fourier Complexe et L'Equation de Convolution. IX, 137 pages. 1973. DM 16,–

Vol. 326: A. Robert, Elliptic Curves. VIII, 264 pages. 1973. DM 22,–

Vol. 327: E. Matlis, 1-Dimensional Cohen-Macaulay Rings. XII, 157 pages. 1973. DM 18,–

Vol. 328: J. R. Büchi and D. Siefkes, The Monadic Second Order Theory of All Countable Ordinals. VI, 217 pages. 1973. DM 20,–

Vol. 329: W. Trebels, Multipliers for (C, α)-Bounded Fourier Expansions in Banach Spaces and Approximation Theory. VII, 103 pages. 1973. DM 16,–

Vol. 330: Proceedings of the Second Japan-USSR Symposium on Probability Theory. Edited by G. Maruyama and Yu. V. Prokhorov. VI, 550 pages. 1973. DM 36,–

Vol. 331: Summer School on Topological Vector Spaces. Edited by L. Waelbroeck. VI, 226 pages. 1973. DM 20,–

Vol. 332: Séminaire Pierre Lelong (Analyse) Année 1971-1972. V, 131 pages. 1973. DM 16,–

Vol. 333: Numerische, insbesondere approximationstheoretische Behandlung von Funktionalgleichungen. Herausgegeben von R. Ansorge und W. Törnig. VI, 296 Seiten. 1973. DM 24,–

Vol. 334: F. Schweiger, The Metrical Theory of Jacobi-Perron Algorithm. V, 111 pages. 1973. DM 16,–

Vol. 335: H. Huck, R. Roitzsch, U. Simon, W. Vortisch, R. Walden, B. Wegner und W. Wendland, Beweismethoden der Differentialgeometrie im Großen. IX, 159 Seiten. 1973. DM 18,–

Vol. 336: L'Analyse Harmonique dans le Domaine Complexe. Edité par E. J. Akutowicz. VIII, 169 pages. 1973. DM 18,–

Vol. 337: Cambridge Summer School in Mathematical Logic. Edited by A. R. D. Mathias and H. Rogers. IX, 660 pages. 1973. DM 42,–

Vol: 338: J. Lindenstrauss and L. Tzafriri, Classical Banach Spaces. IX, 243 pages. 1973. DM 22,–

Vol. 339: G. Kempf, F. Knudsen, D. Mumford and B. Saint-Donat, Toroidal Embeddings I. VIII, 209 pages. 1973. DM 20,–

Vol. 340: Groupes de Monodromie en Géométrie Algébrique. (SGA 7 II). Par P. Deligne et N. Katz. X, 438 pages. 1973. DM 40,–

Vol. 341: Algebraic K-Theory I, Higher K-Theories. Edited by H. Bass. XV, 335 pages. 1973. DM 26,–

Vol. 342: Algebraic K-Theory II, "Classical" Algebraic K-Theory, and Connections with Arithmetic. Edited by H. Bass. XV, 527 pages. 1973. DM 36,–

Vol. 343: Algebraic K-Theory III, Hermitian K-Theory and Geometric Applications. Edited by H. Bass. XV, 572 pages. 1973. DM 38,–

Vol. 344: A. S. Troelstra (Editor), Metamathematical Investigation of Intuitionistic Arithmetic and Analysis. XVII, 485 pages. 1973. DM 34,–

Vol. 345: Proceedings of a Conference on Operator Theory. Edited by P. A. Fillmore. VI, 228 pages. 1973. DM 20,–

Vol. 346: Fučík et al., Spectral Analysis of Nonlinear Operators. II, 287 pages. 1973. DM 26,–

Vol. 347: J. M. Boardman and R. M. Vogt, Homotopy Invariant Algebraic Structures on Topological Spaces. X, 257 pages. 1973. DM 22,–

Vol. 348: A. M. Mathai and R. K. Saxena, Generalized Hypergeometric Functions with Applications in Statistics and Physical Sciences. VII, 314 pages. 1973. DM 26,–

Vol. 349: Modular Functions of One Variable II. Edited by W. Kuyk and P. Deligne. V, 598 pages. 1973. DM 38,–

Vol. 350: Modular Functions of One Variable III. Edited by W. Kuyk and J.-P. Serre. V, 350 pages. 1973. DM 26,–

Vol. 351: H. Tachikawa, Quasi-Frobenius Rings and Generalizations. XI, 172 pages. 1973. DM 18,–

Vol. 352: J. D. Fay, Theta Functions on Riemann Surfaces. V, 137 pages. 1973. DM 16,–

Vol. 353: Proceedings of the Conference on Orders, Group Rings and Related Topics. Organized by J. S. Hsia, M. L. Madan and T. G. Ralley. X, 224 pages. 1973. DM 20,–

Vol. 354: K. J. Devlin, Aspects of Constructibility. XII, 240 pages. 1973. DM 22,–

Vol. 355: M. Sion, A Theory of Semigroup Valued Measures. V, 140 pages. 1973. DM 16,–

Vol. 356: W. L. J. van der Kallen, Infinitesimally Central-Extensions of Chevalley Groups. VII, 147 pages. 1973. DM 16,–

Vol. 357: W. Borho, P. Gabriel und R. Rentschler, Primideale in Einhüllenden auflösbarer Lie-Algebren. V, 182 Seiten. 1973. DM 18,–

Vol. 358: F. L. Williams, Tensor Products of Principal Series Representations. VI, 132 pages. 1973. DM 16,–

Vol. 359: U. Stammbach, Homology in Group Theory. VIII, 183 pages. 1973. DM 18,–

Vol. 360: W. J. Padgett and R. L. Taylor, Laws of Large Numbers for Normed Linear Spaces and Certain Fréchet Spaces. VI, 111 pages. 1973. DM 16,–

Vol. 361: J. W. Schutz, Foundations of Special Relativity: Kinematic Axioms for Minkowski Space Time. XX, 314 pages. 1973. DM 26,–

Vol. 362: Proceedings of the Conference on Numerical Solution of Ordinary Differential Equations. Edited by D. Bettis. VIII, 490 pages. 1974. DM 34,–

Vol. 363: Conference on the Numerical Solution of Differential Equations. Edited by G. A. Watson. IX, 221 pages. 1974. DM 20,–

Vol. 364: Proceedings on Infinite Dimensional Holomorphy. Edited by T. L. Hayden and T. J. Suffridge. VII, 212 pages. 1974. DM 20,–

Vol. 365: R. P. Gilbert, Constructive Methods for Elliptic Equations. VII, 397 pages. 1974. DM 26,–

Vol. 366: R. Steinberg, Conjugacy Classes in Algebraic Groups (Notes by V. V. Deodhar). VI, 159 pages. 1974. DM 18,–

Vol. 367: K. Langmann und W. Lütkebohmert, Cousinverteilungen und Fortsetzungssätze. VI, 151 Seiten. 1974. DM 16,–

Vol. 368: R. J. Milgram, Unstable Homotopy from the Stable Point of View. V, 109 pages. 1974. DM 16,–

Vol. 369: Victoria Symposium on Nonstandard Analysis. Edited by A. Hurd and P. Loeb. XVIII, 339 pages. 1974. DM 26,–

Vol. 370: B. Mazur and W. Messing, Universal Extensions and One Dimensional Crystalline Cohomology. VII, 134 pages. 1974. DM 16,–

Vol. 371: V. Poenaru, Analyse Différentielle. V, 228 pages. 1974. DM 20,–

Vol. 372: Proceedings of the Second International Conference on the Theory of Groups 1973. Edited by M. F. Newman. VII, 740 pages. 1974. DM 48,–

Vol. 373: A. E. R. Woodcock and T. Poston, A Geometrical Study of the Elementary Catastrophes. V, 257 pages. 1974. DM 22,–

Vol. 374 S. Yamamuro, Differential Calculus in Topological Linear Spaces. IV, 179 pages. 1974. DM 18,–

Vol. 375: Topology Conference 1973. Edited by R. F. Dickman Jr. and P. Fletcher. X, 283 pages. 1974. DM 24,–

Vol. 376: D. B. Osteyee and I. J. Good, Information, Weight of Evidence, the Singularity between Probability Measures and Signal Detection. XI, 156 pages. 1974. DM 16.–

Vol. 377: A. M. Fink, Almost Periodic Differential Equations. VIII, 336 pages. 1974. DM 26,–

Vol. 378: TOPO 72 – General Topology and its Applications. Proceedings 1972. Edited by R. Alò, R. W. Heath and J. Nagata. XIV, 651 pages. 1974. DM 50,–

Vol. 379: A. Badrikian et S. Chevet, Mesures Cylindriques, Espaces de Wiener et Fonctions Aléatoires Gaussiennes. X, 383 pages. 1974. DM 32,–

Vol. 380: M. Petrich, Rings- and Semigroups. VIII, 182 pages. 1974. DM 18,–

Vol. 381: Séminaire de Probabilités VIII. Edité par P. A. Meyer. IX, 354 pages. 1974. DM 32,–

Vol. 382: J. H. van Lint, Combinatorial Theory Seminar Eindhoven University of Technology. VI, 131 pages. 1974. DM 18,–

Vol. 383: Séminaire Bourbaki – vol. 1972/73. Exposés 418-435 IV, 334 pages. 1974. DM 30,–

Vol. 384: Functional Analysis and Applications, Proceedings 1972. Edited by L. Nachbin. V, 270 pages. 1974. DM 22,–

Vol. 385: J. Douglas Jr. and T. Dupont, Collocation Methods for Parabolic Equations in a Single Space Variable (Based on C^1-Piecewise-Polynomial Spaces). V, 147 pages. 1974. DM 16,–

Vol. 386: J. Tits, Buildings of Spherical Type and Finite BN-Pairs. IX, 299 pages. 1974. DM 24,–

Vol. 387: C. P. Bruter, Eléments de la Théorie des Matroïdes. V, 138 pages. 1974. DM 18,–

Vol. 388: R. L. Lipsman, Group Representations. X, 166 pages. 1974. DM 20,–

Vol. 389: M.-A. Knus et M. Ojanguren, Théorie de la Descente et Algèbres d' Azumaya. IV, 163 pages. 1974. DM 20,–

Vol. 390: P. A. Meyer, P. Priouret et F. Spitzer, Ecole d'Eté de Probabilités de Saint-Flour III – 1973. Edité par A. Badrikian et P.-L. Hennequin. VIII, 189 pages. 1974. DM 20,–

Vol. 391: J. Gray, Formal Category Theory: Adjointness for 2-Categories. XII, 282 pages. 1974. DM 24,–

Vol. 392: Géométrie Différentielle, Colloque, Santiago de Compostela, Espagne 1972. Edité par E. Vidal. VI, 225 pages. 1974. DM 20,–

Vol. 393: G. Wassermann, Stability of Unfoldings. IX, 164 pages. 1974. DM 20,–

Vol. 394: W. M. Patterson 3rd, Iterative Methods for the Solution of a Linear Operator Equation in Hilbert Space – A Survey. III, 183 pages. 1974. DM 20,–

Vol. 395: Numerische Behandlung nichtlinearer Integrodifferential- und Differentialgleichungen. Tagung 1973. Herausgegeben von R. Ansorge und W. Törnig. VII, 313 Seiten. 1974. DM 28,–

Vol. 396: K. H. Hofmann, M. Mislove and A. Stralka, The Pontryagin Duality of Compact O-Dimensional Semilattices and its Applications. XVI, 122 pages. 1974. DM 18,–

Vol. 397: T. Yamada, The Schur Subgroup of the Brauer Group. V, 159 pages. 1974. DM 18,–

Vol. 398: Théories de l'Information, Actes des Rencontres de Marseille-Luminy, 1973. Edité par J. Kampé de Fériet et C. Picard. XII, 201 pages. 1974. DM 23,–